Introduction to Nuclear Reactions

A typical experimental area for making measurements on the scattering of a beam of particles accelerated by a cyclotron. (Photograph courtesy of E. E. Gross, Oak Ridge National Laboratory.)

INTRODUCTION TO NUCLEAR REACTIONS

G. R. SATCHLER, MA, DSc(Oxon), FAPS

Oak Ridge National Laboratory
Oak Ridge, Tennessee 37831

SECOND EDITION

New York
Oxford University Press
1990

Published in Great Britain by Macmillan Education Ltd

Published in the United States by
Oxford University Press, Inc
200 Madison Avenue, New York 10016

Library of Congress Cataloging-in-Publication Data
Satchler, G. R. (George Raymond)
Introduction to nuclear reactions.
Includes bibliographical references and index.
1. Nuclear reactions. I. Title.
QC794.S26 1990 539.7′5 90-7560
ISBN 0-19-520841-2
0-19-520842-0 (ppr)

Printing (last digit) 9 8 7 6 5 4 3 2 1

Printed in Hong Kong

Contents

Preface to the Second Edition

The past decade has seen a remarkable growth in the extent and variety of experiments being done on nuclear reactions. The range of energies available has increased by two orders of magnitude. Because of the rapid expansion of heavy-ion science, the number of projectiles in use is very much greater and now includes nuclei as heavy as those of uranium atoms. However, the basic physical principles and techniques involved in the description of these reactions has remained unchanged, although the emphasis may have shifted to some degree. For example, the application of semi-classical ideas became more extensive with the increased interest in heavy-ion reactions, while relativistic considerations became more important as the energy increased.

These advances have resulted in some changes being made in the text, together with some modernisation of the references. Again, the reader is urged to consult review journals and the proceedings of recent conferences for the latest developments. The many experimental examples shown in the figures have not been changed; these were mostly drawn from work done at energies lower than those currently available. Nonetheless, the features illustrated remain relevant at these higher energies.

The opportunity has been taken to correct some misprints, errors, omissions and obscurities that were present in the original. I am indebted to numerous people for bringing many of these to my attention, especially R. C. Johnson, S. Kobayashi and A. G. Tibell.

Oak Ridge, Tennessee, 1990 G.R.S.

Preface to the First Edition

This book is aimed primarily at the undergraduate or beginning graduate student, although I believe others will also find it useful. The specialist from another field may find here a summary of the present situation in our understanding of nuclear-reaction phenomena. The established nuclear physicist may find reading it to be helpful in refreshing his memory about areas in which he is not currently working himself.

Chapters 1 and 2 are intended to provide an overview of the subject which can be readily understood by the novice. They also serve as an introduction to the somewhat more serious remainder. Chapter 3 reviews scattering theory with emphasis on the underlying physical ideas. It also provides schematic *entrées* to the more advanced topics. (There are other excellent texts available which expound these more formal and mathematical aspects of scattering theory.) The discussion is not specific to nuclei, so that Chapter 3 may serve equally well as an introduction to the theory of atomic and molecular collisions. The physical models which have been developed to account for the various aspects of nuclear reaction phenomena are described in more detail in Chapter 4, which is the largest section of the book. I believe that this arrangement enables the book to cater to the needs of a variety of readers without sacrificing any coherence of the presentation as a whole.

Some acquaintance with quantum mechanics is assumed, but not to any great depth. In general, the emphasis here is on the word 'introduction' in the title. There are a number of books and many review articles which treat various parts of the subject in detail and at a higher level; some of these are referred to in the text and are listed in the references at the end of each chapter. Many references are made also to original research papers in the belief that the reader should be encouraged to dip into these other sources of material even if generally they do seem to be more technically advanced than he needs.

Nuclear physics is still very much a living, developing field of study. Consequently any book such as this one is in danger of being obsolescent in some respect as soon as it appears. (This is particularly liable to apply to the descriptions of heavy-ion reactions). This is another reason to urge the reader to supplement the material presented here by resorting freely to the current literature such as review journals and books, and reports of the proceedings of conferences.

The instrumental, experimental and technological aspects of studying nuclear reactions are scarcely mentioned in the present volume; the purpose here is to understand the results of measurements rather than to describe how they are made. There are a number of good books available which address themselves to experimental problems and techniques; we mention in particular *Atomic Nuclei and Their Particles* by E. J. Burge (Oxford University Press, 1977) and, at a more advanced level, the excellent *Techniques in Nuclear Structure Physics* by J. B. A. England (Halsted, New York, 1974).

The first draft of this book was written in the summer of 1970 while the author was a guest of the Aspen Center for Physics, Colorado. I am indebted to the Center for providing a climate so conducive to this achievement. Various sections of the text were read by D. M. Brink, K. T. Hecht and D. K. Scott; I am grateful for their helpful comments. I am also grateful to the many colleagues who have permitted me to reproduce their illustrations here. Finally, I am indebted to Mrs. Althea Tate for efficiently, patiently and cheerfully typing and retyping the manuscript.

Oak Ridge, Tennessee, 1980 G.R.S.

1

Some Background Information

The atomic nucleus was discovered as a consequence of a nuclear-reaction experiment and the investigation of its properties relies to a large extent upon measurements made on a variety of nuclear reactions. However, before discussing nuclear reactions, we should answer the question: 'what is a nucleus?'

1.1 DISCOVERY OF THE NUCLEUS

Geiger and Marsden (1909) studied the scattering of *alpha particles* (α-particles)—the nuclei of helium atoms, emitted by radioactive atoms—impinging on foils of gold and silver. They found that a small number, about 1 in 8 000, was deflected through an angle greater than 90°; that is, they were scattered backwards, although the most probable angle of deflection was found to be less than 1°. Now the massive α-particle will not be noticeably deflected by the light electrons of the atoms in the foil, so the scattering must be due to the heavier positive charges. If these were uniformly distributed in a cloud with atomic dimensions (inferred from the kinetic theory of gases to be about 10^{-10} m), the electric field acting on an impinging α-particle would be so weak that the probability of deflection through such large angles would be minute, many orders of magnitude smaller than that observed. In order to obtain the necessary field strength, the positive charges of the atom must be concentrated in some way. Rutherford (1911) then returned to the idea that the positive charges (and hence most of the mass of the atom) were concentrated at the centre of the atom. Further experiments by Geiger and Marsden (1913) confirmed that the distribution in angle of the scattered α-particles was in accord with Rutherford's theory of the scattering from this centre (*see* Chapter 2). One can deduce from Geiger and Marsden's results that this core or *nucleus* of positive charge must have a radius of less than a few times 10^{-14} m. Then the negative electrons must be arranged around this nucleus at distances up to about 10^{-10} m, thus defining the atomic dimensions. Further, in agreement with Barkla's (1911) results, the number of positive charges was found to be equal to the atomic number, or roughly one-half the atomic weight. The application by Bohr (1913) of this nuclear model to explain the spectrum of radiation from the hydrogen atom and hence to lay the foundations of modern atomic theory is outside the scope of the present book. We are here concerned with the nucleus itself and the ways in which we study it by scattering processes.

1.2 CONSTITUTION OF THE NUCLEUS

As we have seen, the nucleus is very small even in relation to the atom itself. A nuclear dimension is of order 10^{-14} m, or some ten thousand times smaller than a typical atomic dimension of 10^{-10} m. To give this figure in a more familiar perspective, if an atom filled a cubic room with sides of 10 m, the nucleus at the centre of the room would be about 1 mm across. All but a very small fraction (one part in a few thousand) of the mass of the atom resides in this nucleus. As a consequence, the density of matter in the nucleus is immensely greater than the densities to which we are accustomed in everyday life; it is about 3×10^{17} kg m^{-3} or 3×10^{14} times that of water.

So far, we have said nothing about the structure of this atomic nucleus. It was evident that the phenomena of radioactivity should be associated with the nucleus, so it was natural to speculate on the possibility of artificially inducing nuclear disintegrations. The first example of this artificial transmutation of matter was obtained by Rutherford (1919) who bombarded nitrogen atoms with the α-particles from radium and observed the emission of *protons* (the nuclei of hydrogen atoms with a charge equal and opposite to that of the electron). At first, it was supposed that nuclei were composed of protons and electrons since electrons are emitted during β-decay, but this hypothesis led to some difficulties. The problem was removed when Chadwick (1932) (*see* also Curie and Joliot, 1932) established the evidence for the *neutron*, a highly penetrating neutral particle with a mass very close to that of the proton. Since the mass of an atom is close to an integer A (called the mass number) times the mass of the hydrogen atom, the discovery of the neutron led immediately to the idea that nuclei were composed of Z protons and $(A-Z)$ neutrons, where Z is the atomic number of the element (and so Ze is the total charge on the nucleus if $-e$ is the charge on an electron). Neutrons and protons, as the constituents of nuclei, are now referred to as *nucleons.*

1.3 THE STUDY OF THE NUCLEUS BY NUCLEAR REACTIONS

We may study nuclei passively by simply observing the radiation from radioactive decay. In other cases we may study the interaction between a nucleus and its electronic environment, which gives rise, for example, to the hyperfine structure in atomic or molecular spectra. However, the most fruitful source of information is obtained by applying an external field. This happens when we make the nucleus interact with another particle by shooting a beam of those particles at some target material containing the nucleus and observing the various products and their distribution in angle and energy, etc. The particles scattered may be photons, electrons, mesons, neutrons, hyperons, protons or other nuclei.

Geiger and Marsden (1909) and Rutherford (1919) used other naturally radioactive material as their source of radiation. However, as well as the discovery of the neutron, the year 1932 saw the first successful nuclear transmutation induced by artificially accelerated particles. Cockcroft and Walton (1932) accelerated protons by applying an electrostatic potential of up to 500 kV. Directing these upon lithium, they observed the emission of two α-particles (helium nuclei) which could be interpreted as the capture of the proton by a lithium nucleus followed by its break-up. One obvious advantage of artificial acceleration is that a much greater intensity of radiation can be obtained; a current as small as 1 μA represents some 6×10^{12} singly charged particles per second, or about the number of α-particles emitted by 160 g of radium. Of course one also has much greater control over the energy of the particles and can direct them in a concentrated beam. Artificial transmutations may also be sources of secondary radiations, such as neutrons or mesons, which do not occur naturally in any abundance.

The study of nuclei by nuclear reactions expanded rapidly. Fermi (1934) showed that nuclei could be transmuted by the capture of secondary neutrons produced by another reaction; in a few years this was followed by Hahn and Strassmann's (1939) observations of the *fission* of heavy nuclei into two lighter ones by neutron capture, the consequences of which are only too familiar. One consequence of all this effort is that, during the intervening forty years, some 1300 radioisotopes have been produced, identified and studied in addition to the 300 stable nuclear species. The search for new nuclei continues; one current interest is in the possibility of producing superheavy nuclei with $A \gtrsim 300$, which according to some theories may be metastable.

Machines to accelerate charged particles have increased in power until it is now possible to produce particles with energies more than ten million (10^7) times greater than those available to Cockroft and Walton in 1932. The study of the collisions of nuclei and their constituents has been extended by the discovery of many new subnuclear particles, *mesons* and *hyperons* (*see*, for example, Hughes, 1971; Kirk, 1986; Close, 1983) until the discipline has evolved into two fields, nuclear physics and particle physics (the latter sometimes called high-energy physics because much of its work is with the accelerators that have the highest energies). While nuclear physicists are primarily concerned with the properties of many-nucleon systems (where 'many' means greater than 2!), there are of course strong links with particle physics. For example, from particle physics must come our understanding of the forces between nucleons which hold nuclei together.

1.4 SOME PRACTICAL APPLICATIONS

The study of nuclear reactions is not only of academic interest, but numerous applications have been found in industry, medicine and other branches of science as well. Nuclear fission, of course, made possible the reactor as an important source of electrical power (*see*, for example, Inglis, 1973 and Murray, 1975).

As nuclear reactors become more sophisticated, precise information about the interactions of neutrons with nuclei becomes vital and may determine the overall direction of a development programme. The theoretical description of these interactions provides supplementary information which may allow one to interpolate the needed parameters into regions where measurements are difficult or impractical. This information is needed not only within the reactor but also for the materials around it which provide radiation shielding. Shielding is important in other areas also; around accelerators for example or as protection from cosmic rays on spaceships. These require knowledge about interactions with charged particles and gamma rays (γ-rays), as well as neutrons.

One of the present-day hopes is that it can be shown feasible to produce power economically from the *fusion* of two very light nuclei. The realisation of this hope requires an understanding of a number of nuclear reactions involving charged particles.

The radiations from artificial radioactive isotopes, produced by nuclear reactions, are used as tools in industry, chemistry, biology and medicine. Different isotopes can be tailor-made for different applications. The activation, by bombardment, of stable isotopes can be used to detect and measure trace elements in concentrations as small as 1 part per million or less, with important applications in metallurgy, archaeology, criminology, etc.

Nuclear reactions also play an important role in astrophysics (Truran, 1984; Reeves, 1968). An important source of stellar energy is nuclear fusion and an understanding of stellar-burning cycles requires the knowledge and understanding of a number of nuclear reactions. Furthermore, the study of nucleogenesis attempts to understand the present relative abundances of the elements in the universe through the sequences of nuclear reactions which have taken place during its history.

1.5 THE ROLE OF MODELS

A *model* is an imitation of the real thing. As we begin to learn some of the properties of nuclei, we invent simple models, based upon our past experience of other systems, which have similar properties. For example, the high density of a nucleus suggests that the neutrons and protons are closely packed, but the nuclear force between nucleons is known to have a short range so that we may expect that only nearest neighbours will interact strongly. These are two properties also possessed by the constituent atoms in liquids and suggest the model of a liquid drop for the nucleus. The next step is to predict other properties which such a system would have and then search for them. For example, such a drop might undergo oscillations in its shape, or its stable shape may not be spherical. If the number of protons is sufficiently large, the Coulomb repulsion may cause the drop to split into two and fission.

As our knowledge grows in detail, so must our models become more detailed. We may find that different sets of properties suggest different models which appear almost self-contradictory. Indeed, the two major models of nuclear structure are the *shell* (or *independent-particle*) model and the *collective* model (a sophisticated form of the liquid drop). The shell model regards the nucleus as a cloud of nucleons moving in more-or-less independent orbits like the electrons of the atom. The collective model seems to treat the nucleus like a drop of fluid in which the motions of individual nucleons are highly correlated and do not need to be discussed in detail. These two pictures appear at first sight to be mutually exclusive; however, it is possible to superimpose an overall slow collective or 'drift' motion upon the independent orbital motions without too much interference between them. When such an understanding is reached we often say we have a 'unified' model.*

An understanding of this kind, however, is often a qualitative one based upon plausibility, not a rigorous derivation from first principles. When we are dealing with the many-body problems posed by such systems as atoms, nuclei, electrons in metals, etc., we know we shall never achieve a complete and exact solution. Even though the electromagnetic forces are known, the positions of the spectral lines of a complex atom can be measured far more accurately than they can be calculated. Nuclear phenomena are much more complicated, and even the forces cannot be described in any great detail. It is clear then that models are not just a feature of the early history of the science, but will always remain as important vehicles of our understanding.

The model here is then a simpler physical system whose properties we can calculate more easily and, in some sense, understand, perhaps even visualise, more easily. It is a first approximation to the nuclear system and schemes may be devised for making it more and more 'realistic' by introducing further refinements. It will never be able to reproduce all nuclear parameters to an unlimited degree of accuracy for this would be equivalent to solving exactly the original many-body problem. Indeed, through a given model we can seldom hope for more than the accurate description of a small category of properties, together with a less detailed understanding of the characteristics of some others. It is not surprising then that we may use several models, each of which emphasises different categories of properties, or perhaps different sets of nuclei. The shell and collective models already mentioned are two examples. If we view each model as a starting point in an approximation scheme (akin to the choice of a

*This fascinating ability of the nucleus to reveal quite different phenomena when studied in different ways, and the consequent desire to understand the apparent complexity within a unified theory, provide one of the dominant intellectual motivations for conducting basic research in nuclear science.

convenient representation in the language of quantum mechanics), each may be appropriate to its own line of enquiry but not necessarily suitable for another.

This emphasis on the use of models does not mean that we abandon any thought of a more fundamental theory. But in general we do not entertain hopes of confronting a fundamental theory directly with the experimental material. Rather, the models serve as manifestations of intermediate phenomenological theories and the understanding of these and the explanation of their relatively few parameters is the task of a fundamental theory. It is as though we must impose some order on the raw data and parameterise the correlations we find before we can ask about the general principles underlying them.

1.6 CONSERVATION LAWS AND SYMMETRY PRINCIPLES

We have already emphasised the complex nature of the many-body problem and the difficulty of solving it. Fortunately there is one kind of information that often we can obtain from very general arguments based upon conservation laws and symmetry principles. The interactions between the particles of the system will possess symmetry properties which lead to conservation laws (*see*, for example, Frauenfelder and Henley, 1975). For example, if the interaction between two particles is invariant under a rotation of the coordinate axes, then their total angular momentum is conserved. (Not the separate angular momentum of each particle because their interaction allows them to transfer angular momentum back and forth.) Similarly, invariance under translation of the coordinates leads to conservation of momentum.

Some applications are almost trivial. Conservation of energy and momentum, for example, can be used to deduce the energy and direction of motion of one particle after a two-particle collision, provided those for the other are measured. The conservation of angular momentum, however, plays a peculiarly important role in atomic and nuclear structure (Brink and Satchler, 1968). This comes about because we are dealing with small, finite systems which are, or almost are, spherically symmetric. Because the system is finite, the centre of the system provides a point about which rotations have a particular physical significance. (An electron moving in a metal, for example, does not have any such point of reference, and angular momentum is not a very useful property in a description of its motion.) Because the system is also small, the average angular momentum of a particle in it is not large compared to Planck's constant. So the discreteness of angular momentum in quantum mechanics is particularly important. A typical nuclear radius is 5×10^{-15} m and a typical average velocity of a nucleon in the nucleus is about $c/5$ (corresponding to a kinetic energy of about 30 MeV). Such a nucleon travelling around the extreme periphery of the nucleus has an angular momentum of about $L = 5\hbar$. On average, a nucleon would be found with only two or three times $\hbar$, which is comparable to the quantum $\hbar$ itself. Hence quantum effects will be important. Further, since atoms and nuclei are approximately spherically symmetric, each electron or nucleon *on average* moves in an almost spherically symmetric environment. Thus it makes sense to use a model such as the shell model in which, to a first approximation, the angular momentum of each particle is conserved.

The smallness of a nucleus ensures that quantum effects are important for ordinary momentum also. The typical nucleon just mentioned has a momentum of about $p = (10^{15}\ m^{-1}) \times \hbar$. It is confined to a region of radius 5×10^{-15} m so Heisenberg's uncertainty relation tells us its momentum is uncertain to

$$\Delta p \geqslant \hbar/\Delta x \approx \hbar/(5 \times 10^{-15}\ m) = \tfrac{1}{5}p$$

Hence this uncertainty is proportionately about as large as the uncertainty ($\Delta L \approx \hbar$) in angular momentum. However, the atom or nucleus being a finite system, an electron or nucleon on average does not find itself in an environment which is translationally invariant, so it does not make sense to speak of its momentum p being conserved even on average. On the other hand, if the total nuclear system (for example, two colliding nuclei) is not in an external field, then its *total* momentum *is* conserved.

For a long time it was believed that the interactions between particles which occurred in nature were invariant under a reflection of the coordinates through the origin ($r \to -r$). This led to the concept of *parity* (*see*, for example, Swartz, 1965). It was then found that some weak interactions, such as that involved in β-decay, are not invariant under space reflection (they are intrinsically left- or right-handed), but nonetheless the conservation of parity remains a powerful principle in nuclear physics. A similar symmetry is invariance under time reversal ($t \to -t$); for all practical purposes, this is a good symmetry for the physics of nuclei.

Unique to nuclear and elementary particle physics is the symmetry represented by *charge independence*. In the nuclear case this means that the form of the specifically nuclear interaction between two nucleons is independent of whether they are both neutrons, both protons or one of each. It leads to the concept of *isospin*, and the conservation of isospin, in direct analogy with angular momentum. Although only approximately valid (if only because of the electromagnetic forces), the idea of charge independence remains a useful tool in nuclear spectroscopy (*see*, for example Wilkinson, 1969).

The Pauli Principle requires us to use wave functions for a system of two or more particles with the appropriate *exchange symmetry*. Exchanging the coordinates of any two identical particles will produce no change (for *bosons*) or just change the sign (for *fermions*) of the wave function. Since nuclei are made of neutrons and protons, which are fermions, their wave functions must change sign; we say they are *antisymmetric*. This symmetry rule may be applied directly; for example, in the collison of two identical particles it requires that the angular distribution of the scattered particles must show fore-and-aft symmetry.

Some uses for symmetry principles have already been mentioned. They may give rise to *selection rules* which determine which transitions are allowed. They provide constraints on the forms which may be assumed by various physical quantities, such as the angular distribution of radiation emitted by a nucleus. If certain symmetries are assumed, they restrict the form that nuclear interactions may take.

In each case we are separating those features of the problem which are consequences of general symmetries or conservation laws but which are independent of the details of the interactions, etc. Some of these symmetries are believed to be absolute (such as conservation of momentum), some are only approximate (such as conservation of isospin). Having an understanding of what consequences follow from symmetry alone, we are in a better position to understand what features will give us specific information about the details of nuclear forces and structure. A more familiar classical example is the motion of a particle in a central field. The central nature or symmetry of the interaction determines that the orbit is planar and that angular momentum is conserved. The exact shape of the orbit will depend upon the detailed form of the force.

1.7 SOME BASIC FACTS ABOUT NUCLEI

Here we review very briefly some of those properties of nuclei which are pertinent to studies of nuclear reactions. For more details *see*, for example, Preston and Bhaduri (1975).

1.7.1 Mass, charge and binding energy

A particular nucleus is denoted symbolically as ${}^A_Z\mathrm{X}$, or just ${}^A\mathrm{X}$, where X is the abbreviation for the chemical element, Z the atomic number and A the mass number. This nucleus consists of A nucleons, the collective name for neutrons and protons; there are Z protons and $N = (A - Z)$ neutrons held together by their mutual attractions. Sometimes, the neutron number N is also displayed, as in ${}^A_Z\mathrm{X}_N$. Nuclei with the same Z but different A (and therefore different N) are called *isotopes*. Since they have the same charge they form similar atoms which have the same chemical properties. Those nuclei with the same N but different A are called *isotones*, while those with the same A are called *isobars*. The electric charge on the nucleus is Ze, where $-e$ is the charge on the electron. The masses of a proton and neutron are approximately equal, so the mass of this nucleus is approximately A times that of one nucleon. It is convenient to express nuclear masses (*see* Appendix C) in ***atomic mass units*** (amu or simply u)

$$1\ \mathrm{u} \equiv 1\ \mathrm{amu} = 1.66057 \times 10^{-27}\ \mathrm{kg} = 931.502\ \mathrm{MeV}/c^2$$

An amu is defined* as 1/12th the mass of the neutral ${}^{12}\mathrm{C}$ atom. (Also, MeV = 10^6 eV, where 1 eV is the energy change for an electron or a proton

*This definition was adopted by the International Union of Physics and Chemistry. Until 1961 nuclear physicists used an amu defined as 1/16 the mass of the neutral ${}^{16}\mathrm{O}$ atom. A convention of this type is adopted because it is much easier to make precise measurements of the *ratio* of the masses of two atoms than to determine each absolutely.

upon traversing an electrostatic potential change of 1 V.) On this scale, the nucleons have masses close to unity

$$m_n = 1.008665 \text{ u}, \qquad m_p = 1.007276 \text{ u}$$

and hence the mass of a nucleus is approximately equal to its mass number A. The nuclear mass is actually somewhat less than the mass of its constituent nucleons because of the relativistic contribution ($\delta m = \delta E/c^2$) from its (negative) energy of binding. It is

$$m_A = N m_n + Z m_p - B/c^2$$

where B is the *binding energy*. (It is conventional to give B as a positive number, so that B is the energy required to break up the nucleus into N neutrons and Z protons.) B varies from nucleus to nucleus, but typically it is about 8 MeV per nucleon for stable nuclei, or $B \sim 8A$ MeV. (This may be compared with the energy of the rest mass of a nucleon, about 930 MeV, or of an electron, about 0.5 MeV. These are much greater than typical atomic energies; for example the binding energy of the H atom is about 14 eV and the energy of a photon from the Na D-lines in the optical spectrum is about 2 eV.) B is larger for nuclei with certain *magic* numbers of neutrons and protons, and this led to the concept of the nuclear shell model. These particularly stable nuclei are those with *closed shells* (Mayer and Jensen, 1955), analogous to the closed electron shells of the noble gas atoms. The attractive nuclear forces give most binding when $N \approx Z$, while the repulsive Coulomb forces favour $N > Z$. The competition between these results in the binding energy surface, $-B(N, Z)$, showing a 'valley of stability' along the $N \approx Z$ line for light nuclei which bends over to $N/Z \sim 1.5$ for the heaviest nuclei. As we move away from this valley in the N, Z plane, the nuclei have progressively smaller binding energies until they become unstable against the emission of one or more nucleons. Even before this, they become unstable to β-decay, which converts a neutron into a proton, or vice-versa, and changes the nucleus into one closer to the stability line.

1.7.2 Size and radial shape

As was described earlier, Rutherford's original experiments on the scattering of α-particles, which established the nuclear atom, also indicated that the nucleus had a size* of less than a few times 10^{-14} m. More sophisticated scattering experiments have yielded more precise information. Analysis of electron-scattering measurements tells us that the protons in a nucleus are distributed like the curves shown in *Figure 1.1*. The radius R of the distribution (i.e. the radius at which it has fallen to one-half its central value) is roughly proportional

*The unit fm ≡ femtometer = 10^{-15} m is thus a natural unit of length for describing nuclei. It was originally called a 'fermi' or 'Fermi' after the physicist of that name and in some publications it will be found denoted by the symbol F. Many nuclear physicists, with a touch of nostalgia, continue to refer to it by its old name.

to $A^{1/3}$, $R \approx r_0 A^{1/3}$, with $r_0 \approx 1.07$ fm. Thus a nucleus like ^{120}Sn has a radius of about 5 fm = 5×10^{-15} m. (For comparison, the Bohr radius of the electron's orbit in the H atom is about 5×10^{-11} m, while that for the most tightly bound (1*s*) electron in the Sn atom is roughly 10^{-12} m. Again, the wavelength of the *D*-lines in the optical spectrum of Na is about 6×10^{-7} m, while the *K* X-rays from Sn have a wavelength of about 5×10^{-11} m.)

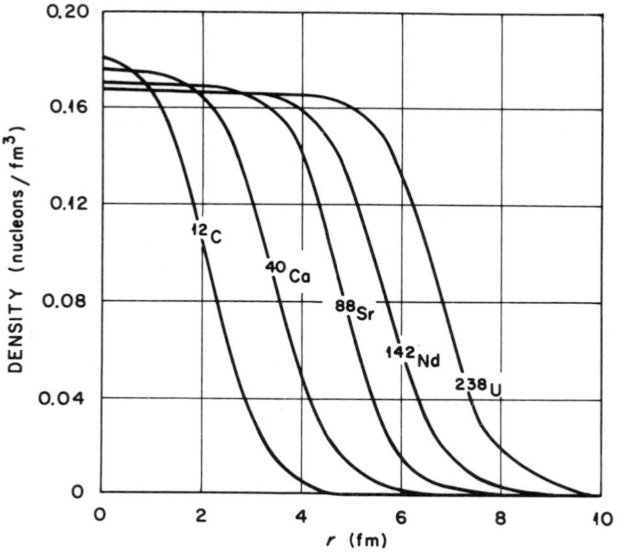

Figure 1.1 The distribution of mass in some typical nuclei as a function of the distance *r* from their centres

The surface rounding of the density distribution is found to be such that the density falls from 90% to 10% of its central value in about 2.5 fm. This rounding is not negligible even for a heavy nucleus and it can mean that a light nucleus like ^{12}C is almost all 'surface' because the radius of ^{12}C is only about 2.5 fm (*see Figure 1.1*). The neutron distribution has to be deduced from measurements of the total matter distribution, e.g. by α-particle scattering. It is similar to the proton distribution, as we would expect because there are strong attractive forces between neutrons and protons. Since $R \approx r_0 A^{1/3}$, the nuclear volume is proportional to A, the number of nucleons, so that the density in the nuclear interior is about the same in all nuclei. Further, each nucleon occupies a volume of about $(4/3)\pi r_0^3$, and on average the nucleons are separated from one another by about $2r_0$.

1.7.3 Spins, parities and moments

A nucleus may have total internal angular momentum, or *spin*, which is non-zero. We know the neutron and proton each has an intrinsic spin of $\frac{1}{2}\hbar$ (usually referred to as spin-$\frac{1}{2}$, the unit $\hbar$ being understood). A nucleus with odd mass number A then has a half-integral value of spin. Nuclei with even-A have integral spins; those with even-N and even-Z invariably have zero spin in their ground states, although their excited states will have non-zero spin in general. This internal angular momentum is the resultant of the intrinsic spins of the A constituent nucleons plus the orbital angular momentum associated with their motion around the centre of the nucleus. Further, each nuclear state is characterised by a definite parity, + or −, corresponding to whether the corresponding nuclear wave function is unchanged or changes sign upon reflection of the coordinates through the origin, $\mathbf{r} \to -\mathbf{r}$.

To a first approximation, nuclei are spherical. When a cloud of particles is held together by strong mutually attractive forces, the spherical shape tends to be the most favoured energetically. However, there are often small departures from the spherical shape and these can be of great importance. They can be described in terms of the *multipole moments* of the system (Ramsey, 1953; Brink and Satchler, 1968). Many nuclei (those with spin-1 or greater) are found to have electric *quadrupole* moments, the majority with a positive value (corresponding to an elongated or prolate shape for the charge distribution) but some are negative (corresponding to a compressed or oblate shape). *Hexadecapole* (2^4-pole) moments have been detected in some nuclei. Except for light nuclei, the deformation away from the spherical shape is not very great; the difference between the major and minor axes seldom exceeds 20 or 30% of the radius (*see Figure 1.2*).

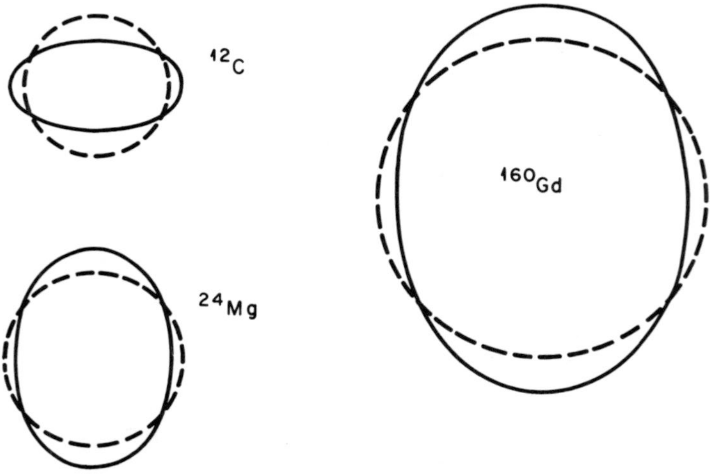

Figure 1.2 The shapes of some strongly deformed (non-spherical) nuclei. The dashed lines represent spheres with the same volume. ^{12}C is oblate (flattened), the other two are prolate (elongated), each with an axis of symmetry in the plane of the figure running from top to bottom

There are some very heavy nuclei for which the nearly spherical state is not the most stable but which gain binding energy by deforming to such an extent that they eventually fly apart into two separate fragments. This process is known as *fission* (Preston and Bhaduri, 1975).

1.7.4 Excited states

Just like atoms, nuclei can exist in excited states, which can be excited as a consequence of a nuclear reaction. The spacing between the lowest excited states ranges in order from 1 MeV in light nuclei to 100 keV or less in the heaviest

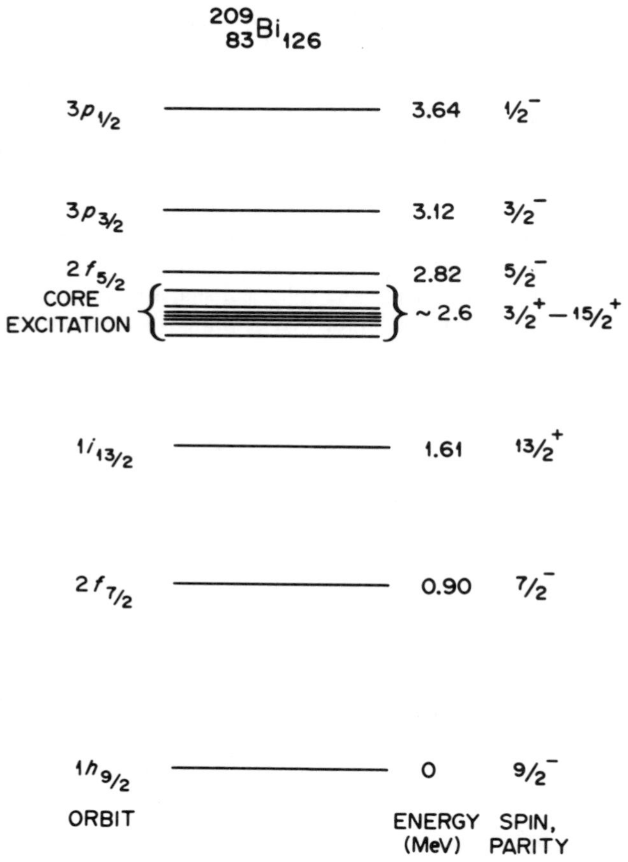

Figure 1.3 Energy levels of ^{209}Bi, interpreted as the single-particle (or shell model) states of the 83rd proton moving in a potential well due to the ^{208}Pb core. The group of states near 2.6 MeV correspond to this proton remaining in the $1h_{9/2}$ orbit but with the ^{208}Pb core excited to its 3^- state at 2.6 MeV. Vector addition of these two angular momenta results in states with spins from 3/2 to 15/2 with positive parity

nuclei. The density of these excited levels increases rapidly (exponentially) as the excitation energy is increased. Two particular kinds of excited states are of especial interest. If we have a nucleus with one 'valence' nucleon plus closed shells (in analogy with the single valence electron of the alkali atoms), there will be low excited states which correspond to moving this odd nucleon into higher shell model orbits without disturbing the inner closed shells. Such a spectrum is shown by ^{209}Bi (*Figure 1.3*). The excitation and identification of these single-particle levels is important for understanding the shell model of nuclear structure (Mayer and Jensen, 1955). A nucleus with a closed shell minus one nucleon will

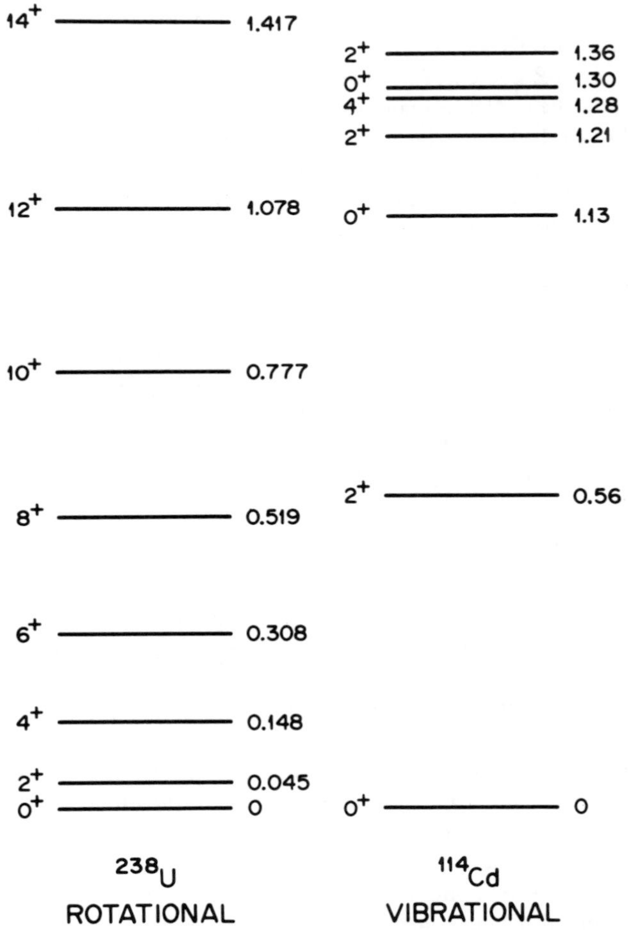

Figure 1.4 The energy levels of a typical rotational and a typical vibrational nucleus. The 0^+, 2^+, 4^+ triplet of states corresponding to the excitation of two quadrupole phonons appears just above 1 MeV in ^{114}Cd; additional 0^+ and 2^+ states due to a different kind of excitation are found nearby

exhibit analogous single 'hole' levels as the vacancy is moved into more deeply bound occupied orbits. These single-particle states correspond to the quantum orbits of a nucleon moving in an attractive potential 'well' whose shape corresponds roughly to the shape of the nuclear density distribution (*Figure 1.1*).

Other nuclei, between closed shells, may be sufficiently deformed away from the spherical shape that they have excited states corresponding to a well-defined ***rotational*** motion in which the deformed nucleus as a whole slowly rotates (like a diatomic molecule). This is characterised by an energy spectrum

$$E(I) = (\hbar^2/2\mathscr{I})\,[I(I+1) - I_0(I_0+1)]$$

where I is the spin of the level with excitation energy $E(I)$, I_0 is the spin of the ground state with $E(I_0) = 0$, and $\mathscr{I}$ is the moment of inertia of the nucleus. For an even-N, even-Z nucleus, $I_0 = 0$ and in most cases only even-I exist in the lowest band. The nuclei of many rare earth atoms and the very heavy atoms display this kind of spectrum (*Figure 1.4*). These states are called collective because many nucleons participate in the rotational motion. For the same reason, the radiative decay of one of these levels into a lower one is much faster than the decay between single-particle states, and the transitions themselves are also called collective.

Between the extremes of the single-particle states and the collective rotational states there is another category of collective states called ***vibrational***. These are visualised as harmonic oscillations in shape about a spherical mean; the quanta of these oscillations are called phonons and have an energy $\hbar\omega_\lambda$. They are characterised by the angular momentum they carry, primarily $\lambda = 2\hbar$ and $3\hbar$ for quadrupole and octupole phonons, respectively, with parity $(-1)^\lambda$. In an even-N, even-Z nucleus, the corresponding energy spectrum consists of evenly spaced levels with

$$E_{n(\lambda)} = n(\lambda)\hbar\omega_\lambda$$

where $n(\lambda) = 0, 1, 2, \ldots$ is the number of 2^λ-pole phonons. The $n(\lambda) > 1$ states are degenerate multiplets with more than one spin I; for example, two quadrupole phonons, $n(2) = 2$, each with an angular momentum of 2 units, can couple to resultant angular momenta $I = 0, 2$, and 4. Transitions between vibrational states are strong, but not as strong as in the rotational case. However, well-defined vibrational spectra, corresponding to a successively increasing number of phonons, are not observed for nuclei like they are for molecules. Almost all even-N, even-Z nuclei have a 2^+ first excited state and a low-lying 3^- state, both of which show strong or collective transitions to the ground state; except in the deformed nuclei whose spectra have the rotational form just discussed, these are often referred to as one-phonon vibrational states. In some nuclei, a triplet of states is observed having some of the properties of the two-quadrupole-phonon triplet (*Figure 1.4*); no states corresponding to two octupole phonons have been positively identified. It appears that these higher phonon states have become mixed with other kinds of excitation which occur with similar energies. Nonetheless, the vibrational pnonon picture remains a convenient first approximation in describing nuclear dynamics.

1.7.5 Time scales

A particle with mass M (in amu) and a kinetic energy E (in MeV) has a velocity of about

$$1.4 \times 10^7 \, (E/M)^{1/2} \text{ m s}^{-1}$$

A typical kinetic energy for a nucleon within a nucleus is 30 MeV, hence its velocity is 7.6×10^7 m s^{-1}. The circumference of the nucleus of an Sn atom is about 3×10^{-14} m, hence such a nucleon would make a complete orbit in 4×10^{-22} s or less. Thus a characteristic time for nucleon motions in the nucleus is 10^{-22} s. (This is much shorter than characteristic times for atomic motion. For example, the time for an electron to complete one Bohr orbit in the H atom is about 1.5×10^{-16} s, while the period of the D-lines in the optical spectrum of Na is about 2×10^{-15} s.)

A phonon for a quadrupole vibration of a nucleus has a typical energy of $\hbar\omega_2 \approx 1$ MeV, or a period of about 4×10^{-21} s. This is an order of magnitude slower than the characteristic nucleon orbiting time. A typical excitation energy for the lowest 2^+ state in the heavier even-N, even-Z rotational nuclei is 100 keV, corresponding to a rotational period of about 4×10^{-20} s which is another order of magnitude slower. Thus we can understand that after averaging over the rapid individual nucleon motions we may still be left with relatively slow drift motions which correspond to collective oscillations or rotations.

Nuclear excited states are unstable against decay, by emission of γ-rays or other radiation, to other states of lower energy. Such unstable states do not have a precise energy associated with them, but rather a small spread of energies determined by the Heisenberg uncertainty principle in the form

$$\Delta E \, \Delta t \geqslant \hbar$$

We may take for Δt the lifetime τ of the state and then ΔE will be the width Γ of the energy distribution. Numerically

$$\Gamma(\text{MeV}) = (6.6 \times 10^{-22})/\tau(\text{s})$$

A nuclear state which only lived long enough for a nucleon to make one orbit would have a large uncertainty in energy, $\Gamma \sim 1$ MeV, whereas a width of, say, $\Gamma = 1$ eV corresponds to $\tau = 6.6 \times 10^{-16}$ s, time enough for 10^6 orbital revolutions. We shall see later that states of both types can be observed in scattering measurements.

1.7.6 Relativity and nuclear physics

Often it is necessary to take into account the effects of the theory of relativity when considering atomic and electronic phenomena. How about in the description of nuclear physics? A typical kinetic energy of a nucleon within a nucleus is about 30 MeV or 1/30 of its rest mass; that is, $(v/c)^2 \simeq 1/15$ where v is its velocity. Hence relativistic corrections of a few per cent can be anticipated. Generally these are ignored because they are smaller than various experimental and theoretical uncertainties which are usually present; for example the forces acting between nucleons are not known very precisely, also the difficulty of solving the many-body problem leads to the use of various approximations which introduce errors into the calculations. Consequently a non-relativistic treatment is usually adequate for most nuclear-physics phenomena. There are some exceptions. In β-decay, the emitted electron may have an energy well in excess of its rest mass and the neutrino always travels with the speed of light; these leptons must be described relativistically using Dirac's equation but even so we may still use non-relativistic equations for the nucleons involved in the β-decay. Also, if in a nuclear-scattering experiment we have bombarding energies of several hundred or more MeV per nucleon of the projectile, we should describe the kinematics of the collision in a Lorentz-invariant way. In some cases, relativistic models of the reaction itself have been introduced.

REFERENCES

Barkla, C. G. (1911). *Phil. Mag.* Vol. 21, 648, and earlier papers

Bohr, N. (1913). *Phil. Mag.* Vol. 26, 1

Brink, D. M. and Satchler, G. R. (1968). *Angular Momentum*, 2nd edn. Oxford; Oxford University Press

Chadwick, J. (1932). *Nature*, Vol. 129, 312; *Proc. Roy. Soc.* Vol. A136, 692

Close, F. (1983). *The Cosmic Onion*. New York; American Institute of Physics

Cockcroft, J. D. and Walton, E. T. S. (1932). *Proc. Roy. Soc.* Vol. A137, 229

Curie, I. and Joliot, F. (1932). *Compt. Rend.* Vol. 194, 273

DeVries, R. M. and Clover, M. R. (1975). *Nucl. Phys.* Vol. A243, 528

Fermi, E. (1934). *Nature*. Vol. 133, 898

Frauenfelder, H. and Henley, E. M. (1975). *Nuclear and Particle Physics, A.* Reading, Mass.; W. A. Benjamin

Geiger, H. and Marsden, E. (1909). *Proc. Roy. Soc.* Vol. A82, 459

Geiger, H. and Marsden, E. (1913). *Phil. Mag.* Vol. 25, 604

Hahn, O. and Strassmann, F. (1939). *Naturwissenschaften.* Vol. 27, 11

Hughes, I. S. (1971). *Elementary Particles*. London; Penguin Books

Inglis, D. R. (1973). *Nuclear Energy: Its Physics and Its Social Challenge.* Reading, Mass.; Addison–Wesley

Kirk, W. T. (1986). *Elementary Particle Physics*. Washington, D.C.; National Academy Press

Mayer, M. G. and Jensen, J. H. D. (1955). *Elementary Theory of Nuclear Shell Structure.* London; Chapman and Hall
Murray, R. I. (1975). *Nuclear Energy*. Oxford; Pergamon
Preston, M. A. and Bhaduri, R. K. (1975). *Structure of the Nucleus*. Reading, Mass.; Addison-Wesley
Ramsey, N. F. (1953). *Nuclear Moments.* London; Chapman and Hall
Reeves, H. (1968). *Stellar Evolution and Nucleosynthesis*. London; Gordon and Breach
Rutherford, E. (1911). *Phil. Mag.* Vol. 21, 669
Rutherford, E. (1919). *Phil. Mag.* Vol. 37, 537
Truran, J. W. (1984). *Ann. Rev. Nucl. Part. Sci.* Vol. 34, 53
Wilkinson, D. H. ed. (1969). *Isospin in Nuclear Physics*. Amsterdam; North-Holland

EXERCISES FOR CHAPTER 1

1.1 Calculate the energy in MeV and the de Broglie wavelength in fm for the following cases:

(i) a car of mass 1000 kg moving with a speed of 100 km h^{-1};
(ii) a ball of mass 1/5 kg thrown at a speed of 50 km h^{-1};
(iii) a proton of mass 1.67×10^{-24} g accelerated to a speed of 4.4×10^7 m s^{-1}.

What energy would a proton have if it travelled at the same speed as the car or the ball?

1.2 A beam carrying a current of 1 mA of (i) doubly charged ^{16}O ions, (ii) α-particles, or (iii) protons, each with an energy of 140 MeV, is fully stopped in a target. The kinetic energy of the particles is converted to heat. Calculate in each case the rate at which heat is produced. What force is exerted on the target?

1.3 There is an accelerator which produces a beam of ^{32}S ions each of which has a kinetic energy of 200 GeV per nucleon. What velocity would a mass of 1 mg have if it had this kinetic energy? If this velocity were directed upwards, how high would the mass rise against the earth's gravitational pull? (This exercise illustrates the possibility of a single ion having a macroscopic amount of energy.)

1.4 Two $^{238}_{92}U$ nuclei are placed with their centres 15 fm apart so that they are almost touching. Calculate the potential energy due to the Coulomb repulsion of their electric charges. What force, expressed in kg weight, is required to keep them toge :r? If released, what would their acceleration be in terms of g, the acceleration due to gravity at the earth's surface ($g = 9.81$ m s^{-2})?

1.5 Derive an expression for the potential energy V_c (in MeV) of a point charge $Z_1 e$ in the Coulomb field due to a nucleus with a charge $Z_2 e$ distributed uniformly throughout a sphere of radius R. Plot this as a function of the distance r from the centre of the sphere. How is $V_c(r = 0)$ related to $V_c(r = R)$? What values does this Coulomb energy assume for a proton and a Pb nucleus ($Z_2 = 82$; assume $R = 7$ fm)? What is the average value of this Coulomb energy for a proton which may be found with equal probability throughout the sphere?

Discuss how the expression for the potential energy is modified if the particle with charge $Z_1 e$ is not a point but also a sphere with a finite radius. (For numerical results, *see*, for example, DeVries and Clover, 1975.)

1.6 What is the potential energy (in MeV) of two protons separated by the average distance between two adjacent nucleons in a nucleus (take this to be $r = 2$ fm) due to their electric charges? Compare this to the potential energy due to their gravitational attraction. (Take the gravitational constant to be $G = 6.7 \times 10^{-10}$ erg m g^{-2}.)

What would be the energy due to a nuclear potential of the form

$$-V\frac{e^{-r/a}}{r/a}$$

with $V = 40$ MeV, $a = 1.4$ fm?

How do these various energies change when the separation between the protons is reduced to 1 fm? Calculate the corresponding forces and compare them.

Compare these forces, in order of magnitude, to the forces between atoms in a solid. (Hint: typical atomic energies are of the order of eV and typical separations are of the order of Å = 10^{-10} m.)

1.7 By measuring their deflection by applied magnetic and electric fields, Rutherford deduced that the α-particles emitted by radium (actually Ra C′ or ^{214}Po) had a velocity of $v = 1.9 \times 10^7$ m s^{-1} and a charge-to-mass ratio $q/m = 4.8 \times 10^4$ C g^{-1}. He also observed that α-particles were emitted at the rate of 3.4×10^{10} s^{-1} g^{-1} of radium and when the charge was collected it was found to accumulate at the rate of 31.6 esu s^{-1} g^{-1}. Deduce the charge and the mass of these α-particles. How do these values compare with our present knowledge (*see* Appendix C)? Calculate the energy of the α-particles in MeV units.

1.8 According to the kinetic theory of gases, the mean kinetic energy of a molecule in a gas at a temperature T is $3/2\, k_B T$, where k_B is the Boltzmann constant (*see* Appendix C). What is the mean kinetic energy (in ergs and in eV) and the corresponding velocity of an He atom in a helium gas at room temperature? (Take room temperature to be 20°C.) At what temperature would this mean energy be the same as for the α-particles emitted from radium (*see* previous question)?

1.9 The incompressibility of nuclear matter is defined as

$$\chi = 9\rho^2 \frac{\partial^2 E}{\partial \rho^2}$$

evaluated at the normal density, $\rho = \rho_0$ say, for which E is a minimum, $\partial E/\partial\rho = 0$. Here E is the energy per nucleon of nuclear matter at a density of ρ nucleons per unit volume. This χ has a value of about 200 MeV. It is related to the usual bulk modulus of elasticity K by $K = (1/9)\rho\chi$.

Calculate K for nuclear matter and compare it with the bulk moduli of familiar materials such as steel. (Normal nuclear matter has a density of about 0.17 nucleon fm^{-3}.)

1.10 The concept of quadrupole and higher multipole moments appears in the classical theory of electrostatic potentials. If a given charge distribution $\rho(\mathbf{r})$ has axial symmetry, the potential ϕ outside the distribution can be expanded in terms of Legendre polynomials (*see* Appendix A3)

$$\phi(r, \theta) = \frac{1}{r} \sum_{n=0}^{\infty} \frac{a_n}{r^n} P_n(\cos\theta)$$

The constant a_0 is simply the total charge, a_1 is the dipole moment, a_2 is one-half the quadrupole moment Q, etc. The dipole moment a_1 vanishes for a system in a stationary state of definite parity. The quadrupole moment is given by (Brink and Satchler, 1968)

$$Q = 2a_2 = \int (3z^2 - r^2)\,\rho(\mathbf{r})d\mathbf{r}$$

Deduce an expression for Q for a total charge of Ze uniformly distributed over an ellipsoid with axial symmetry and major semi-axis a, minor semi-axis b.

Assuming the nucleus $^{160}_{64}\text{Gd}$ has an ellipsoidal shape and a quadrupole moment $Q = 760\ e\ \text{fm}^2$, deduce the ratio of its major and minor axes. (Assume that $R = 1.2\,A^{1/3}$ fm is the radius of a sphere with the same volume as the ellipsoid.)

(i) Take $(a/b) = 1 + \delta$, assuming $\delta \ll 1$.

(ii) Compare with the more precise value obtained using the relations $a = (1 + \epsilon)R$, $b = (1 - \frac{1}{2}\epsilon)R$, as is appropriate for a quadrupole shape.

1.11 The special theory of relativity says that the mass m of a free particle with velocity v and rest mass m_0 is $m = m_0\gamma$, while its total energy (rest energy m_0c^2 plus kinetic energy K) is given by mc^2. Here $\gamma = (1 - \beta^2)^{-1/2}$ with $\beta = (v/c)$. Show that to order β^2 we have

$$\frac{\delta m}{m_0} = \frac{K}{m_0c^2} \approx \tfrac{1}{2}\beta^2$$

where $\delta m = m - m_0$, the increase in mass due to the motion.

Show to order β^2 that K is related to the classical value of the kinetic energy, $K_{\text{class}} = \frac{1}{2}m_0v^2$, by

$$K = K_{\text{class}}\left[1 + \tfrac{3}{2}\frac{K_{\text{class}}}{m_0c^2}\right]$$

Alternatively, show that to order β^2 the relativistic velocity for a given kinetic energy K is less than what would be deduced classically by the factor

$$\left[1 - \frac{3}{4}\frac{K}{m_0c^2}\right]$$

At what kinetic energy does a particle have a mass equal to twice its rest mass?

2

Introduction to Nuclear Reactions

2.1 INTRODUCTION

When we wish to observe an object, we usually illuminate it with a beam of light. The light is then reflected, refracted, diffracted, absorbed, in various ways. By interpreting our measurements on the scattered light, we learn about the shape and size of the object and its other properties. For example, we may learn whether the object is transparent or opaque; if the former, we may measure its refractive index, if the latter, we can find its absorptivity, and so on. If we study further by using monochromatic light, we find these various properties vary with the wavelength (in particular, we may find the object is coloured; i.e. it preferentially scatters light of a particular wavelength). In some cases we may find more than just scattering or absorption of the light. The object may continue to re-emit light after the incident beam is removed (phosphorescence), it may emit radiation of a different wavelength from that used to illuminate it (fluorescence) or it may emit radiation of a different kind (e.g. electrons in the photo effect).

So it is with nuclear reactions. The light we use to illuminate a nucleus may also be electromagnetic radiation. More usually, it consists of the matter waves associated with a beam of energetic particles such as electrons, mesons, hyperons, neutrons, protons or other nuclei. These waves may also be reflected, refracted, diffracted and absorbed. The various processes which can occur are investigated by making scattering experiments.

If we wish to see the details of an object, it is necessary to illuminate it with radiation of wavelength λ which is shorter than the size of the object. Thus an optical microscope operating with visible light ($\lambda \sim 10^{-7}$ m) is useless for studying objects much smaller than about 1 μm (10^{-6} m). The de Broglie wavelength of the matter waves associated with a beam of particles with momentum p is $\lambda = h/p$. Electrons with a kinetic energy of $E = 1$ keV have $\lambda = h/(2m_eE)^{1/2} \sim 0.3 \times 10^{-10}$ m, so an electron microscope operating with 1-keV electrons can see much finer details than an optical microscope and can even resolve features of atomic or molecular dimensions. Nuclei have radii of about 5 fm and particles like neutrons and protons are about 10 times smaller; thus it is necessary to use radiation with a very short wavelength and hence a high energy to study details of nuclear or particle structure. A photon with a reduced wavelength $\lambda\!\!\!\!-= \lambda/2\pi =$ 1 fm has an energy of 197 MeV. The reduced de Broglie wavelength for a massive particle at non-relativistic energies is $\lambda\!\!\!\!- \approx 4.5/[m(\text{amu})E(\text{MeV})]^{1/2}$ fm, where m is its mass and E its kinetic energy, so that a proton with $\lambda\!\!\!\!- = 1$ fm has an energy of about 20 MeV. *Table 2.1* lists values of $\lambda\!\!\!\!-$ for several particles commonly used in scattering experiments.* A heavy particle has a smaller de Broglie wavelength than a light particle with the same kinetic energy E provided it is non-relativistic ($E \ll mc^2$, the rest energy). At very relativistic energies ($E \gg mc^2$) the particle momentum $p \approx E/c$ and the corresponding wavelength is more or less independent of the particle's rest mass, $\lambda\!\!\!\!- \approx 197/E(\text{MeV})$ fm.

TABLE 2.1 *Reduced de Broglie wavelengths* $\lambda\!\!\!\!-$, *in fm, for various particles and energies*

Energy	*Photon*	*Electron*	*Pion*	*Proton*	*α-Particles*	^{16}O	^{40}Ar	^{208}Pb
1 MeV	197	140	12	4.5	2.3	1.14	0.72	0.32
10 MeV	19.7	18.7	3.7	1.4	0.72	0.36	0.23	0.10
100 MeV	2.0	2.0	1.0	0.45	0.23	0.11	0.072	0.032
1 GeV	0.20	0.20	0.17	0.12	0.068	0.035	0.023	0.010

If the wavelength of the incident radiation is very short (and hence the energy or mass or both is very high), the correspondence principle of wave mechanics tells us that we may describe the collision in the particle language of classical mechanics. This classical description is often useful because of the physical insight it provides. However, it is well to remember that over the energy domain of most interest for nuclear physics, 0–1000 MeV say, quantal or wave effects are generally very important and quantum mechanics is required for a quantitative description of the phenomena.

2.2 THE CENTRE-OF-MASS COORDINATE SYSTEM

In the usual nuclear-reaction experiment we have target nuclei A at rest being bombarded by projectiles a. Usually, but not necessarily, the target A is heavier than the projectile a. Anyway, the same reactions could be induced by firing nuclei A at a target containing particles a. Hence it is convenient to use a reference frame which reflects this inherent symmetry of the scattering process. One

*Note that neutrons with the very low energy of, say, 1/40 eV, which is typical of the energies of the thermal neutrons obtainable from many nuclear reactors, have a wavelength $\lambda\!\!\!\!- \simeq 0.3 \times 10^{-10}$ m = 0.3 Å. Hence diffraction will occur on an atomic scale, such as Bragg reflection by crystals. Indeed, such slow neutrons are an important tool for the study of the structure of materials.

quantity we know to be conserved in a collision between two systems is their *total* momentum; our symmetry requirement is then satisfied by choosing a reference system in which this total momentum is zero. Referred to a coordinate system fixed in the laboratory (the LAB system) this means a system moving with the same velocity V_{CM} as the centre of mass of the colliding pair (hence called the CMS or centre-of-mass system). The kinetic energy, $\frac{1}{2}(M_A + M_a)V_{CM}^2$, associated with this centre-of-mass motion in the laboratory system is conserved and thus is not available for producing nuclear excitations, etc. In the CMS, this energy has been transformed away, the centre of mass is at rest and the two nuclei approach one another with equal and opposite momenta, $\mathbf{p}_A^{CM} = -\mathbf{p}_a^{CM}$ (*see Figure 2.1*). Similarly, if there are only two products, b and B, of a reaction, these separate in the CMS with equal and opposite momenta, $\mathbf{p}_B^{CM} = -\mathbf{p}_b^{CM}$. The kinematic consequences of this transformation from the LAB to the CM system are described in Appendix B. If the target A is much heavier than the projectile a, there is little difference between the two systems but for light targets the differences may be considerable. For example if a and A have equal masses, V_{CM} is one-half the velocity of the projectile in the LAB and one-half the bombarding energy is taken up in motion of the centre of mass. Further, the way the products are distributed in angle will be different in the two systems. In the elastic scatter-

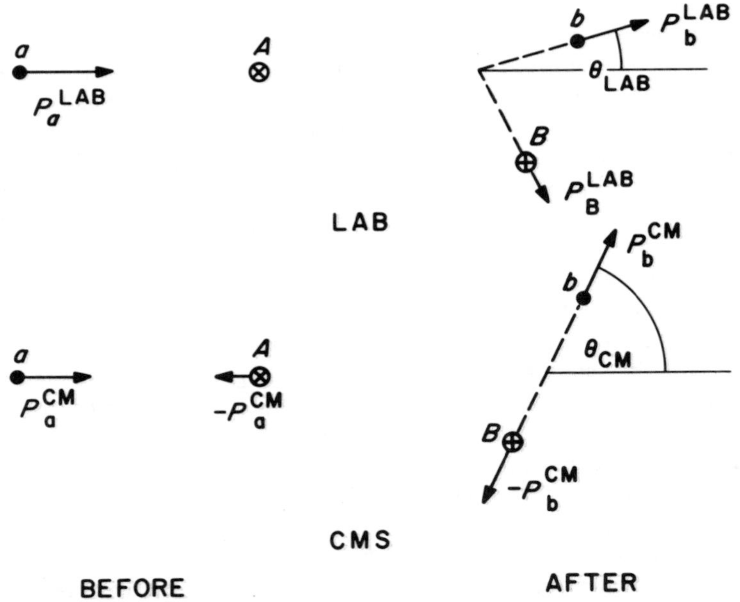

Figure 2.1 LAB and CMS coordinates for the reaction A(a, b)B. In the LAB system, the target A is at rest and the projectile a is incident with momentum p_a^{LAB}. The products b and B are shown moving forward. In the CMS, a and A approach one another with equal but opposite momenta and the products b and B separate with equal but opposite momenta

ing of two equal masses, conservation of momentum and energy ensures that the scattering angle θ_{LAB} (*Figure 2.1*) can never exceed 90°, whereas in the CMS all angles θ_{CM} up to 180° are possible. Indeed in the special case where the scattering intensity was proportional to cos θ_{LAB} in the LAB system we should find it to be isotropic, that is of equal intensity at all angles θ_{CM}, in the CMS (*see* Appendix B).

The CMS can also be used for scattering experiments at relativistic energies although it is often more convenient to use relativistic invariants, such as the square of the four-momentum transfer, which have the same value in all coordinate systems.

2.3 TYPES OF REACTION

Many different processes may take place when two particles collide. A typical nuclear reaction may be written

$$A + a \rightarrow B + b + Q \tag{2.1}$$

If A is the symbol for the target nucleus, a that for the projectile, while B is the residual nucleus and particle b is observed, this reaction would be written A(a, b)B. When specific isotopes are intended, the mass number is written as a superscript to the left of the chemical symbol in this expression. Special symbols are used for 'elementary' particles and the lightest nuclei; for example: e = electron, π = pion, p = proton, n = neutron, d = deuteron or ^2H, t = triton or ^3H, and α = alpha particle or ^4He. A photon or gamma ray is signified by γ. Further, either B or b or both may be left in excited states; sometimes the excitation energy is given as a subscript, or a state of excitation may be simply indicated by an asterisk, B*, etc.

The symbol Q in equation 2.1 refers to the energy released during the reaction; if the residual particles b and B are in their ground states, this is denoted Q_0. Since the total energy is conserved in all reactions, $Q \neq 0$ means that kinetic energy has been converted into internal excitation energy (or rest energy) or vice-versa, so that $Q = E_f - E_i$, where E_f is the total kinetic energy of the particles in the final state and E_i is the corresponding quantity in the initial state. If the Q-value, as it is called, is positive, the reaction is said to be exoergic (or exothermic), while a reaction with a negative Q-value is endoergic (or endothermic). In the latter case, a bombarding energy above a definite threshold value is required in order for the reaction to take place; in the CMS an energy E_i greater than $-Q$ is needed for $E_f > 0$.

Of course, although we shall refer mostly to these, we are not confined to just two particles b and B in the final state; three or more may result from the collision. If sufficient energy is available, it is possible for the two colliding systems to be broken up entirely into their constituents, although such an event is rather unlikely. When there is an appreciable number of reaction products, the collision is often called a spallation reaction.

There are several major classes of reactions.

(i) *Elastic scattering*: here b = a and B = A. The internal states are unchanged so that $Q = 0$ and the kinetic energy in the CMS is the same before and after the scattering. We have a + A → a + A; for example

$$n + {}^{208}Pb \rightarrow n + {}^{208}Pb, \quad \text{or} \quad {}^{208}Pb\,(n, n)\,{}^{208}Pb \qquad (2.2)$$

(ii) *Inelastic scattering*: this term most often means collisions in which b = a but A has been raised to an excited state, B = A* say, Consequently $Q = -E_x$, where E_x is the excitation energy of this state. Since a is then emitted with reduced energy, it is commonly written a′, $A + a \rightarrow A^* + a' - E_x$; for example

$$\alpha + {}^{40}Ca \rightarrow \alpha' + {}^{40}Ca^* \quad \text{or} \quad {}^{40}Ca\,(\alpha, \alpha')\,{}^{40}Ca^* \qquad (2.3)$$

If a is itself a complex nucleus, it may be left in an excited state instead of the target, or both may be excited through a *mutual excitation* process. An example of the latter is

$${}^{12}C + {}^{208}Pb \rightarrow {}^{12}C^* + {}^{208}Pb^* \quad \text{or} \quad {}^{208}Pb\,({}^{12}C, {}^{12}C^*)\,{}^{208}Pb^* \qquad (2.4)$$

(iii) *Rearrangement collision or reaction*: here b ≠ a and B ≠ A so that there has been some rearrangement of the constituent nucleons between the colliding pair (a transmutation). There are many possibilities; A + a → $B_1 + b_1 + Q_1$ or $\rightarrow B_2 + b_2 + Q_2$, etc. Some examples are:

$$p + {}^{197}Au \rightarrow d + {}^{196}Au^* \quad \text{or} \quad {}^{197}Au\,(p, d)\,{}^{196}Au^* \qquad (2.5a)$$

$$\alpha + \alpha \rightarrow {}^{7}Li + p \quad \text{or} \quad {}^{4}He\,(\alpha, p)\,{}^{7}Li \qquad (2.5b)$$

$${}^{32}S + {}^{54}Fe \rightarrow {}^{28}Si + {}^{58}Ni \quad \text{or} \quad {}^{54}Fe\,({}^{32}S, {}^{28}Si)\,{}^{58}Ni \qquad (2.5c)$$

$$\text{or} \quad {}^{32}S\,({}^{54}Fe, {}^{28}Si)\,{}^{58}Ni$$

The process 2.5a, for example, represents many reactions for there are many different energy levels in ^{196}Au which may be excited. The process 2.5c may be written several ways; the two shown correspond firstly to a beam of ^{32}S ions incident on a ^{54}Fe target and secondly to a beam of ^{54}Fe ions incident on a ^{32}S target; in both cases the emerging ^{28}Si is observed. This kind of symmetry is especially likely to occur in *heavy-ion reactions* where the projectile and target may be nuclei of comparable mass (the term 'heavy ion' conventionally denotes a projectile nucleus with a mass number greater than 4).

(iv) *Capture reactions*: this is a special case of class (iii); the pair A + a coalesce, forming a compound system in an excited state; this excitation energy is then lost by emitting one or more γ-rays, $A + a \rightarrow C + \gamma + Q$. For example

$$p + {}^{197}Au \rightarrow {}^{198}Hg + \gamma \quad \text{or} \quad {}^{197}Au\,(p, \gamma)\,{}^{198}Hg \qquad (2.6)$$

(v) *Other reactions*: we mean here reactions not included in equation 2.1 because there are more than two particles in the final state, such as

$A + a \rightarrow B + b + c + Q$. For example

$$\alpha + {}^{40}\text{Ca} \rightarrow p + \alpha' + {}^{39}\text{K} \quad \text{or} \quad {}^{40}\text{Ca}\,(\alpha, \alpha' p)\,{}^{39}\text{K} \tag{2.7}$$

The designations we have just described are not always applied rigidly. The term inelastic is sometimes used for any reaction other than elastic scattering; these processes may also be referred to as non-elastic. The name nuclear reaction may also be applied to any scattering process involving nuclear particles.

Each possible combination of particles may be called a *partition*. Each partition is further distinguished by the state of excitation of each nucleus and each such pair of states may be called a *channel*. The initial partition A + a, both in their ground states, constitutes the incident, or entrance, channel; the various possible sets of products in their various possible energy states become the exit channels. Thus there is an inelastic channel like equation 2.3 for each excited state of the target A and a reaction channel like equation 2.5 for each pair of excited states of the residual nuclei B and b. If a channel cannot be reached because there is not enough energy available ($Q < 0$ and $E_i < -Q$), it is said to be *closed. Open channels* are those which are energetically available.

2.4 ENERGY AND MASS BALANCE

The Q-value was defined above as the energy released in the reaction. Hence it is equal to the change in the sum of the kinetic energies of the colliding particles, $Q = E_f - E_i$. The Q-value can also be related to the rest masses of the particles through the relativistic relation $E = mc^2$. Consider the reaction A(a, b)B. If the rest mass of particle i is m_i

$$m_A + m_a = m_B + m_b + Q/c^2 \tag{2.8}$$

Alternatively, Q is equal to the change in the binding energies B_i of the particles

$$B_A + B_a = B_B + B_b - Q \tag{2.9}$$

(If i is an elementary particle, such as a nucleon, we regard $B_i = 0$ in this equation. Also Q appears here with a sign opposite of that in equation 2.8 because of the convention that binding energies are positive—*see* Chapter 1.) Hence an exoergic (positive Q) reaction results in systems more tightly bound than in the entrance channel, and an endoergic (negative Q) reaction results in less tightly bound systems.

The Q-value can be deduced from tables of masses or binding energies (e.g. Mattauch *et al.*, 1965); often, a measurement of the Q-value of a reaction is used to give the mass or binding energy of one of the particles if the others involved are known. As an example, consider the reaction 2.5b, $\alpha + \alpha \rightarrow {}^7\text{Li} + p$, leaving ${}^7\text{Li}$ in its ground state:

$$Q_0 = B_{\text{Li}} - 2B_\alpha = 39.245 - 2 \times 28.297 = -17.35 \text{ MeV}$$

Alternatively,

$$Q_0 = 2m_\alpha - m_{\text{Li}} - m_{\text{p}} = 2 \times 4.001506 - 7.014357 - 1.007276$$
$$= -0.018621 \text{ amu}$$
$$= -17.35 \text{ MeV}$$

This is an endoergic reaction and will not be initiated unless the sum of the kinetic energies of the two α-particles is greater than 17.35 MeV *in the CMS.* (It is not sufficient to have a ^{4}He target at rest bombarded with 17.35-MeV α-particles in the LAB system. Since the total momentum must be conserved, the two residual nuclei, ^{7}Li and p, will always recoil and thus carry some kinetic energy. In the CMS the total momentum is zero by definition and all the kinetic energy is available for excitation. In this simple case of two identical particles the CMS energy is exactly one-half the LAB bombarding energy; thus a LAB energy of at least 34.7 MeV is required for this reaction. At the threshold energy of 34.7 MeV, the ^{7}Li and p would be formed at rest in the CMS.) If we wish to leave ^{7}Li in an excited state, the Q-value will be even more negative and additional energy will have to be supplied.

2.5 OTHER CONSERVED QUANTITIES

Besides the total energy and momentum, various other quantities are conserved during a nuclear reaction. The total electric charge always remains constant. In the usual nuclear reaction in the absence of β-decay, the total number of neutrons and the total number of protons are separately conserved. (This last rule may be violated if mesons or hyperons are involved, either as products or as bombarding particles, for then nucleons may be transmuted into other particles or into each other. For example, the capture of μ-mesons by nuclei, which is analogous to the inverse of β-decay, may transmute a neutron into a proton or vice-versa. The capture of a π^--meson by a proton may produce a K-meson and a hyperon or a π^0 and a neutron. If the bombarding energy in a reaction is sufficiently high, mesons or hyperons may be produced. For example, the threshold for charged pion production is about 140 MeV in the CMS. An energy of about 2 GeV enables one to create a nucleon–antinucleon pair. However, in all these processes another quantum number, the total *baryon* number, remains constant.)

Two other important conserved quantities are *parity* and the total *angular* momentum. Any change in the total (vector sum) of the internal angular momenta (spins) of the nuclei must be compensated for by a corresponding change in the orbital angular momentum of their relative motion. Similarly, any change in the product of their intrinsic parities must be reflected in a change in the parity of their relative motion. We return to these matters in Chapter 3.

2.6 CROSS-SECTIONS

We need a quantitative measure of the probability that a given nuclear reaction will take place. For this we introduce the concept of a *cross-section* which we define in the following way. Consider a typical reaction A(a, b)B. If there is a flux of I_0 particles of type a per unit area incident on a target containing N nuclei of type A, then the number of particles b emitted per unit time is clearly proportional to both I_0 and N. The constant of proportionality is the cross-section, σ, and has the dimensions of area. Then the cross-section for this particular reaction will be

$$\sigma = \frac{\text{number of particles b emitted}}{\text{(number of particles a incident/unit area) (number of target nuclei within the beam)}}$$

A convenient unit of area for nuclear physics is the *barn* (symbol b: 1 barn = 10^{-28} m^2 = 100 fm^2) and cross-sections are usually given in barns or the sub-units millibarn, 1 mb = 10^{-3} b, and microbarn, 1 μb = 10^{-6} b, etc.

If we ask for the number of particles b emitted per unit time within an element of solid angle $d\Omega$ in the direction with polar angles (θ, ϕ) with respect to the incident beam, clearly this is proportional to $d\Omega$ as well as I_0 and N (*see Figure 2.2*). The constant of proportionality in this case is the *differential* cross-section, $d\sigma/d\Omega$. Since solid angles are dimensionless, this also has the dimensions of area.

In general, the probability of emission of b, and hence the differential cross-section, will depend upon the angles θ and ϕ. Only in special cases will the angular distribution be isotropic. To emphasise this, the differential cross-section $d\sigma/d\Omega$ is sometimes written as $d\sigma\,(\theta, \phi)/d\Omega$. However (unless the spins of one or more of the particles are polarised—*see* section 2.13), the scattering process is quite symmetrical about the direction of the incident beam which means that the differential cross-section cannot depend on the azimuthal angle ϕ. In that case, we may write it simply as $d\sigma\,(\theta)/d\Omega$.

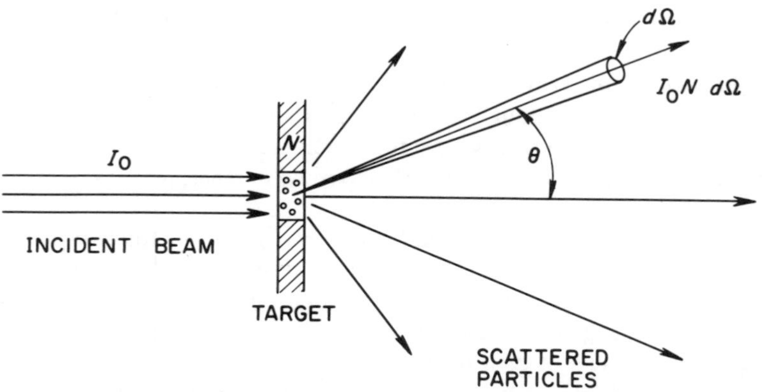

Figure 2.2 Diagram for the definition of differential cross-section. Usually the size of the irradiated area of the target is small, very much smaller than the distance to the detector, so that the scattering angle θ is well defined

Clearly the two kinds of cross-sections are related by

$$\sigma = \int_0^{4\pi} (\mathrm{d}\sigma/\mathrm{d}\Omega)\mathrm{d}\Omega \tag{2.10}$$

or, since the solid angle $\mathrm{d}\Omega = \sin\theta\,\mathrm{d}\theta\,\mathrm{d}\phi$

$$\sigma = \int_0^{\pi} \sin\theta\,\mathrm{d}\theta \int_0^{2\pi} \mathrm{d}\phi(\mathrm{d}\sigma/\mathrm{d}\Omega)$$

If there is no spin polarisation so that $\mathrm{d}\sigma/\mathrm{d}\Omega$ is independent of ϕ, this becomes

$$\sigma = 2\pi \int_0^{\pi} (\mathrm{d}\sigma/\mathrm{d}\Omega) \sin\theta\,\mathrm{d}\theta$$

Cross-sections are measures of probabilities. At a given bombarding energy, we may define a cross-section for each available set of states of each possible set of residual nuclei; that is, for each open channel. Since different channels correspond to nuclei in different energy states, there is no quantum interference between the corresponding probability amplitudes. We may simply add the cross-sections for different reaction channels. The sum of all these cross-sections for non-elastic processes is then called the *reaction* or *absorption* cross-section for the pair at that energy. When the elastic cross-section is added also, we have the *total* cross-section; it is a measure of the probability that *something* will happen during the collision.

To make the concept of cross-section more physical, consider the collision of two classical spheres (*Figure 2.3*). Let sphere 1 be at rest and let sphere 2 be impinging upon it. The two spheres will not collide unless their impact distance b is less than the sum of their radii, $b \leqslant R_1 + R_2$. The effect is the same as for the collision of a point particle with a disc of radius $R_1 + R_2$. The area of this disc, $\pi(R_1 + R_2)^2$, is the cross-section for the collision.

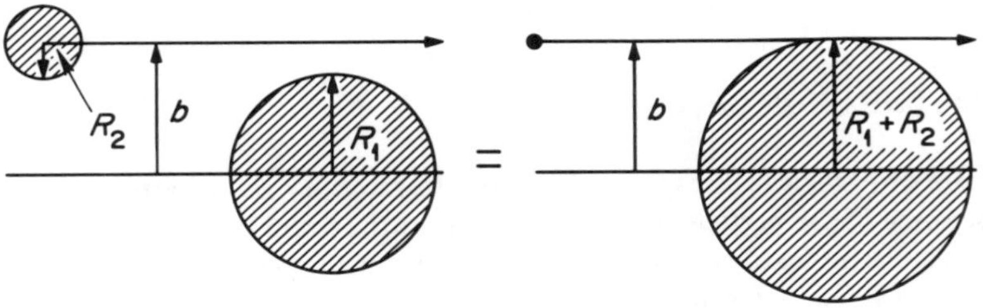

Figure 2.3 The collision of two spheres (left) has the same cross-section as a point particle incident upon a sphere whose radius is the sum $R_1 + R_2$ (right)

We learn one important feature from this picture. The cross-section is not a property of the target alone, but reflects properties of the projectile also. In our classical example, it is the *sum* of the radii which enters. A different projectile will give a different cross-section for the same target if its radius R_2' is different, and this must be taken into account if we wish to extract the radius of the target from the results of a scattering measurement. In addition, physical systems like nuclei do not have sharp edges. Their surfaces are diffuse and, as they approach one another, there is a transition region in which they are only partly in contact. The situation is further complicated by the fact that the forces acting between the systems have non-zero ranges. The interaction assumed in our classical picture is a contact interaction but actual nuclear forces act over finite distances. This has the same kind of blurring effect as the diffuse surface of the density distributions. The two effects make it impossible to completely characterise the interaction between two systems simply in terms of a radius. At least one other parameter is required, such as a range or a surface diffuseness. (The electrostatic or Coulomb force, which is inversely proportional to the square of the distance between the two systems, has a very long range and requires special treatment. As we shall see later (Chapter 3), its effects can be taken into account explicitly so that we can separate out the effects due to the short-ranged nuclear forces.)

As we have mentioned before, wave effects such as refraction and diffraction are important for actual nuclear scattering. Even with contact interactions and in the absence of surface diffuseness, these wave effects would modulate the cross-section so that only in an average sense (over a region of bombarding energy, say) would it be $\pi(R_1 + R_2)^2$. Nuclear structure effects may also make the cross-section vary with bombarding energy and even to exhibit sharp resonances—*see* section 2.7 below. Further, the cross-section for a particular reaction need not be comparable to $\pi(R_1 + R_2)^2$. Although the *total* cross-section for two strongly interacting systems is usually close to this geometric area, the probability of the interaction leading to a particular final state may be very much smaller—even zero—because of nuclear structure or other effects. If the systems are not strongly interacting, the cross-section can also be very small; for example, the cross-section for neutrinos scattering from a nucleus is almost vanishingly small because nuclei are almost transparent to neutrinos. At the opposite extreme, in a strongly resonant situation, such as sometimes occurs for slow neutrons on nuclei, the cross-section may become much greater than $\pi(R_1 + R_2)^2$.

2.7 ATTENUATION OF A BEAM

As a beam of particles passes through matter, its intensity will be attenuated because some of the particles are scattered out of the beam or induce reactions. Since even elastic scattering removes particles from the beam, it is the *total* cross-section which determines this attenuation. Consider a beam of intensity I_0 particles per unit time per unit area (*Figure 2.4*) incident upon a slab of material. At a depth x, the intensity is I. Let there be N target particles per unit volume in the material. The number of collisions per unit time per unit area in the slice of thickness dx is then proportional to the number of incident particles I and the number of target particles $N\,dx$. By definition the constant of proportionality is the total cross-section, σ_T. Then

$$dI = -\sigma_T I N\, dx$$

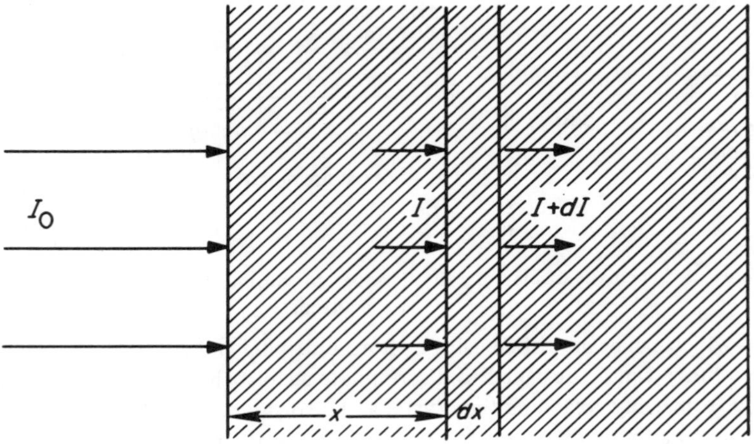

Figure 2.4 Diagram for the attenuation of a beam passing through a slab

Integrating and applying the initial condition $I = I_0$ at $x = 0$

$$I = I_0 \exp(-N\sigma_T x) = I_0 \exp(-x/\Lambda) \tag{2.11}$$

where $\Lambda = 1/N\sigma_T$ is the 'mean free path' between collisions. Clearly Λ is the distance over which the beam intensity falls by e^{-1}. The attenuation 2.11 may be measured in order to give a value for Λ and hence σ_T if N is known. This is known as a transmission experiment.

2.8 NUCLEAR SIZES FROM NEUTRON SCATTERING AND A SIMPLE TRANSMISSION EXPERIMENT

As probes of nuclei, neutrons have the advantage of not being charged so that there is no repulsive Coulomb potential to hinder their approach to a target nucleus. Neutrons with a kinetic energy of 10 MeV have a reduced wavelength $\lambda\!\!\!^{-} \approx 1.5$ fm, which is a fraction of a typical nuclear radius. The scattering or absorption of such fast neutrons is then a suitable way to study nuclear sizes. As we shall see later, the total cross-section σ_T for particles of wavelength $\lambda\!\!\!^{-}$ incident upon a strongly absorbing sphere of radius R is approximately

$$\sigma_T = 2\pi (R + \lambda\!\!\!^{-})^2 \tag{2.12}$$

Measurement of σ_T in a transmission experiment will then give a measure of the nuclear radius R. An experiment of this kind (suitable for an undergraduate laboratory) has been described by Fowler *et al.* (1969). The basic apparatus is shown in *Figure 2.5*. The Pu–α–Be source contains Pu and Be; the α-particles from the decay of the Pu induce an (α, n) reaction with the Be. This provides neutrons (typically $\sim 10^6$ s^{-1}) with a spread of energies with a maximum at about 10.7 MeV. Neutrons which pass through the target collide with protons in the organic material of the detector. The recoiling protons produce light

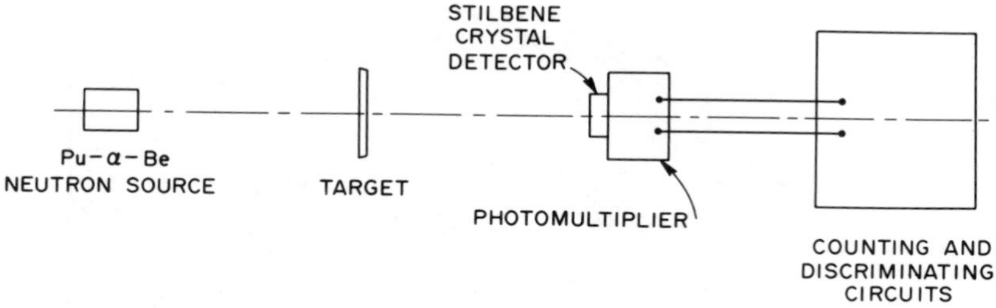

Figure 2.5 Schematic display of a simple apparatus for measuring the total cross-section for neutrons

pulses which can be amplified using a photomultiplier and counted electronically. The circuitry can be designed to discriminate against the background of pulses arising from γ-rays reaching the crystal and also can be biased to select only pulses from a known segment at the upper end of the neutron energy spectrum. Counting neutrons reaching the detector with and without the target in place tells us the attenuation produced by the target. If we know the target thickness and the number of nuclei per unit volume, equation 2.11 then tells us σ_T.

Some results from more sophisticated experiments (*see*, for example, England, 1974) with 14-MeV neutrons ($\bar{\lambda} \approx 1.2$ fm) are shown in *Figure 2.6*. The quantity $(\sigma_T/2\pi)^{1/2}$ is plotted against $A^{1/3}$, where A is the mass number of the target nucleus. According to equation 2.12, this is $(R + \bar{\lambda})$ so we see evidence that nuclear radii are proportional to $A^{1/3}$. The straight line is drawn for $R = 1.4 \times A^{1/3}$ fm; the measured values show some oscillations about this line. As we shall

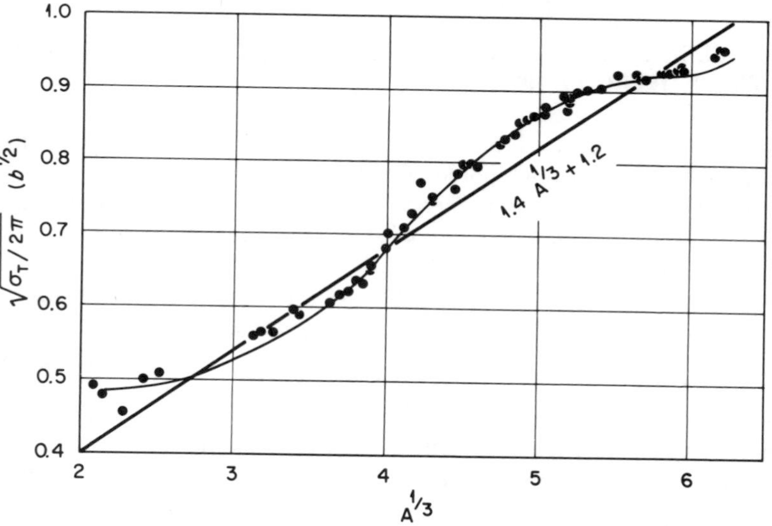

Figure 2.6 Total cross-sections σ_T for 14-MeV neutrons plotted against $A^{1/3}$ where A is the mass number of the target nucleus. The curve which follows the data was obtained from an optical model (*see* Chapter 4) which allows for nuclei to be partially transparent to neutrons

see later (Chapter 4), these oscillations arise because the nuclei are not perfectly black to neutrons. The proportionality constant of about 1.4 fm is appreciably larger than the value 1.1 fm found for the nuclear charge distribution from electron scattering, but this is due to the diffuse surface of the nuclei (the simple expression, equation 2.12, is for a sphere with a sharp edge) and the finite range of the nuclear forces.

2.9 A TYPICAL ACCELERATOR EXPERIMENT

There are many types of phenomena associated with nuclear reactions that one may measure, and many experimental arrangements for doing so (*see*, for example, England, 1974; also Cerny, 1974). We shall describe one such arrangement which is often used in order to illustrate some of the features of such measurements. It is shown in a very schematic form in *Figure 2.7*.

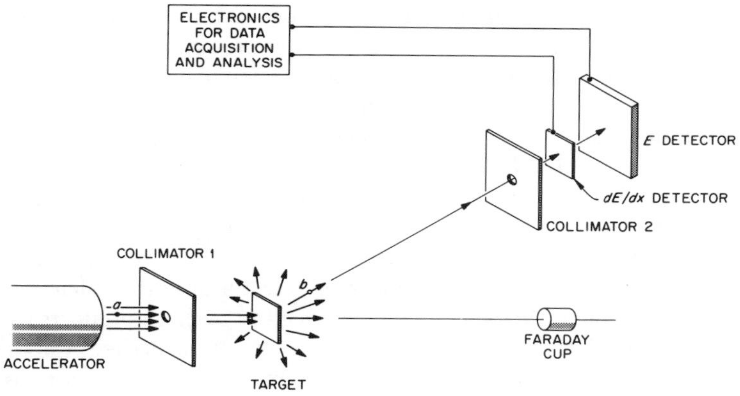

Figure 2.7 Schematic layout of the apparatus for a typical experiment using an accelerator. Compare with an actual set-up as shown in the frontispiece!

First, the accelerator provides a beam of charged particles of type a. (It is possible these have been passed through a bending magnet first in order to select ions of the required energy, etc.) Baffles and shielding (represented by the screen 1) help collimate the beam into a small spot on the target, and to remove stray particles. The target may be a thin film of the required material containing the nuclei A, or this may be deposited on a thin backing of some other material. The target must be thin if we are concerned with precise energy measurements. The beam ions can be scattered by the electrons of the atoms in the target and passage through a thick target will cause an undue spread in the energies of the beam particles. On the other hand, the thicker the target, the more target nuclei there are and the larger is the scattered intensity. Most experiments involve a compromise between these two requirements.

We must know the incident beam intensity as well as the scattered intensity.

For this purpose, and to monitor the uniformity of the beam, we need at least one device. One simple device shown is a Faraday cup in which the beam is collected for a known time, the charge is measured, and hence the current can be estimated.

The points discussed so far are common to most experiments. The mode of detection of the reaction products varies widely. We may wish to distinguish definite types of emitted particles b and to measure the distribution of their energies. One way to do this is shown. The scattered particles are collimated by screens and baffles (represented by screen 2), partly so as to define the scattering angle θ and partly to shield the detectors from background radiation. The scattered particle b first passes through a thin '$\mathrm{d}E/\mathrm{d}x$' counter. The degree of ionisation produced in this distinguishes the type of particle b. The second detector stops the particle b; the ionisation and hence the size of pulse produced tells us its kinetic energy. By registering these two types of pulses *in coincidence*, i.e. by only recording pairs of pulses which occur within a short predetermined time interval, we obtain an energy distribution or spectrum for each type of particle emitted from the target. By moving the detector system around, these measurements can be made on particles emitted at different angles to the incident beam direction and hence angular distributions can be obtained.

Obviously this is an oversimplified account of an actual experiment. There may also be many practical limitations. For example, it may not be possible to distinguish easily between two or more types of particle with the detector arrangement described; ^{3}He and ^{4}He ions of similar energy yield similar ionisation, for example. The length of time that a measurement can be continued may be limited by the stability of the accelerator or detector or both, which limitation may render difficult the accurate measurement of reaction products with low intensities. Invariably there will be nuclei other than the desired A in the target, both from impurities and from other elements necessary for the construction of the target. Some of these, notoriously carbon and oxygen, may give rise to very strong groups of emitted particles which may obscure the radiations one is attempting to measure.

2.10 COULOMB SCATTERING AND RUTHERFORD'S FORMULA

One of the most important cross-sections in physics is the differential cross-section for the scattering of two charged particles. The Coulomb force will cause scattering even in the absence of other, specifically nuclear, forces. Since all nuclei and the majority of elementary particles are charged, Coulomb scattering is a common experience. In a classic paper, Rutherford (1911) gave the formula for the differential cross-section for Coulomb scattering in the non-relativistic case. He used it to interpret his early experiments on the scattering of α-particles which provided evidence for the nuclear atom. Rutherford used classical mechanics, but it was found subsequently that the formula remains true in non-relativistic quantum mechanics (*see*, for example, Mott and Massey, 1965).

2.10.1 Classical derivation

The Coulomb force between a projectile with a mass m and a charge $Z_1 e$ ($Z_1 = 2$ for the α-particle) and a target nucleus with a charge $Z_2 e$ is $Z_1 Z_2 e^2/r^2$, where r is the distance between them. The force is repulsive if the two particles have charges of the same sign. The corresponding potential is $Z_1 Z_2 e^2/r$. If the target is much heavier than the projectile, we may consider it to remain at rest while the projectile describes an orbit which is one branch of a hyperbola (*Figure 2.8*).

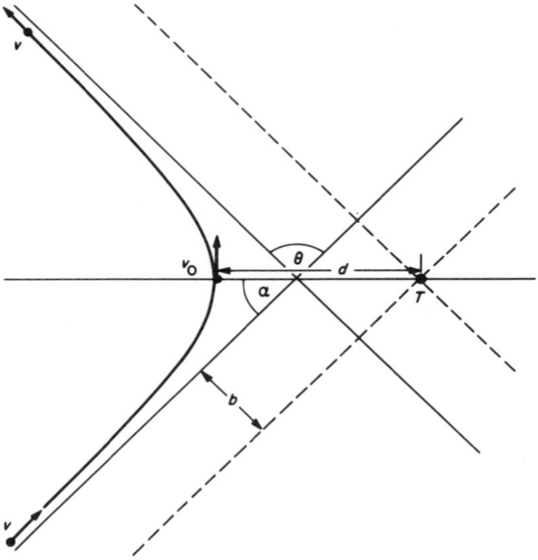

Figure 2.8 Coordinates for describing Rutherford scattering of a charged particle by a target T with a charge of the same sign

Let b be the distance of the target T from the asymptote of the hyperbola; b is called the impact parameter. Let d be the distance of closest approach of the orbit to the target. Further, let the projectile velocity be v at a very great distance from the nucleus where the potential is negligible and let its velocity be v_0 at the point of closest approach, $r = d$. Then conservation of energy gives

$$\tfrac{1}{2} mv^2 = \tfrac{1}{2} mv_0^2 + \frac{Z_1 Z_2 e^2}{d}$$

or

$$\left(\frac{v_0}{v}\right)^2 = 1 - \frac{d_0}{d}, \; d_0 = \frac{2Z_1 Z_2 e^2}{mv^2} = \frac{Z_1 Z_2 e^2}{E} \qquad (2.13)$$

where $E = \frac{1}{2} mv^2$ is the bombarding energy. (Note that $e^2 = 1.44$ MeV fm.) We see that d_0 is the distance of closest approach for which $v_0 = 0$, i.e. a head-on collision ($b = 0$). Conservation of angular momentum implies

$$mvb = mv_0 d$$

Hence with equation 2.13 we have a relation between b and d

$$b^2 = d(d - d_0) \tag{2.14}$$

It is a property of the hyperbola that

$$d = b \cot\left(\frac{\alpha}{2}\right);$$

with equation 2.14 we soon find $\tan \alpha = 2b/d_0$ or, since $\theta = \pi - 2\alpha$

$$\cot\left(\frac{\theta}{2}\right) = \frac{2b}{d_0} \tag{2.15}$$

Equation 2.15 gives the scattering angle θ as a function of the impact parameter b; note that θ increases as b decreases, until $\theta = \pi$ as $b = 0$, the head-on collision (*Figure 2.9*).

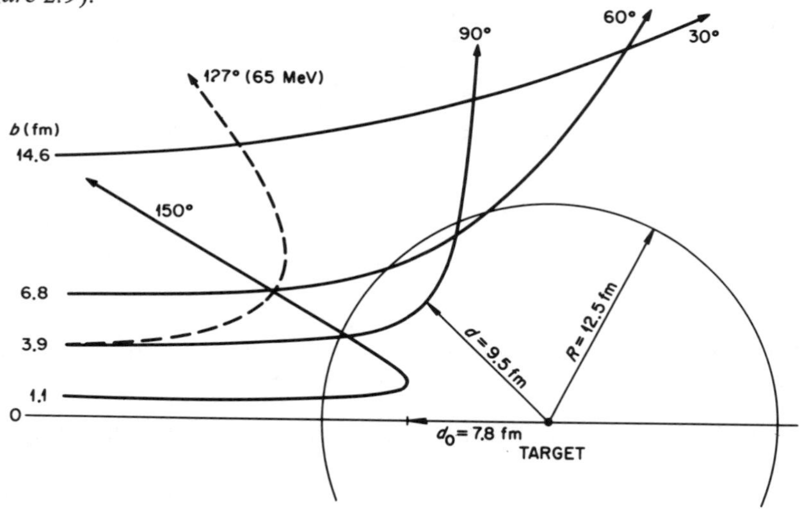

Figure 2.9 Some typical Rutherford (classical) orbits for a charged particle scattered by a charged target for different impact parameters b. The curves are drawn for ^{16}O ions with energy 130 MeV incident on ^{208}Pb, except that the dashed curve for $b = 3.9$ fm is for an energy of 65 MeV. The circle has a radius $R = 12.5$ fm which is equal to the distance of closest approach at which the nuclear forces begin to act for this system. We see that the orbit with $b = 6.8$ fm barely touches this region but orbits with $b < 6.8$ fm approach more closely. In practice the nuclear forces would distort these latter orbits and the scattering would deviate from Rutherford for scattering angles $\theta \gtrsim 60°$

Consider a flux of I_0 particles per unit area and unit time crossing a plane perpendicular to the beam (*Figure 2.10*). The flux passing through the annulus with radii b, $b + \mathrm{d}b$ is

$$\mathrm{d}I = 2\pi I_0 b \mathrm{d}b$$

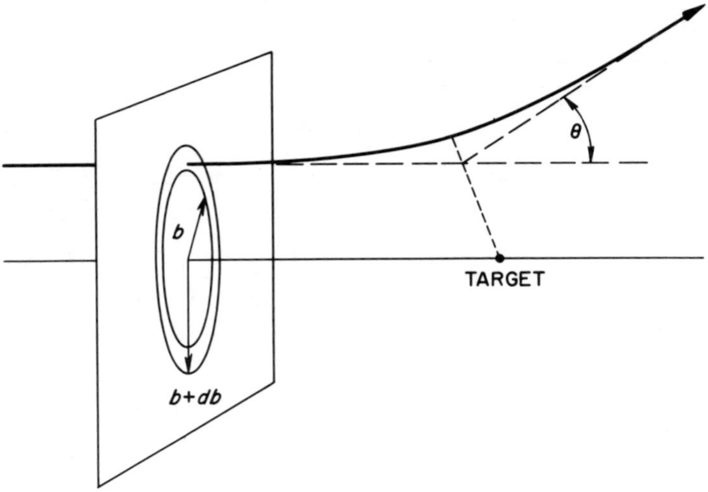

Figure 2.10 Coordinates for defining the Rutherford cross-section

which from equation 2.15 may be written

$$\mathrm{d}I = \tfrac{1}{4}\,\pi I_0\, d_0^2 \frac{\cos\left(\dfrac{\theta}{2}\right)}{\sin^3\left(\dfrac{\theta}{2}\right)}\,\mathrm{d}\theta$$

This is the flux scattered between the cones with angles θ and $\theta + \mathrm{d}\theta$, which enclose a solid angle $\mathrm{d}\Omega = 2\pi \sin\theta\,\mathrm{d}\theta$. Thus the differential cross-section (section 2.6) is

$$\frac{\mathrm{d}\sigma}{\mathrm{d}\Omega} \equiv \frac{1}{I_0}\frac{\mathrm{d}I}{\mathrm{d}\Omega} = \left(\frac{d_0}{4}\right)^2 \frac{1}{\sin^4\dfrac{\theta}{2}} = \left(\frac{Z_1 Z_2 e^2}{4E}\right)^2 \frac{1}{\sin^4\dfrac{\theta}{2}} \tag{2.16}$$

This is Rutherford's formula for Coulomb scattering. We see that the *angular distribution* has a universal form* (*Figure 2.11*), while the *magnitude* of the cross-section depends only upon the product $Z_1 Z_2$ of the charges of the two particles and the bombarding energy $E = \frac{1}{2}mv^2$.

The condition that the target mass be much greater than that of the projectile may be relaxed. It turns out that equation 2.16 is still true if we take account of the recoil of the target; we only have to reinterpret E, the energy, and θ, the scattering angle, as being measured in the centre-of-mass system, E_{CM} and θ_{CM} respectively (section 2.2 and Appendix B), then equation 2.16 gives the cross-section in the CMS.

*An apparatus and procedure for demonstrating this result which is suitable for an undergraduate teaching laboratory has been described by Ramage *et al.* (1975).

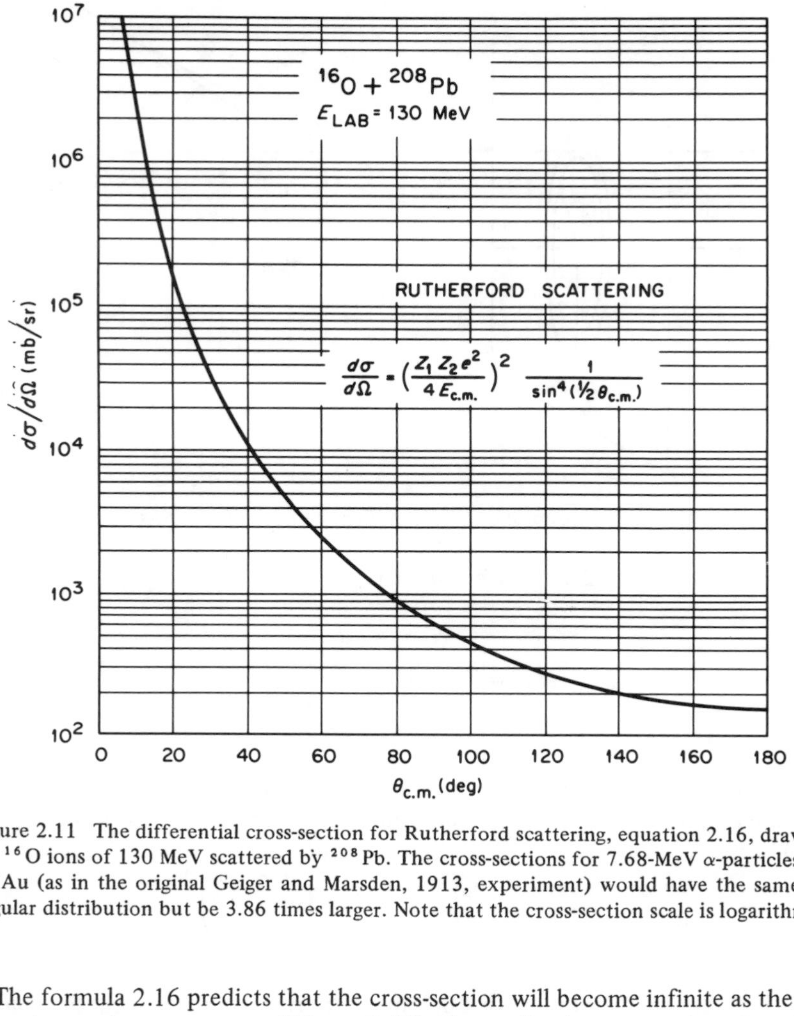

Figure 2.11 The differential cross-section for Rutherford scattering, equation 2.16, drawn for ^{16}O ions of 130 MeV scattered by ^{208}Pb. The cross-sections for 7.68-MeV α-particles on ^{197}Au (as in the original Geiger and Marsden, 1913, experiment) would have the same angular distribution but be 3.86 times larger. Note that the cross-section scale is logarithmic

The formula 2.16 predicts that the cross-section will become infinite as the scattering angle goes to zero (*Figure 2.11*). Physically this means that the Coulomb potential has a very long range so that even particles with very large impact parameters are deflected slightly by it. If we integrate equation 2.16 to obtain the total cross-section, this is also infinite. In practice, the differential cross-section cannot be measured for *very* small angles; the beam of incident particles has a finite width so that there is an upper bound on the impact parameter b, and those particles scattered in the most forward direction cannot be distinguished from the beam itself. Further, the charge of the target nucleus will usually be screened by atomic electrons so that the potential felt at distances of the order of atomic dimensions is no longer Coulombic.

2.10.2 *Quantum and relativistic effects*

The derivation outlined above is based upon classical mechanics. It is valid if wave effects can be neglected; that is provided the wavelength $\bar{\lambda}$ is small enough. The precise condition is that $\bar{\lambda}$ should be small compared to half the distance of closest approach for a head-on collision

$$\bar{\lambda} \ll \tfrac{1}{2} d_0 = \frac{Z_1 Z_2 e^2}{2E_{\mathrm{CM}}} \tag{2.17}$$

Usually, because of wave effects, the scattering formulae derived using classical and quantum mechanics are different even for the same interaction potential, but by some mathematical chance Rutherford's formula for the scattering of two electrically charged particles remains valid in non-relativistic quantum mechanics.

At sufficiently high energies, where the kinetic energy E becomes comparable to the energy of the rest mass, mc^2, relativistic effects must be taken into account. This is very often the case for electrons whose rest energy, $m_e c^2$ is small, $\simeq 0.5$ MeV. The cross-section for scattering high-energy electrons has to be obtained from Dirac's relativistic wave equation (*see*, for example, Uberall, 1971). There is a simple expression for the scattering from a point charge Ze, provided Z is small compared to the fine structure constant, $Z \ll (\hbar c/e^2) = 137$. Then

$$\frac{\mathrm{d}\sigma}{\mathrm{d}\Omega} = \left(\frac{Ze^2}{2E\beta^2}\right)^2 \left(1 - \beta^2 \sin^2 \frac{\theta}{2}\right) \frac{1}{\sin^4\left(\frac{\theta}{2}\right)}$$

where $\beta = v/c$, the ratio of the electron's speed to the speed of light. Note that here E represents the relativistic expression for the kinetic plus the rest energy of the electron

$$E = \frac{m_e c^2}{(1 - \beta^2)^{1/2}}$$

In the limit of low energy, $\beta \to 0$ and $E\beta^2 \to m_e v^2$, so we regain the Rutherford formula 2.16. At extreme relativistic energies $\beta = 1$ and we have the *Mott formula*

$$\left(\frac{\mathrm{d}\sigma}{\mathrm{d}\Omega}\right)_{\mathrm{Mott}} = \left(\frac{Ze^2}{2E}\right)^2 \frac{\cos^2\left(\frac{\theta}{2}\right)}{\sin^4\left(\frac{\theta}{2}\right)} \tag{2.18}$$

The additional angle dependence, $\cos^2(\theta/2)$, compared to the Rutherford expression which arises from the relativistic treatment, is not negligible.

2.10.3 Extended particles

Strictly, Rutherford's formula is only true for the scattering of *point* charges. However, it also describes the scattering of particles with extended charge distributions provided they do not approach one another too closely. Consider the scattering of a nucleus with radius R_1 and charge Z_1e by another with radius R_2 and charge Z_2e. The interaction potential when their centres are separated by r is

$$\frac{Z_1Z_2e^2}{r} \qquad \text{if } r \geqslant R_1 + R_2 = R$$

and is often represented by (see Exercise 1.5) (2.19)

$$\frac{Z_1Z_2e^2}{2R}\left(3 - \frac{r^2}{R^2}\right) \qquad \text{if } r \leqslant R_1 + R_2 = R$$

If the distance of closest approach for a head-on collision, equation 2.13, is greater than the sum of the two radii $d_0 > R$, then only the Coulomb force will be experienced and the Rutherford formula will hold. This means the bombarding energy should be less than $Z_1Z_2e^2/R$; this critical energy is often referred to as the *Coulomb barrier*. At energies greater than this, the scattering will deviate from equation 2.16. The scattering angle θ and the distance of closest approach d are related by equations 2.13–2.15

$$\cot\left(\frac{\theta}{2}\right) = 2\left[\frac{d}{d_0}\left(\frac{d}{d_0} - 1\right)\right]^{1/2} \qquad (2.20)$$

Deviations will occur for those θ for which $d < R$. Since the smaller d result in the larger θ (*Figure 2.9*), the deviations first appear at $\theta = 180°$ and then at progressively smaller angles as the energy is increased.

In Geiger and Marsden's original experiment of 1913, α-particles with $E = 7.68$ MeV from an RaC (or ^{214}Po) source were scattered from gold. Their results agreed with the Rutherford formula for all scattering angles. Since $d_0 = 3 \times 10^{-14}$ m = 30 fm in this case ($Z_1 = 2, Z_2 = 79$), they concluded that the gold nucleus had a radius of less than 30 fm. (It is now known that the radius of the gold nucleus is about 7 fm; the α-particles would need an energy of over 30 MeV to penetrate to $d_0 = 7$ fm.)

If there are also nuclear forces acting between the two particles, the potential 2.19 will be further modified. The nuclear forces are known to have a short range δ, of the order of 1 fm, so they are not felt for separations r which are much greater than $R + \delta$. At larger values of r, the potential 2.19 is still valid; at smaller values it is modified by the addition of the (attractive) nuclear potential. Again the scattering will be described by Rutherford's formula if $d_0 > R + \delta$, but deviations will occur if d_0 is less than this amount. (This discussion assumes that the particles can be precisely localised; it is valid if the condition 2.17 is satisfied. Otherwise wave effects will modify the results.) *Figure 2.12* shows the differential cross-sections for O and C ions scattering from various target nuclei,

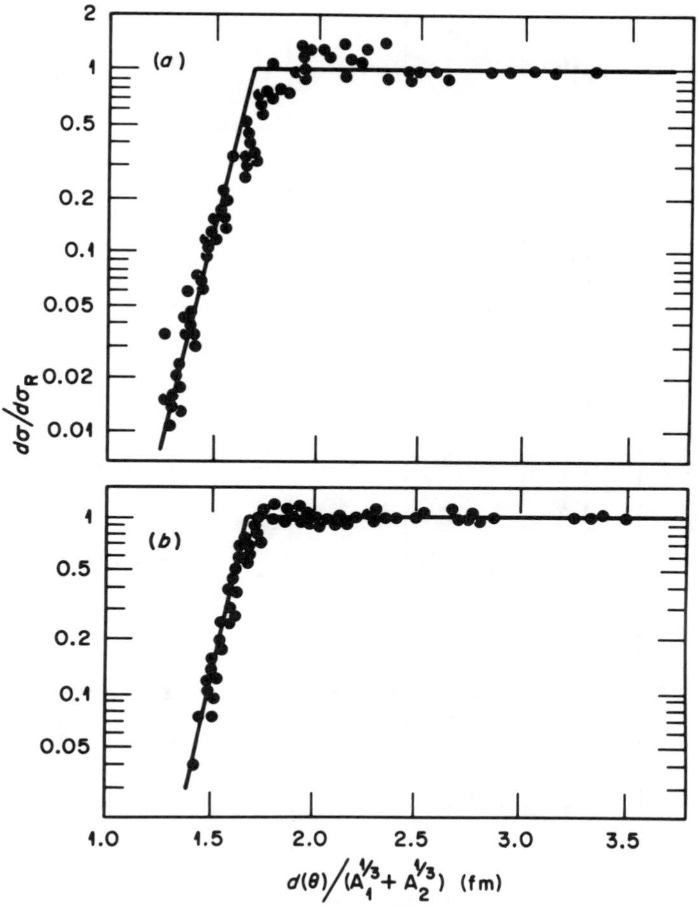

Figure 2.12 The differential cross-sections for O and C ions, in ratio to the Rutherford ones, scattering from various targets plotted against the distance of closest approach d, instead of the scattering angle θ, by using equation 2.20. The distance d has been divided by $(A_1^{1/3} + A_2^{1/3})$, where A_i is the mass number of nucleus i. The measured cross-sections then fall on a universal curve, showing that nuclear radii are approximately proportional to $A^{1/3}$. (a) $^{16}O + {}^{40,48}Ca$ at 49 MeV, $^{16}O + {}^{40,48}Ca$, ^{50}Ti, ^{52}Cr, ^{54}Fe, ^{62}Ni at 60 MeV and $^{18}O + {}^{60}Ni$ at 60 MeV; (b) $^{12}C + {}^{96}Zr$ at 38 MeV, $^{16}O + {}^{96}Zr$ at 47, 49 MeV, $^{16}O + {}^{88}Sr$, ^{93}Zr at 60 MeV and $^{18}O + {}^{90}Zr$ at 60, 66 MeV. (After Christensen *et al.*, 1973)

plotted versus the distance of closest approach d instead of the scattering angle, using the relation 2.20. The cross-section is given by the Rutherford value until the two nuclei approach to within a distance of about $1.7(A_1^{1/3} + A_2^{1/3})$ fm, when the cross-section begins to fall rapidly to zero. This decrease occurs because the two nuclei have surmounted their mutual Coulomb barrier and come under the influence of the nuclear forces. These induce non-elastic reactions and the

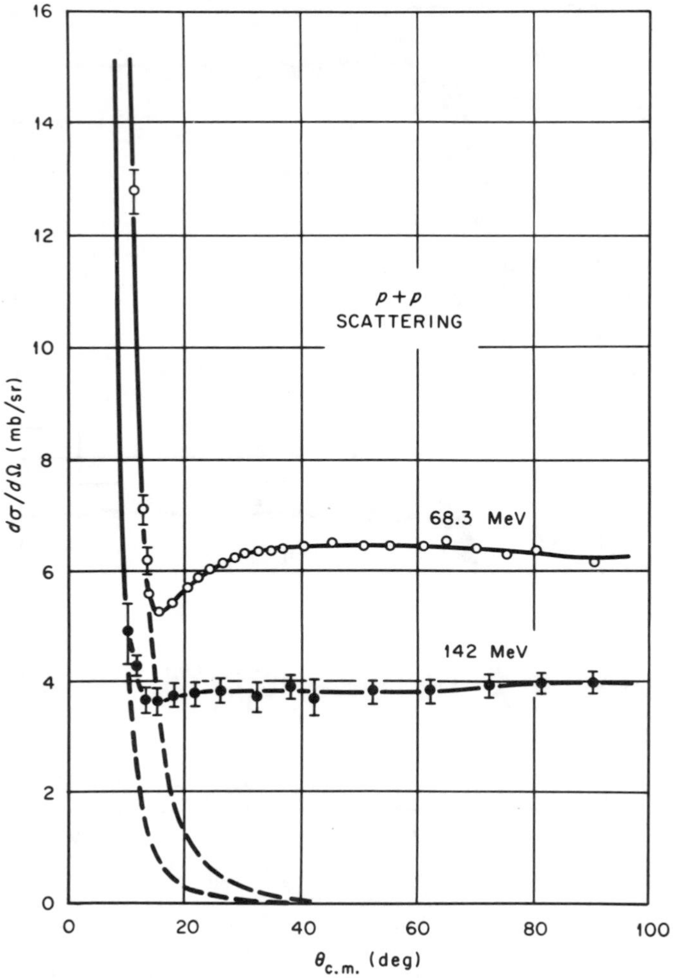

Figure 2.13 Differential cross-sections for protons with LAB energies of 68 and 142 MeV scattering from protons. The dashed curves show the Rutherford cross-sections. The measured values follow the Rutherford curves at small angles but show large deviations at larger angles where the nuclear force is important. The bars attached to the data points indicate the experimental uncertainties. (After Breit *et al.*, 1960)

nuclei are removed from the elastic channel; they are said to be ***absorbed***, in analogy to the absorption of light by a black object.

The differential cross-section for the elastic scattering of charged particles is always dominated by Rutherford scattering at small angles. ***Figure 2.13*** shows the cross-sections for the scattering of protons by protons with CMS energies of 34 and 71 MeV. The cross-sections predicted by the Rutherford formula are shown as dashed lines. For small angles the measured values become large and agree with equation 2.16, but for larger angles the cross-section becomes bigger than the Rutherford value because of the nuclear force. (The preceding semi-classical arguments cannot be used to deduce the 'size' of the proton, i.e. the range of this force, because the condition 2.17 is not satisfied; at 34 MeV, for example, $\lambda\!\!\!^{-} \approx 1$ fm while $d_0 \approx 0.04$ fm.) Because of this importance of the Coulomb scattering, measured differential cross-sections are often expressed in ratio to the Rutherford values, equation 2.16, instead of being given absolutely.

2.10.4 *Classical relations for Coulomb orbits*

It is often useful to think of heavy charged particles following classical orbits in the Coulomb field of the target even when quantum mechanics is required for a detailed description of the scattering. This is a valid picture if condition 2.17 is satisfied, which it often is, especially for heavy ions which are massive and carry large charges. We give here some simple and useful relations between the quantities involved; although Planck's constant h appears, this simply represents a convenient choice of units and does not imply the use of quantum mechanics.

It has been found convenient to define a quantity called the *Sommerfeld parameter*

$$n = \frac{Z_1 Z_2 e^2}{\hbar v} \approx \frac{Z_1 Z_2}{6.3}\left(\frac{m(\mathrm{u})}{E(\mathrm{MeV})}\right)^{1/2} \tag{2.21}$$

where $\hbar = h/2\pi$. The second, numerical, form holds if the projectile mass is measured in amu and the bombarding energy in MeV. In terms of n, the condition 2.17 becomes

$$n \gg 1 \tag{2.22}$$

Also

$$d_0 = 2n\lambda\!\!\!^{-} = 2n/k, \quad k = 1/\lambda\!\!\!^{-} \tag{2.23}$$

where

$$k = 1/\lambda\!\!\!^{-} = (2mE/\hbar^2)^{1/2} \approx 0.22\,(m(\mathrm{u})E(\mathrm{MeV}))^{1/2}\ \mathrm{fm}^{-1} \tag{2.24}$$

The angular momentum of the projectile about the target is

$$L\hbar = mvb, \quad \text{or} \quad L = b/\lambda\!\!\!^{-} = kb \tag{2.25}$$

Then equation 2.15 relates the scattering or deflection angle of the orbit to the angular momentum (in units of $\hbar$) of the particle following that orbit

$$L = n \cot\left(\frac{\theta}{2}\right) \tag{2.26}$$

With equation 2.14 we also find a relation between the angular momentum and the distance of closest approach d of the orbit

$$L^2 = \rho(\rho - 2n), \quad \text{or} \quad \rho = n + (L^2 + n^2)^{1/2} \tag{2.27}$$

where $\rho = kd = d/\lambda\!\!\!^{-}$. Of course, if $L = 0$, then $\rho = 2n$ or $d = d_0$. Combining equations 2.26 and 2.27, we also have

$$\rho = n\left(1 + \operatorname{cosec}\frac{\theta}{2}\right), \quad \text{or} \quad \sin\frac{\theta}{2} = \frac{n}{\rho - n} \tag{2.28}$$

2.11 ELECTRON SCATTERING

In section 2.10.3 we discussed the departures from the Rutherford formula when the charges of two heavy particles scattering from one another are not concentrated at points but are distributed over finite regions. Such deviations also occur for electrons scattering from nuclei. By measuring these, we obtain information about the distribution of charge within the nucleus. (It is sufficient to regard the *electron* as a point charge.) Because the electron rest mass is small ($m_e c^2 \approx 0.5$ MeV), wave effects are generally important for electron scattering, so we must have electrons whose de Broglie reduced wavelength $\lambda\!\!\!^{-}_e$ is smaller than or at least comparable to the nuclear size. This occurs for energies $E \gtrsim 50$ MeV since

$$\lambda\!\!\!^{-}_e \approx [197/E(\text{MeV})] \text{ fm}$$

when $E \gg m_e c^2$. In order to see fine details in the charge distribution, which we expect to occur over distances of order 1 fm, we need $\lambda\!\!\!^{-} \lesssim 1$ fm or $E \gtrsim 200$ MeV.

The observed differential cross-section may be expressed as the product of the Mott formula 2.18 for scattering by a point charge, times the square of a quantity called the *form factor*

$$\frac{d\sigma}{d\Omega} = \frac{d\sigma}{d\Omega}_{\text{Mott}} [F(q)]^2 \tag{2.29}$$

where q represents the momentum transferred when the electron is scattered through an angle θ, $q = 2k_e \sin(\theta/2) = (2/\lambda\!\!\!^{-}_e)\sin(\theta/2)$. Thus $F(q)$ is a measure of the deviation from point charge scattering and carries the information about the actual charge distribution. If it is adequate to use the Born approximation (*see* Chapter 3), $F(q)$ becomes the Fourier transform of the charge distribution $\rho_{ch}(r)$ of the target nucleus

$$F(q) = \frac{4\pi}{qZ}\int_0^\infty \rho_{ch}(r) \sin(qr) r \, dr \tag{2.30}$$

Here $\rho_{ch}(r)$ is normalised so that $F(q = 0) = 1$, namely

$$4\pi \int \rho_{ch}(r) r^2 \, dr = Z \tag{2.31}$$

For small momentum transfers, $\sin(qr)$ in equation 2.30 may be expanded in powers of qr, so that equation 2.30 may be written

$$F(q) = 1 - \tfrac{1}{6} q^2 \langle r^2 \rangle + \ldots \tag{2.32}$$

where $\langle r^2 \rangle$ is the mean square radius of $\rho_{ch}(r)$. Hence we see that measurements with small q (that is, small E or small θ) will not tell us more than the value of $\langle r^2 \rangle$. This is an example of the rule that high energies are required to reveal the detailed shape of the charge distribution.

If we knew $F(q)$ accurately for a large range of values of q, we could invert the Fourier transform 2.30 and obtain $\rho_{ch}(r)$ directly. In practice, usually we have to content ourselves with choosing some simple functional form for $\rho_{ch}(r)$ and adjusting its parameters to fit the experimental data. The most commonly used form is called a Fermi distribution (*see Figure 2.14*)

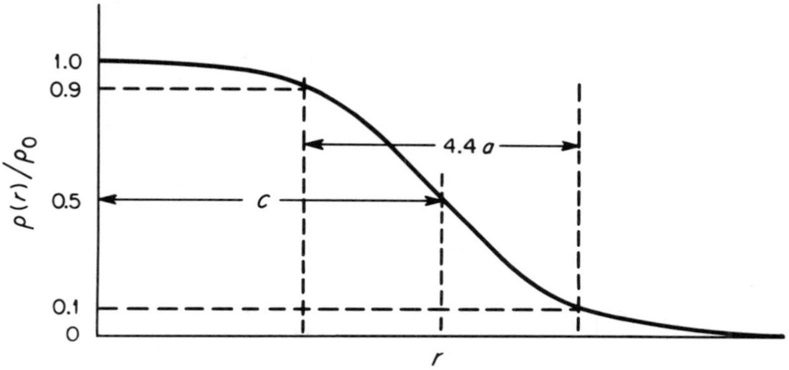

Figure 2.14 The Fermi shape, equation 2.33, used for describing the charge distribution in a nucleus. The ratio c/a used to make the drawing corresponds to a nucleus with $A \sim 20$

$$\rho_{ch}(r) = \frac{\rho_0}{1 + \exp\left(\dfrac{r - c}{a}\right)} \tag{2.33}$$

This reproduces the principal result of electron-scattering measurements that the distribution $\rho_{ch}(r)$ has an almost flat region in the nuclear interior with a rounded edge in which the density falls from 90% to 10% of its central value in about 2.5 fm. It is found that typically $c \approx 1.1 \times A^{1/3}$ fm, $a \approx 0.55$ fm and $\rho_0 \approx 0.07$ fm^{-3}. The constant ρ_0 is chosen so that $\rho_{ch}(r)$ is normalised according to equation 2.31; its value corresponds to about 1 proton per 14 fm^3. (The fact that ρ_0 is roughly the same for all nuclei (*see Figure 1.1* in Chapter 1) follows from the finding that the nuclear radius c is approximately proportional to $A^{1/3}$.)

Careful measurements involving large q values and detailed analyses of them show that the form 2.33 is only approximately true and that, for example, there are slight oscillations in the density in the nuclear interior. *Figure 2.15* shows the results of measurements on the scattering from ^{40}Ca and ^{208}Pb while *Figure 2.16* shows charge distributions which have been deduced from them. These variations from a uniform density can be understood in terms of the shell model orbits occupied by the protons in a particular nucleus.

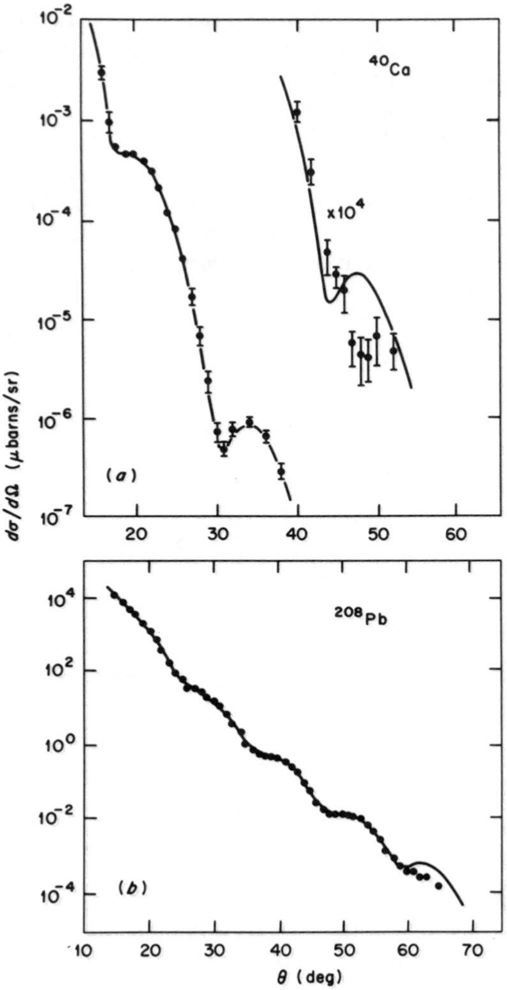

Figure 2.15 The differential cross-sections for electrons scattering elastically from (upper) ^{40}Ca at 750 MeV (after Bellicard *et al.*, 1967) and (lower) ^{208}Pb at 502 MeV. The curves are theoretical fits to the data (after Heisenberg *et al.*, 1969)

The charge distribution of the simplest nucleus of all, the proton, has been measured by electron scattering and found to have a root mean square radius of about 0.8 fm. This size can be related to the cloud of virtual mesons which surround the 'bare' proton. For the same reason, the neutron itself is found to have a charge distribution of finite extent; although its total charge is zero, it has a short-ranged distribution of positive charge and a longer-ranged distribution of negative charge, with a net root mean square radius of 0.36 fm.

In addition to the charge distribution we may also determine the distribution

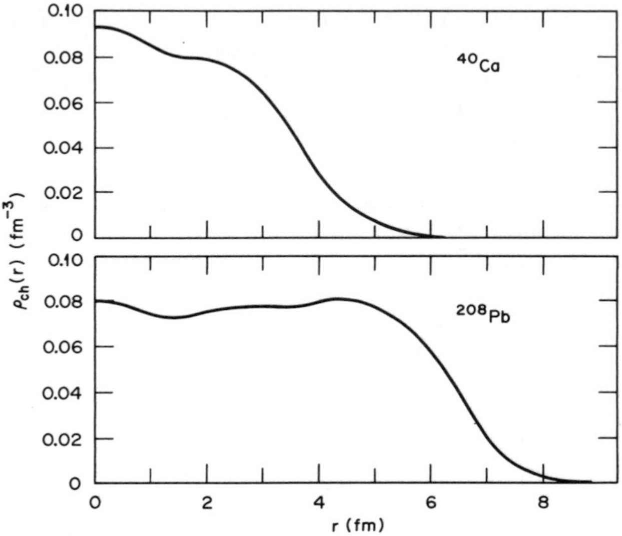

Figure 2.16 The charge distributions of ^{40}Ca and ^{208}Pb nuclei deduced by theoretical fits to the measurements such as those shown in *Figure 2.15*. The shapes at small radii are obtained by fitting the data for the larger angles (that is, for the larger momentum transfers.) (After Friar and Negele, 1973; Sinha *et al.*, 1973)

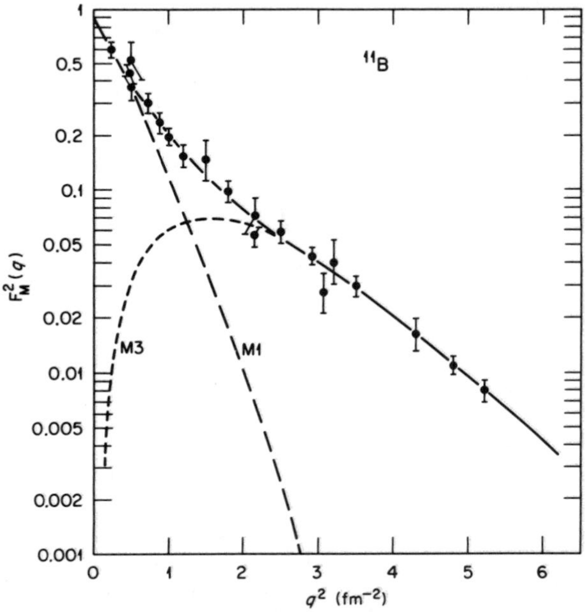

Figure 2.17 Dipole (M1) and octupole (M3) magnetic form factors for ^{11}B deduced from electron-scattering measurements. (After Rand *et al.*, 1966)

of *magnetisation* within the nucleus by observing the scattering of electrons. The nuclear magnetisation arises both from the current due to the orbital motion of the protons and from the intrinsic magnetic moments of the neutrons and protons. (In contrast to charge scattering, neutrons contribute importantly to magnetisation scattering because of their intrinsic magnetic moments.) Magnetic scattering of the electrons dominates for large scattering angles nea: 180° (note that the Mott formula, equation 2.18, predicts zero cross-section for charge scattering through 180°). Not only magnetic dipole, but magnetic octu-

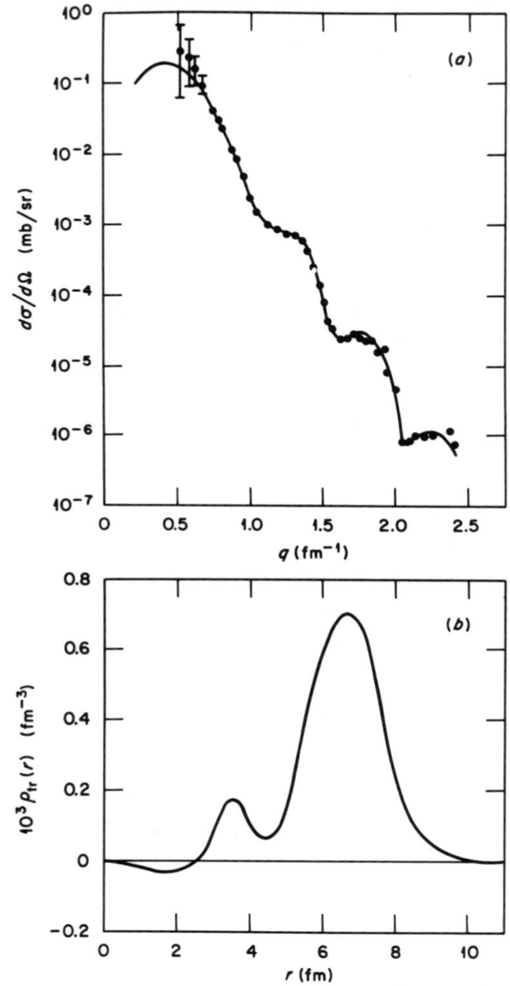

Figure 2.18 (a) Differential cross-sections plotted against momentum transfer q for the inelastic scattering of 248-MeV electrons from ^{208}Pb, exciting the 3^- state with 2.63-MeV excitation energy; the curve is a theoretical fit to the data. (b) The transition density deduced from theoretical fits to these data and other data for 502-MeV electrons.

(After Heisenberg and Sick, 1970)

pole and higher multipole distributions can be measured in this way. Magnetic form factors, $F_M(q)$, are defined in analogy with the charge form factor, equation 2.30. *Figure 2.17* shows some results for scattering from ^{11}B where both dipole (M1) and octupole (M3) moments contribute.

The incident electrons can also excite the target nucleus, transferring energy to it and thus exhibiting inelastic scattering. A simple case which can be visualised classically is the excitation of rotational states in a deformed nucleus (*see* section 1.7.4). The deformed (non-spherical) nucleus has an electric quadrupole moment; the non-uniform electric field due to the passing electron exerts a torque upon this quadrupole moment which tends to set it rotating. In quantum mechanics, interactions of this kind cause transitions between quantum states, i.e. inelastic scattering. The interaction strength is not distributed uniformly throughout the nucleus and we define a *transition density* $\rho_{tr}(r)$ in analogy to the charge distribution $\rho_{ch}(r)$ which produces elastic scattering. The transition density, especially for the strong collective transitions, tends to be concentrated near the nuclear surface. *Figure 2.18* shows the results of measurements on electron inelastic scattering which excites the first excited state of ^{208}Pb, together with the transition density which has been deduced from them.

A comprehensive account of electron scattering has been given by Uberall (1971).

2.12 COULOMB EXCITATION

As just remarked, the Coulomb force between an electron and a nucleus may excite the nucleus when the electron is scattered from it. Similarly the scattering of any other charged particle can excite the nucleus. However, the expression 'Coulomb excitation' is usually only applied to the scattering of protons or heavier nuclei. If the bombarding energy of such particles is below the Coulomb barrier (*see* section 2.10.3), they do not approach closely enough for the short-ranged nuclear forces to act, the long-ranged and well-known Coulomb force acts by itself. The interpretation of measurements under these conditions is then unambiguous and much valuable information on transition rates to excited states can be obtained. If the bombarding energy is then raised above the Coulomb barrier, both types of force act. This also is a useful situation since there are *interference* effects between the repulsive Coulomb and attractive nuclear forces; since the Coulomb force is known, observation of these effects can lead to information about the nuclear force.

If the wavelength of the projectile is sufficiently short (*see* section 2.10.2), such as is the case with heavy ions, the collision can be visualised semi-classically. The projectile moves along a classical Coulomb (or Rutherford) orbit and at some point near the distance of closest approach emits a virtual photon, losing kinetic energy and thereby slowing down. The virtual photon is absorbed by the target nucleus which is thus excited. The emission and absorption process have to be treated quantum mechanically; the excitation of the target nucleus is like an inverse γ-decay, the 'photon' being provided by the field of the scattered projectile.

Detailed expositions of Coulomb excitation can be found in Alder and Winther (1975) and Biedenharn and Brussard (1965); *see* also McGowan and Stelson (1974).

2.13 POLARISATION

The existence of nuclear spins adds a further dimension to the description of a nuclear reaction. A beam of particles with spin may be polarised, that is to say their spin angular momenta may be oriented preferentially in a particular direction. The orientation need not be complete and usually is only partial. This orientation is relevant, in part because the force between two nucleons depends upon the relative orientation of their spins (Preston and Bhaduri, 1975), and in part because the reaction mechanism itself may favour one orientation over another. One simple way that this may occur is through one or both nuclei having a nonspherical shape (Johnson, 1985). Kondo *et al.* (1986) provide many examples of polarisation phenomena.

The spins of the atomic nuclei in a piece of material, or the spins of the particles in the beam from an accelerator, are unpolarised or randomly oriented unless some specific action is taken. One way of polarising nuclei is to apply a uniform magnetic field **B**, because the spin angular momentum vector **I** has associated with it a magnetic moment $\mathbf{u} = g_I \mathbf{I}$. The magnetic moments will tend to align themselves with the magnetic field, while this tendency will be opposed by the thermal agitation. If the component of **I** along the direction of **B** is M, and the probability of finding the value of M is $w(M)$, we define the polarisation P of an assembly of such nuclei as the average value of M/I

$$P = \langle \mathrm{M/I} \rangle \tag{2.34}$$

$$= I^{-1} \sum_M M w(M)$$

where the total probability $\sum_M w(M) = 1$. Since $-I \leqslant M \leqslant I$, we must have $-1 \leqslant P \leqslant 1$. If $P = 1$, there is 100% polarisation and all the nuclei have $M = I$. For the special case of $I = \frac{1}{2}$ such as for protons and neutrons, equation 2.34 becomes

$$P = \frac{w(+\frac{1}{2}) - w(-\frac{1}{2})}{w(+\frac{1}{2}) + w(-\frac{1}{2})}$$

When the particles have spin $I > \frac{1}{2}$, more complicated kinds of non-random orientation may occur. For example, suppose $w(-M) = w(M)$, so that P, equation 2.34, vanishes. Still, the particles may not be randomly oriented. For example, the quantity

$$\langle M^2 - \tfrac{1}{3} I(I+1) \rangle = \sum_M w(M) \, [M^2 - \tfrac{1}{3} I(I+1)]$$

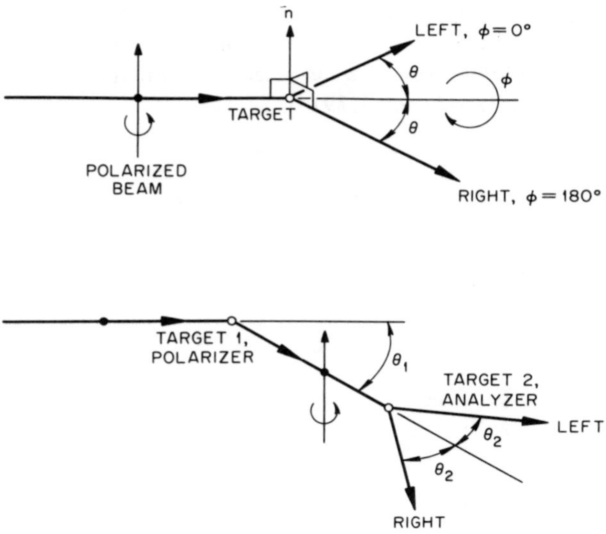

Figure 2.19 Schematic display of the experimental arrangements for measuring (above) the left–right asymmetry in the probability of scattering for a polarised beam and (below) the double scattering of an unpolarised beam

will not be zero unless the probabilities $w(M)$ are independent of M. This and other higher moments of the spin orientation are described as *polarisation tensors*; then P is referred to as the *vector polarisation.*

The energies $\mathbf{u}\cdot\mathbf{B}$ associated with nuclear magnetic moments $\mathbf{u}$ in the magnetic fields $\mathbf{B}$ that one can produce in the laboratory are very small. Then this method requires very low temperatures T so that $\mathbf{u}\cdot\mathbf{B} > k_B T$, where k_B is the Boltzmann constant, otherwise thermal agitation will destroy the polarisation (*see* Daniels, 1965). However, more sophisticated techniques are available, especially for constructing ion sources which will produce polarised particles in an accelerator (*see* England, 1974). Sometimes the polarised products from a previous reaction are used to initiate a secondary reaction, although this method suffers from the low intensities which the first reaction makes available compared to those obtainable from an accelerator.

The most common polarisation measurement on nuclear reactions is the following. We polarise a beam of particles so that their spins are preferentially oriented along a direction $\mathbf{n}$ perpendicular to the beam which is incident on some target. We then measure the scattered intensity (or the intensity of some reaction product) in the plane perpendicular to $\mathbf{n}$ (*see Figure 2.19*). In the absence of any polarisation of the particles in the incident beam or of the target, there is no preferred direction except that of the beam and the symmetry of the arrangement ensures that the scattered intensity does not depend upon the angle ϕ of azimuth about the incident beam, but only on the scattering angle θ. However, using a polarised beam has introduced an asymmetry into the situation. Provided the forces responsible for the scattering or reaction are spin-dependent, there will now be a 'left–right' asymmetry in the detected intensity, $I(\theta)$. That is (*see Figure 2.19*)

$$I_{\text{left}}(\theta) \neq I_{\text{right}}(\theta)$$

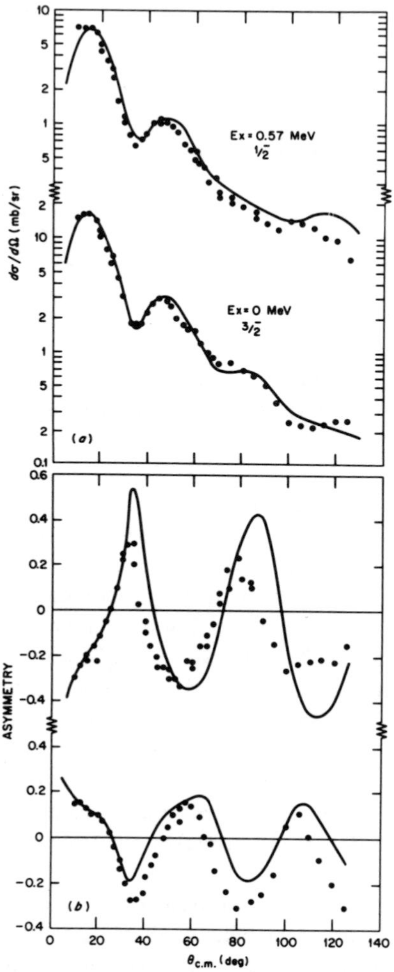

Figure 2.20 The differential cross-sections (a) and the left–right asymmetries (b) of the protons produced by ^{52}Cr (d, p) reactions induced by polarised deuterons of 10 MeV. The upper results are for excitation of a state with spin and parity $1/2^-$, the lower for one with $3/2^-$. Although the differential cross-sections have similar angular distributions, the asymmetries are very different and enable us to distinguish between the two spins. The curves correspond to calculations of deuteron stripping using the distorted-wave theory (*see* section 4.7), assuming direct capture of a neutron into a $p_{1/2}$ and a $p_{3/2}$ orbit, respectively. (After Kocher and Haeberli, 1972)

The asymmetry is defined as

$$A(\theta) = \frac{I_{\text{left}}(\theta) - I_{\text{right}}(\theta)}{I_{\text{left}}(\theta) + I_{\text{right}}(\theta)}$$

and is proportional to the polarisation P of the incident beam. *Figure 2.20* shows as an example the left-right asymmetry from a (d, p) reaction produced by polarised deuterons.

Even when an unpolarised beam is used to initiate a reaction, the products of the reaction, if they have non-zero spin, may be polarised. If a particle is scattered at some angle θ_1, it may be shown that its spin will be polarised perpendicular to the scattering plane (the plane containing both the emitted particle and the incident beam). This induced polarisation may then be measured by using a second scattering at another target as an analyser, and observing the consequent left-right asymmetry (*see Figure 2.19*). Such experiments are difficult because the intensity from the first scattering will be low and hence the number of particles following the second scattering will be very small.

Occasionally experiments have been done in which the spin of the target nucleus itself was oriented by applying a magnetic field or by utilising the magnetic or electric fields arising from the atomic electrons in a crystal (Daniels, 1965). Measurements have also been made on the scattering of polarised neutrons by polarised targets. By comparing intensities when the two spins are polarised in the same and in opposite directions, one may deduce the angular momentum of compound nucleus resonances induced by slow neutrons.

2.14 ANGULAR CORRELATIONS

Angular correlations are closely related to the spin polarisations just discussed. For example, one of the products, B say, of a nuclear reaction A(a, b)B may be left in an excited state which then decays by emitting some further radiation c, a particle or a γ-ray. The direction (θ_c, ϕ_c) of this secondary radiation will be correlated, in general, with the direction (θ_b, ϕ_b) of the other product of the reaction. If the excited nucleus has a non-zero spin, its spin will be polarised by the reaction, and the degree and direction of this polarisation will depend upon the direction of the other emitted particle, b. Further the radiation pattern emitted from such a polarised nucleus is not isotropic in general but shows an angular distribution with respect to the direction of polarisation (just as the radiation from a radio antenna depends upon its angle of emission relative to the orientation of the antenna) (*see* Appendix A). Consequently this radiation intensity is in turn related to the direction of the other reaction product and this angular correlation between b and c may be measured by observing the two in coincidence and seeing how the coincidence rate varies as the two directions are changed. Analysis of such measurements may give information on, for example, the nuclear spins involved, the angular momentum carried away by the radiation and the detailed mechanism of the nuclear reaction.

We set up two counters to detect b and c (*see Figure 2.21*) and measure them *in coincidence*, that is count those events in which b and c are detected simultaneously and hence both come from the same reaction event. The coincidence rate will then depend upon both directions (θ_b, ϕ_b) and (θ_c, ϕ_c). The angular correlation may be observed by varying the position of one or both counters.

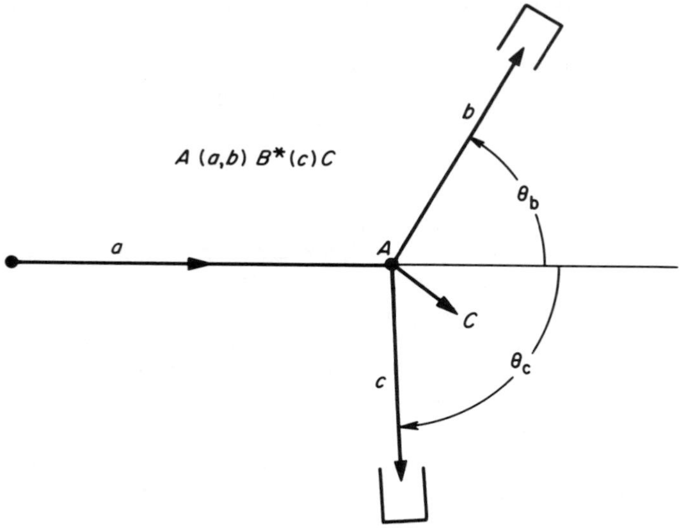

Figure 2.21 The experimental arrangement for a measurement of the angular correlation between particles b and c in a reaction A(a, b)B* (c)C. ('Particles' c may be γ-ray photons.) The directions of emission of b and c in general will not be coplanar with the incident beam. The recoiling nucleus C is not observed

Since the directions of b and c need not be coplanar with the direction of the beam of a particles, this angular correlation depends not only upon θ_b and θ_c, but also upon the azimuth angle ϕ_b-ϕ_c between the (a, b) and (a, c) planes. The results may be expressed as a double-differential cross-section, $d^2\sigma/d\Omega_b d\Omega_c$, the natural generalisation of the differential cross-section $d\sigma/d\Omega$ introduced in section 2.6. (*See*, for example, Ferguson, 1974; Gill, 1975; also Frauenfelder and Steffen, 1966).

2.15 PARTIAL WAVES AND THE WAVE MECHANICS OF SCATTERING

Wave effects are significant in most nuclear-scattering processes, and become very important for low incident energies when the de Broglie wavelength associated with the relative motion of a projectile and target is comparable with or larger than the range of their mutual interaction. Consider a beam of particles each with mass m moving with velocity v impinging on a fixed scattering centre at the origin. If the particles move parallel to the z-axis then the beam can be represented by a plane wave, $\psi = \exp(ikz)$, with wave number $k = mv/h$. The square modulus of the wave function $|\psi|^2 = 1$ everywhere and we can interpret this as representing a density of 1 particle per unit volume or equivalently a flux of v particles per unit area per unit time. The scattering centre interacts with the incident particles to produce an outgoing scattered wave with axial symmetry about the incident beam. At a large distance from the scatterer this wave has the form $\psi_{\text{scatt}} \sim f(\theta) \exp(ikr)/r$. The wave crests form concentric spheres, but their amplitudes $f(\theta)/r$ vary with scattering angle θ as well as decreasing like $1/r$ (so that the intensity obeys the inverse square law). The number of particles in the scattered wave crossing an element of area dS perpendicular to the radius vector in unit time is

$$v\,|\psi_{\text{scatt}}|^2\,\mathrm{d}S = v\left|\frac{f(\theta)\mathrm{e}^{ikr}}{r}\right|^2 \mathrm{d}S = v\,|f(\theta)|^2\,\mathrm{d}\Omega$$

where $\mathrm{d}\Omega = \mathrm{d}S/r^2$ is the solid angle subtended by dS at the origin. Therefore, according to the definition in section 2.6, the differential cross-section is

$$\frac{\mathrm{d}\sigma}{\mathrm{d}\Omega} = |f(\theta)|^2 \tag{2.35}$$

The quantity $f(\theta)$ is called the scattering amplitude. To find it theoretically it is necessary to solve the time-independent Schrödinger equation for the wave function $\psi(r)$ with a boundary condition

$$\psi(\mathbf{r}) \sim \mathrm{e}^{ikz} + f(\theta)\,\frac{\mathrm{e}^{ikr}}{r} \tag{2.36}$$

at large distances from the scatterer.

The above discussion is applicable only to the elastic scattering of a projectile without spin by a spherical target that cannot be excited. It is generalised to include the possibility of reactions in Chapter 3. The restriction to a fixed target can be removed by working in the centre-of-mass coordinate system (section 2.2 and Appendix B).

When a particle without spin is scattered by a spherically symmetric potential $V(r)$ (i.e. one which depends only upon the radial coordinate r and not on the polar angles θ and ϕ) then its orbital angular momentum is conserved. The scattering wave function can be expanded as a sum of angular momentum eigenfunctions, the Legendre functions $P_\ell(\cos\theta)$ (Appendix A)

$$\psi(r,\theta) = \sum_{\ell=0}^{\infty} u_\ell(r)\,P_\ell(\cos\theta) \tag{2.37}$$

If we write the radial wave function as $u_\ell(r) = r^{-1} w_\ell(r)$, then $w_\ell(r)$ satisfies the radial part of the Schrödinger equation in the form

$$\frac{\mathrm{d}^2 w_\ell(r)}{\mathrm{d}r^2} + \left(k^2 - U(r) - \frac{\ell(\ell+1)}{r^2}\right) w_\ell(r) = 0 \tag{2.38}$$

where

$$k^2 = \frac{2mE}{\hbar^2}, \quad U(r) = \frac{2mV(r)}{\hbar^2}$$

and m is the mass of the particle. The expansion 2.37 is called the ***partial wave*** expansion for the wave function and the ℓth term is the component of the wave function with orbital angular momentum ℓh. (More precisely, the square of the angular momentum is $\ell(\ell + 1)\hbar^2$.) In spectroscopic terminology, the term with $\ell = 0$ is called the s-wave component of the wave function, the terms with $\ell = 1$, 2, 3, . . . are the p, d, f, . . . partial waves, etc. The partial wave expansion for the scattering amplitude which corresponds to the expansion 2.37 for the wave function is

$$f(\theta) = \frac{1}{2ik} \sum_{\ell=0}^{\infty} (2\ell + 1) P_\ell (\cos \theta) (\eta_\ell - 1) \tag{2.39}$$

The quantity η_ℓ is a partial wave amplitude which can be calculated from the radial wave function $u_\ell(r)$. The $u_\ell(r)$ itself is obtained by integrating equation 2.38; for a few special forms of $V(r)$ this can be done analytically, but more often the equation must be solved numerically. When there is no potential, $V = 0$ and the wave function is simply the plane wave, $\psi = \exp(ikz)$. The corresponding radial functions are proportional to the *spherical Bessel functions* (section 3.4.2)

$$u_\ell(r) = i^\ell (2\ell + 1) j_\ell(kr) \quad \text{if} \quad V = 0$$

At large distances these have the form

$$i^\ell(2\ell + 1) j_\ell(kr) \to \frac{(2\ell + 1)}{2ikr} \left[e^{ikr} - (-)^\ell e^{-ikr} \right] \tag{2.40}$$

This shows that each partial wave component of a plane wave is the sum of an incoming ($\exp(-ikr)$) and an outgoing ($\exp(ikr)$) spherical wave with equal amplitudes. Provided $V(r)$ is well behaved (falls to zero faster than r^{-1} at large distances), the wave function must have a form similar to equation 2.40 even in the presence of a potential. The incoming and outgoing parts must still have equal magnitudes because the number of particles is conserved. However the effect of the potential $V(r)$ is to shift the phase of the outgoing wave so that asymptotically the radial function is

$$u_\ell(r) \to \frac{(2\ell + 1)}{2ikr} \left[\eta_\ell e^{ikr} - (-)^\ell e^{-ikr} \right] \tag{2.41}$$

where the outgoing wave amplitude is now

$$\eta_\ell = e^{2i\delta_\ell} \tag{2.42}$$

The angle δ_ℓ is called the ***phase shift*** for the ℓth partial wave. Equation 2.36 shows that the scattering amplitude $f(\theta)$ is the excess in amplitude of the outgoing spherical wave caused by the potential V over what was already there in the plane wave which describes the incident beam. Consequently, the difference between equations 2.41 and 2.40 immediately gives the expression 2.39.

A real potential $V(r)$ can only give rise to elastic scattering. When other reactions are possible, such as inelastic scattering and rearrangement collisions (section 2.3), the number of elastically scattered particles observed will be less than the number incident upon the target; some have been diverted (or 'absorbed') into the other reaction channels. Then the outgoing wave amplitude η_ℓ is no longer unimodular and we will have $|\eta_\ell| \leqslant 1$. It can no longer be written in the form 2.42 unless we care to generalise the phase shifts δ_ℓ to be complex.

One reason why the partial wave expansion for the scattering amplitude is useful is that often the series 2.37 is rapidly converging so that only a few terms in the expansion need be considered. According to classical mechanics the relative orbital angular momentum of the projectile and target is pb, where b is the classical impact parameter (*Figures 2.3, 2.8*) and $p = \hbar k$ is the relative momentum. The projectile is scattered by the target only if the impact parameter b is less than the range R of their interaction. Thus we expect that only partial waves with

$$\hbar\ell \lesssim pR \quad \text{or} \quad \ell \lesssim kR \tag{2.43}$$

should contribute to the scattering amplitude.

If $kR \ll 1$ the predominant interaction must occur in states with $\ell = 0$, i.e. in s-states. The wave number $k = 2\pi/\lambda$ where λ is the de Broglie wavelength of the waves associated with the relative motion of the target and projectile. Thus, s-wave scattering predominates if the de Broglie wavelength is much larger than the range of the projectile–target interaction. When the s-wave term dominates, the scattering amplitude is simply

$$f(\theta) \approx \frac{1}{2ik}(\eta_0 - 1)$$

and the differential cross-section

$$\frac{d\sigma}{d\Omega} = |f(\theta)|^2 = \frac{|\eta_0 - 1|^2}{4k^2}$$

is independent of scattering angle.

We can apply these ideas to estimate the range of nuclear forces. *Figure 2.22* shows the differential cross-section for neutron–proton scattering at several energies. The differential cross-section is almost independent of scattering angle for neutrons incident on protons with a laboratory energy of 14 MeV and 18 MeV, suggesting that only s-waves are scattered at these energies. Significant departures from isotropic scattering are developing at 27 MeV and 42 MeV indicating that p- or d-waves are starting to contribute to the scattering amplitude at these energies. The relative wave number k is given by

$$k = \left(\frac{mE_{\mathrm{LAB}}}{\hbar^2}\right)^{1/2} \begin{cases} = 0.41\ \mathrm{fm}^{-1} & \text{when } E_{\mathrm{LAB}} = 14\ \mathrm{MeV} \\ = 0.71\ \mathrm{fm}^{-1} & \text{when } E_{\mathrm{LAB}} = 42\ \mathrm{MeV} \end{cases}$$

where m is the mass of a nucleon (see section 3.1.2 and Appendix B). These numbers suggest a range of less than 2 fm for the neutron–proton interaction.

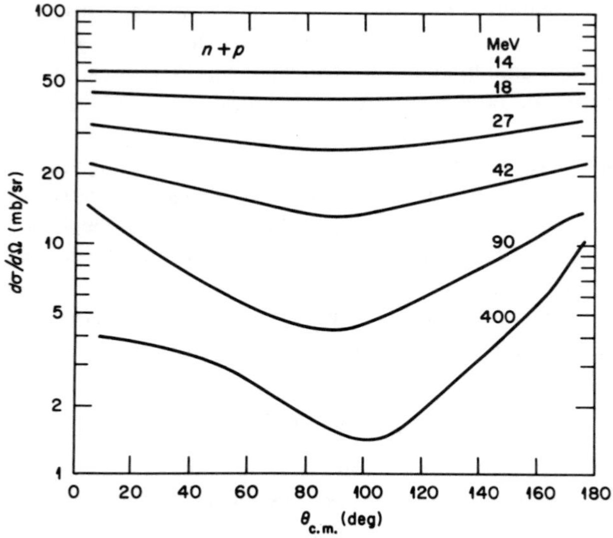

Figure 2.22 The differential cross-sections for neutrons of several energies scattering from protons. (After Hess, 1958)

If $|\eta_\ell| < 1$, other reaction channels are open and diverting or absorbing flux from the elastic channel. If $|\eta_\ell| \approx 0$, this is *strong absorption*. By analogy with the scattering of light by a black sphere we can imagine a target nucleus of radius R which is so strongly absorbing that any particle which touches it suffers a non-elastic collision of some kind and does not return to the elastic channel. If $kR \gg 1$, we would then have

$$\begin{aligned} \eta_\ell &= 0, \quad \ell \leqslant kR \\ &= 1, \quad \ell > kR \end{aligned} \tag{2.44}$$

This simple picture is explored further in Chapter 4, section 4.1. (Another approach to strong absorption also discussed in Chapter 4 uses an analogy to the classical theory of the diffraction of light by a black sphere.) The conditions 2.44 are closely approximated in the scattering of two heavy ions.

The radial wave equation, equation 2.38, contains a term $\ell(\ell+1)\hbar^2/2mr^2$ which represents the rotational kinetic energy of the particle. It appears on the same footing as the potential energy $V(r)$; that is, a particle with angular momentum ℓ obeys the same equation as one with $\ell = 0$ except for the addition of this term which looks like a repulsive potential. It is often referred to as the ***centrifugal potential*** or ***centrifugal barrier***. Its repulsive effect keeps a particle with non-zero ℓ away from the origin; the condition 2.43 is one result of this.

2.16 SCATTERING OF IDENTICAL PARTICLES

The wave function describing a quantal system has to be symmetric or anti-symmetric under the interchange of the coordinates of any two indistinguishable particles contained in it, according to whether the particles are bosons or fermions, respectively. This remains true for the wave function describing the scattering of two nuclear systems. The simplest example of the consequences of this symmetrisation occurs for the scattering of a particle by another, identical, particle when it requires that the differential cross-section (in the CMS) must be symmetric about $\theta = 90°$. This is expected classically if the individual particles cannot be followed by the observer. Then if the two particles are indistinguishable, one would not know which one was being observed emerging from the scattering, and one could not distinguish between the two scattering events shown in *Figure 2.23*. Classically, the observed cross-section would be the sum of the cross-sections for the two possibilities

$$\sigma(\theta) + \sigma(\pi - \theta) \tag{2.45}$$

Quantum theory introduces ***interference***, because we have to add the two corresponding ***amplitudes***. The properly symmetrised wave function includes both events pictured. The cross-section observed is then†

$$|f(\theta) \pm f(\pi - \theta)|^2 = \sigma(\theta) + \sigma(\pi - \theta) \pm 2Re f(\theta) f(\pi - \theta)^* \tag{2.46}$$

Figure 2.23 Two indistinguishable events which may occur when two identical particles scatter

†Equation 2.46 is incomplete for the scattering of two indistinguishable fermions. A fermion must have a non-zero spin and the orientation of this spin must be considered explicitly (*see* Exercise 2.13). The *spatial* part of the scattering wave function for two such fermions then can be separated into a symmetric part and an antisymmetric part. The former contributes terms to the cross-section like equation 2.46 with a plus sign while the latter gives terms with a negative sign. (*See*, for example, Mott and Massey, 1965.)

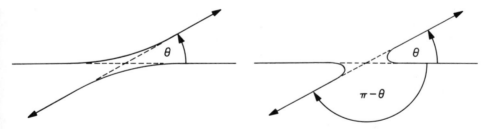

An example of this is shown in *Figure 2.24* for the scattering of ^{12}C nuclei on other ^{12}C nuclei. Classically, one would expect the solid curve, which is just equation 2.45 for Rutherford scattering. The interference term in equation 2.46 introduces the oscillations shown as the dashed curve. The measurements are in very good agreement with the quantum theory.

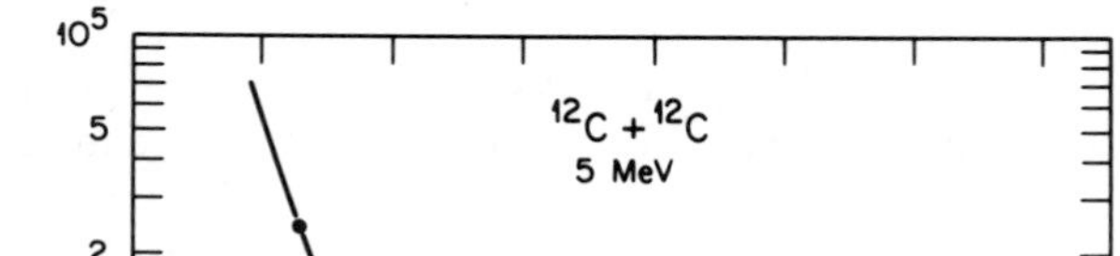

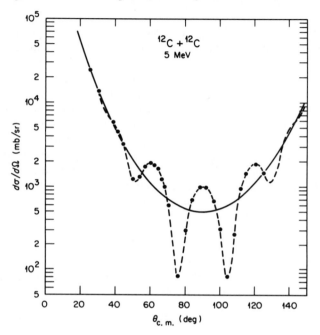

Figure 2.24 The elastic scattering of ^{12}C by ^{12}C at an energy of 5 MeV. The solid curve is the sum, equation 2.45, of the Rutherford cross-sections while the dashed curve includes the quantal interference terms of equation 2.46. The dots are the measured cross-sections. Note that the quantal cross-section is exactly twice the classical one at $\theta = 90°$, as expected from equation 2.46. (After Bromley *et al.*, 1961)

2.17 INVERSE REACTIONS

If the equations describing the reaction process

$$A + a \rightarrow B + b + Q$$

are invariant under time-reversal (changing the sign of the time variable), then they also describe the process

$$B + b \rightarrow A + a - Q$$

For a given total energy, the corresponding cross-sections $\sigma(a \rightarrow b)$ and $\sigma(b \rightarrow a)$ are not the same but they are simply related by the phase space available in the exit channel in each case; that is, the density of final states. The number of states available for momenta between p and $p + dp$ is proportional to p^2, hence $\sigma(a \to b)$ is proportional to p_b^2 if p_b is the momentum of b relative to B, and $\sigma(b \to a)$ is proportional to p_a^2 if p_a is the momentum of a relative to A. Then we have that

$$\frac{\sigma(b \to a)}{p_a^2} = \frac{\sigma(a \to b)}{p_b^2} \tag{2.47}$$

This is known as the ***reciprocity theorem***. It holds for differential as well as integrated cross-sections.

When the particles have spins, I_A, i_a, I_B and i_b say, we must also take into account the associated statistical weights; there are $(2i + 1)$ states of orientation available for a particle with spin i and the relation 2.47 becomes*

$$\frac{\sigma(b \to a)}{(2I_A + 1)(2i_a + 1)p_a^2} = \frac{\sigma(a \to b)}{(2I_B + 1)(2i_b + 1)p_b^2} \tag{2.48}$$

The reciprocity theorem has been tested in a number of experiments. An example of the results is shown in *Figure 2.25* where the differential cross-sections for the two reactions ^{24}Mg(α, p) ^{27}Aℓ and ^{27}Aℓ (p, α) ^{24}Mg, connecting the ground states of ^{24}Mg and ^{27}Aℓ, and measured at the same total energy and same CMS angle are compared. The agreement is excellent and demonstrates, to this accuracy, the time-reversal invariance of the underlying equations.

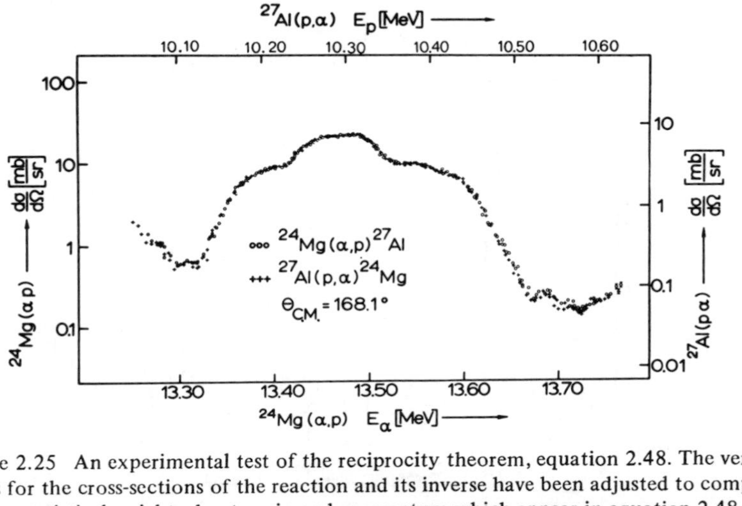

Figure 2.25 An experimental test of the reciprocity theorem, equation 2.48. The vertical scales for the cross-sections of the reaction and its inverse have been adjusted to compensate for the statistical weights due to spin and momentum which appear in equation 2.48. (From von Witsch *et al.*, 1968)

*We are not considering here reactions in which one or more particles is polarised or whose polarisation is measured (*see* section 2.13). More complicated relations hold for these reactions.

2.18 QUALITATIVE FEATURES OF NUCLEAR REACTIONS

Nuclear reactions are found in astonishing variety. Nevertheless, some general characteristics can be discerned and some broad categories defined. For example, not all reactions to all possible final states proceed with anything like equal probability. On the contrary, there is often a high degree of ***selectivity***. *Figure 2.26* illustrates just two examples of this. The reactions $^{12}C\ (^{12}C, \alpha)\ ^{20}Ne$ and $^{16}O\ (^{7}Li, t)\ ^{20}Ne$ lead to the same final nucleus ^{20}Ne, but *Figure 2.26* shows that they do not excite the same excited states of ^{20}Ne with equal probability. Further, the levels of ^{20}Ne in this region of excitation are known to be many times more numerous than the few excited in either of these reactions. Such selectivity will also vary with bombarding energy. It often allows us to deduce much interesting information about nuclear structure and about the mechanisms of the various reactions.

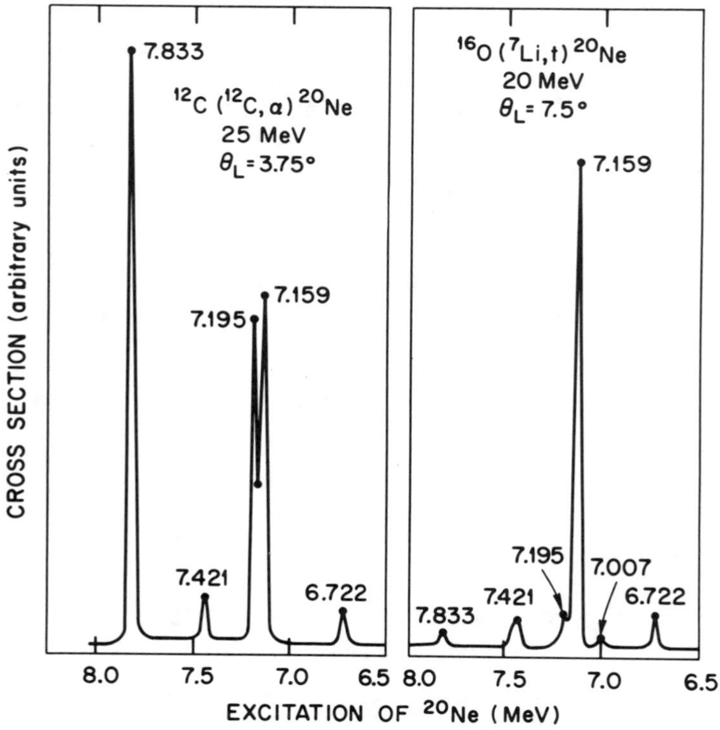

Figure 2.26 A comparison of the energy spectra for the products from the reactions $^{12}C\ (^{12}C, \alpha)\ ^{20}Ne$ and $^{16}O\ (^{7}Li, t)\ ^{20}Ne$ which lead to the same final nucleus. This shows that the two reactions do not excite the states of ^{20}Ne in the same way; both reactions are very 'selective'. The peaks are labelled by the excitation energy of the corresponding states in ^{20}Ne. (After Bromley, 1974)

What one learns from a measurement on a nuclear reaction often depends upon the energy resolution available. It might be supposed that it was best to have very sharp energy resolution; in many cases this is true, for example if we wish to resolve two nuclear levels which have almost the same excitation energy. However, in some instances interesting structure may be observed in an experiment with poor resolution which would be obscured by the complexity of the results from a high-resolution study. Of course, the high-resolution results may be deliberately smoothed later by averaging them over a small energy interval. The point is that such smoothing is easy to do once one realises that it is useful to do it. It is the cruder experiment that points the way to this.* *Figure 2.27*

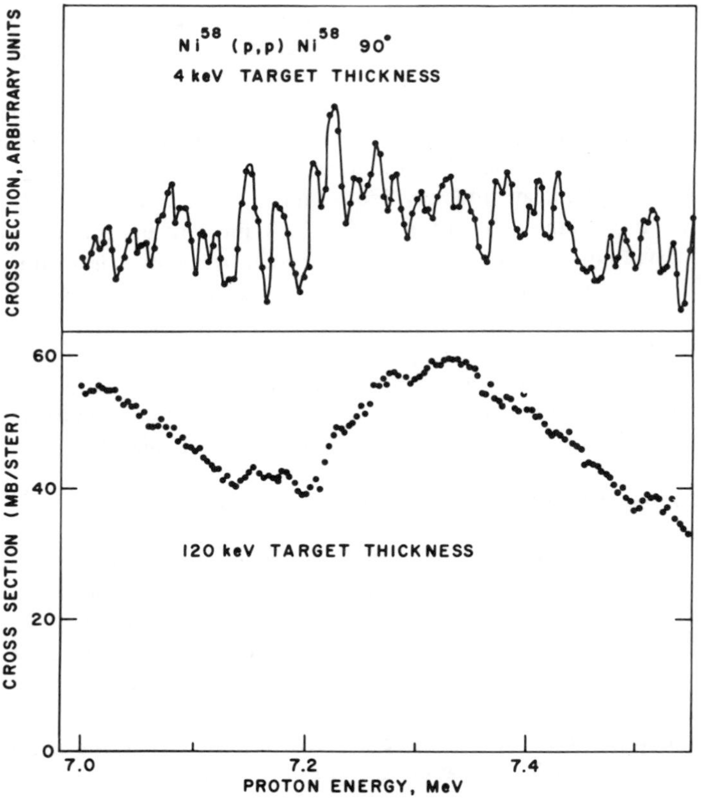

Figure 2.27 Excitation functions (cross-sections as a function of bombarding energy) for protons elastically scattered from ^{58}Ni. The upper portion shows results obtained using a thin target, the lower portion is for a thick target. (After Lee and Schiffer, 1963)

*Fortunately the early stages of a science are usually characterised by poor resolution measurements. Imagine how little progress would have been made systematising the line spectra emitted by atoms if all the fine and hyperfine structure had been observed at the very beginning!

illustrates this for protons scattering elastically from ^{58}Ni to an angle of $\theta_L = 90°$. The upper portion shows the results of high-resolution measurements with a thin target; there are considerable fluctuations as the proton energy is varied. The lower portion shows the result of averaging these results over about 120 keV by using a thick target. The fluctuations are largely removed but some broad 'intermediate' structure remains. (Another example is shown in *Figure 4.38*, Chapter 4.)

The type of information available from reaction measurements also depends upon the nature of the projectile and the bombarding energy. As a particular example, suppose we bring into a nucleus hundreds of MeV of energy via a high-energy proton. This energy is highly concentrated (on one nucleon), and the result is likely to be characteristic of nucleon–nucleon collisions. A target nucleon may be knocked out, a meson or hyperon may be produced. The same amount of energy carried by a heavy ion, such as ^{40}Ar or ^{84}Kr, is diffused over a large volume (many nucleons) and will produce quite different behaviour such as large-scale collective motions of the compound system as a whole and perhaps exciting shock waves in which the local density is much higher than in a normal nucleus. Further, the heavy ion may deposit much larger amounts of angular momentum. For example, a proton of 400 MeV incident upon a nucleus of radius 6 fm will not strike the nucleus if its angular momentum is greater than about $25\hbar$, while an ^{84}Kr ion with 400 MeV can interact with the same target nucleus when their relative angular momentum is as much as $250\hbar$.

The following sections describe the characteristics of some broad categories of nuclear reactions, in particular the two extremes of direct reactions and compound nucleus formation. The detailed models which have been constructed to describe the various types of reactions are discussed in Chapter 4.

2.18.1 Compound nucleus formation and direct reactions

Two extremes have been recognised when two nuclear systems collide, and both are of importance for our understanding of nuclear reaction phenomena.

(i) The two may coalesce to form a highly excited compound system. This is called *fusion* when two heavy ions collide. The *compound nucleus* stays together sufficiently long for its excitation energy to be shared more or less uniformly by all its constituent nucleons. Then, by chance, sufficient energy is localised on one nucleon, or one group of nucleons, for it to escape and in this way the compound nucleus decays (*see Figure 2.28*). Schematically

$$A + a \rightarrow C^* \rightarrow B^* + b$$

If sufficient excitation energy remains in B*, further particle emissions may occur. Otherwise, it will de-excite by β- or γ-decay.

The picture of a nucleus as a liquid drop is perhaps helpful in visualising these processes. In the compound nucleus reaction, the two colliding droplets combine to form a single compound drop which, because it is excited, is at a high tem-

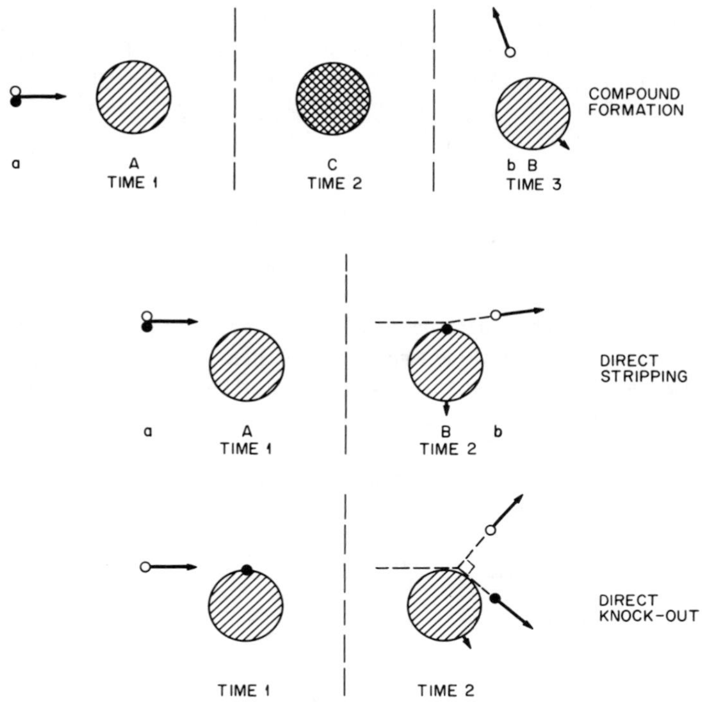

Figure 2.28 Illustrating schematically the two limiting kinds of nuclear reaction, compound nucleus formation and decay, and direct reactions. The latter are represented by stripping and knock-out occurring in the nuclear surface

perature. The decay, or cooling, of this drop can then be thought of in terms of the *evaporation* of one or more of its constituent particles.

Because of the delay between formation and decay, and the many complicated nucleon motions that take place during that period, the system C* may be said to have lost memory of the particular channel A + a by which it was formed, and the probabilities of the various decay modes B + b will be independent of each other and of the entrance channel. If this *independence hypothesis* holds, the cross-section for the reaction will factor into parts depending separately on the entrance and exit channels

$$\sigma = \sigma^{\mathrm{C}}_{\mathrm{Aa}}(E)\, G^{\mathrm{C}}_{\mathrm{Bb}}(E) \tag{2.49}$$

where $\sigma^{\mathrm{C}}_{\mathrm{Aa}}(E)$ is the cross-section for forming the compound nucleus C* from A + a with the total energy E, and $G^{\mathrm{C}}_{\mathrm{Bb}}(E)$ is the relative probability of C at this energy then decaying into B + b. The significance of this is that if the same compound nucleus C* with the same total energy were formed in some other way, say by bombarding target D with projectiles d, then the relative probability for its decay into the B + b channel will be governed by the *same* factor G^{C}_{Bb}. Then the cross-section for the process

$$D + d \rightarrow C^* \rightarrow B^* + b$$

would be

$$\sigma(E) = \sigma^{C}_{Dd}(E)\, G^{C}_{Bb}(E)$$

The classic experiment for verifying the independence hypothesis was made by Goshal (1950) by forming the compound nucleus ^{64}Zn by proton bombardment of ^{63}Cu and by α-bombardment of ^{60}Ni. The decay by emission of the compound nuclei formed in these two ways was then compared as a function of bombarding energy and found to be very similar. The results are shown in *Figure 2.29*; the energy scales have been adjusted so that the excitation of the compound nucleus ^{64}Zn is the same in both cases. It is clear from the similarity of the pairs of excitation curves that the cross-section ratios σ(x, pn)/σ(x, n) or

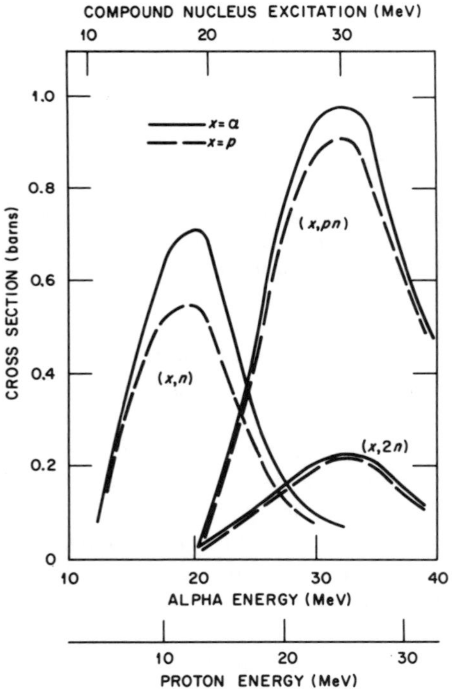

Figure 2.29 Experimental evidence for the independence of formation and decay in a compound nucleus reaction. The same compound nucleus ^{64}Zn, with the same excitation energy, is formed in two different ways, but the excitation curves are very similar. (After Goshal, 1950)

$\sigma(x, 2n)/\sigma(x, n)$, etc., are approximately the same at a given excitation of the compound nucleus whether x is an α-particle or a proton, thus verifying the factorisation in equation 2.49.

(ii) At the other extreme, the two systems may make just glancing contact and immediately separate. Their internal states may be unchanged (elastic scattering), one (or both) may be excited by the contact (inelastic scattering), or one or a few nucleons may be transferred across from one nucleus to the other (rearrangement collision or transfer reaction) (*see Figure 2.28*). Because these reactions occur quickly and proceed directly from initial to final states without forming an intermediate compound state, they are called ***direct reactions*** or ***direct interactions***. Sometimes they are called ***peripheral reactions.*** Clearly then we will not find any sort of independence between entrance and exit channels such as expressed in equation 2.49. Rather, the outcome of such a reaction depends intimately on the way it is initiated and therein lies its importance (*see* below).

Three types of direct reaction are especially important. The first occurs for inelastic scattering. The liquid drop model is again helpful here. We can visualise the momentary, glancing collision of the projectile with the target as setting the target droplet into a state of oscillation, or, if it is initially non-spherical, making it rotate. This suggests, as is indeed the case, that a direct mode of inelastic scattering is particularly effective in exciting collective states (section 1.7.4) of the vibrational or rotational type.

The second important case is the ***stripping reaction*** (or its inverse, known as a ***pick-up reaction***. These are also referred to as ***transfer reactions***). *Figure 2.28* illustrates this; one (or a few) nucleons is stripped from the projectile as it passes the periphery of the target. The prototype of these reactions is the deuteron stripping reaction A(d, p)B in which a neutron is stripped from the deuteron and transferred to the target.

The third concerns knock-out reactions, in which a nucleon or light nucleus is ejected from the target by the projectile, which itself continues freely (*see Figure 2.28*). We then have three particles in the final state. These reactions are also referred to as ***quasi-free scattering*** since the picture is of the projectile and struck particle colliding almost as though the latter was free and the rest of the target nucleus was simply a spectator. Here the prototype is the (p, 2p) reaction. This and other reactions which eject a single nucleon, (p, pn), (e, e′p), etc., are nuclear analogues of the early experiments of Franck and Hertz and others on the ionisation of atoms under electron bombardment. The ejection of other particles, e.g. a (p, p′α) reaction, may tell us about the probability of finding preformed clusters, such as α-particles, within the nucleus.

A multiple-scattering picture can also give us insight into the relation between direct and compound reactions, especially if the projectile is a nucleon. We can consider this incident nucleon as making successive collisions with the various target nucleons. At each collision, the incident nucleon usually will not lose a large fraction of its energy. Hence after one such collision, the target nucleus will not be highly excited. If the incident particle then escapes, we will then have a direct-reaction inelastic scattering, leaving the target in one of its low-lying excited states. If the struck nucleon also escapes we have a direct knock-out or, if it emerges bound to the incident nucleon, a pick-up reaction. If instead the incident nucleon suffers another collision, it will tend to lose some more energy. After a number of such collisions, the original energy has been shared with many target nucleons, the incident particle no longer has sufficient energy to escape, and a compound nucleus has been formed.

Direct and compound nuclear reactions are not mutually exclusive; both types of process may contribute to a given reaction leading to a particular final state. For example one may visualise a glancing, peripheral collision resulting in a direct reaction and a more head-on collision leading to fusion into the compound nucleus and its subsequent decay. Their relative importance can also be expected to depend upon the bombarding energy. *Figure 2.30* shows ***excitation functions*** (cross-section versus bombarding energy) for the inelastic scattering of nucleons exciting the first excited state of a medium-weight nucleus. At the lowest energies the compound nucleus formation dominates but its importance falls off as the energy increases and more open channels become available (*see* section 2.18.5). Meanwhile the direct component increases steadily until it begins to dominate at energies above 10 MeV or so.

In addition, we must recognise that in a given reaction there will usually be a

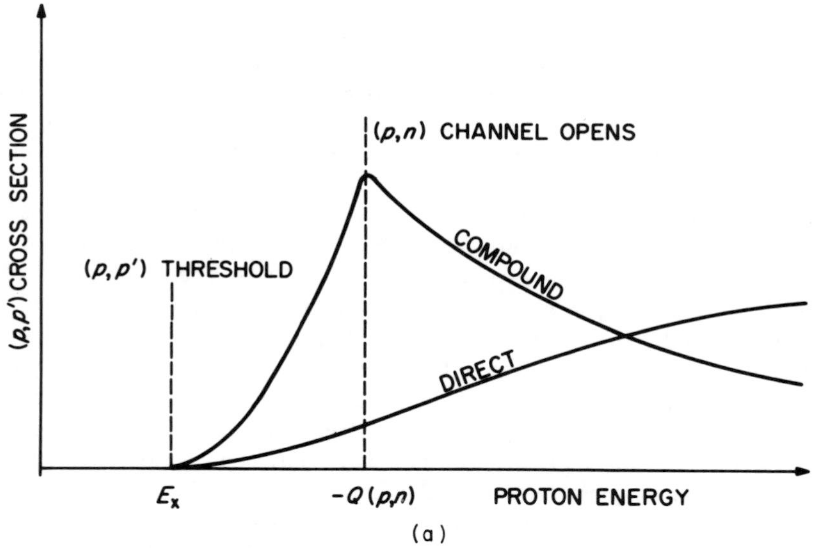

Figure 2.30 (a) Excitation functions for exciting the lowest excited state of a medium-weight nucleus by the inelastic scattering of nucleons: the cross-section for proton scattering is predominantly due to compound nucleus formation at the lowest energies until the threshold for the (p, n) reaction is reached, then neutron decay competes successfully until the direct reaction becomes most important

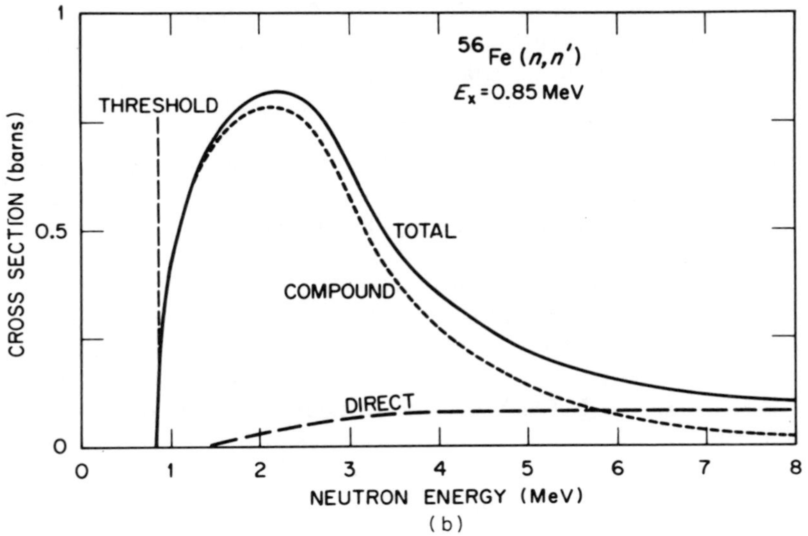

Figure 2.30 (b) Excitation functions for exciting the lowest excited state of a medium-weight nucleus by the inelastic scattering of nucleons: neutron scattering is also dominated by compound nucleus reactions at low energies until the increasing energy makes available many competing (n, n') reactions to higher excited states. (Based upon data and analyses of Kinney and Perey, 1970)

continuous spectrum of reaction processes between the two extreme ones we describe as direct and compound; there will be intermediate processes that do not clearly fall into either category. Until recently there had not been a wide interest in these intermediate processes when they occurred in reactions induced by light-ion projectiles, where they are referred to as *pre-equilibrium reactions* (*see*, for example, Blann, 1975; Hodgson, 1987, for a theoretical description of them), because they did not seem to exhibit any qualitatively new nuclear phenomena. This was not true for collisions between heavy-ion projectiles and heavy-nucleus targets (*see* section 2.18.12).

2.18.2 Compound resonances

A compound nucleus which is formed with a relatively low excitation energy may exhibit discrete quantum states even though it is unstable against particle emission. These states are merely an extension to higher energies of the discrete states observed at lower excitation energies below the threshold for particle emission. They have definite spin and parity, for example. Although enough energy is available for the emission of a particle, most of the time the energy is shared by several particles, none of which has by itself enough energy to escape. Hence the compound system may survive for a time long compared to a typical orbital period of a nucleon, 10^{-22} s. However it will decay eventually. The effect of the instability is to give such a decaying state an imprecise energy. The energy is distributed in probability with a characteristic width Γ related to the lifetime τ of the state by the uncertainty principle

$$\Gamma \simeq \hbar/\tau$$

The requirement that $\tau \gg 10^{-22}$ s implies that $\Gamma \ll 1$ MeV (*see* section 1.7.5). (Even the lower, so-called 'bound', states are unstable to β- and/or γ-decay, but the widths due to these instabilities are usually very small.)

If the bombarding energy in channel a + A is just such as to match the energy E_r of one of these quasi-stationary quantum states of the compound nucleus C, there will be a resonance; the cross-section for formation of the compound nucleus shows a marked increase (*see Figure 2.31*). The shape of the peak in the curve of cross-section versus bombarding energy is the same as that of the response of a resonant electrical circuit to an external stimulus or driving potential, namely

$$\sigma(E) = \sigma_0 \frac{\frac{1}{4}\Gamma^2}{(E - E_r)^2 + \frac{1}{4}\Gamma^2} \qquad (2.50)$$

Here σ_0 is the peak height (at $E = E_r$) and Γ is the full width at half maximum (FWHM). The shape 2.50 is referred to as a *Breit–Wigner resonance*.

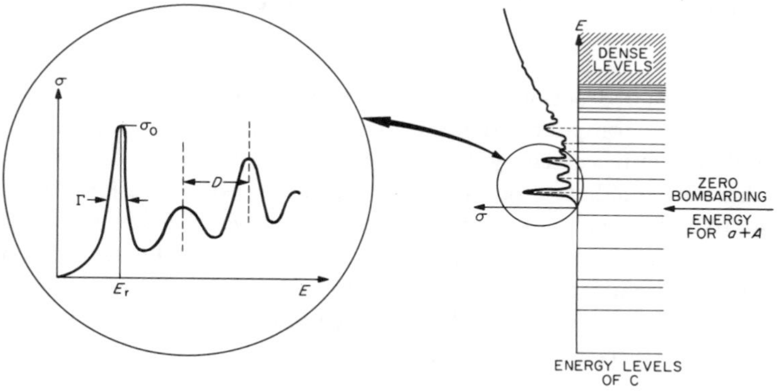

Figure 2.31 Excitation function for a reaction obtained by bombarding nucleus A with ions a. Each peak corresponds to a state of the compound system C = A + a, until the density of such levels becomes sufficiently high that individual ones cannot be resolved

Wave-guides provide an analogy to these resonant states of the compound nucleus. A closed cavity with perfectly reflecting walls has definite resonant frequencies for electromagnetic radiation within it which are determined by its shape and dimensions. These are like the bound levels of a nucleus. If we make a hole in the wall of the cavity, the radiation can escape; this is like the unbound nucleus. However if the hole is small so that the rate of loss is low, the cavity will still resonate; the effect of the hole will be to change the sharp resonance frequency to a narrow distribution so that the response of the cavity will have the form 2.50.

As the bombarding energy (and hence the corresponding energy of the compound system) increases, the levels of the compound nucleus become more closely spaced. With the extra energy available and more decay channels open, their widths also increase, until eventually they overlap sufficiently that a smooth dependence on energy results for the cross-section. When this happens, we can no longer describe the reaction in terms of the quantum numbers of individual resonance levels but we must resort to using average properties. This leads to the statistical model of compound reactions that we shall describe later.

The compound nucleus resonances observed in nuclear reactions have a wide variety of widths Γ and spacings D, ranging from $\Gamma \sim 1/10$ eV, $D \sim 1$ eV for resonances in the interaction of slow neutrons with heavy nuclei, to hundreds of keV for both Γ and D for reactions involving light nuclei.

2.18.3 Reaction times

One essential feature distinguishing the two types of reaction, direct and compound, is the time required for each; the compound reaction is slow, the direct reaction is fast. The time scale by which we judge slow or fast is given by the motion of nucleons within the nucleus. A typical nucleon orbital period is a few times 10^{-22} s (corresponding to a kinetic energy of about 20 MeV) so that it will traverse the nucleus in about 10^{-22} s. If the collision is completed within this time or less, there is no time for appreciable sharing of energy between a struck nucleon and the other nucleons in the target. Formation of a compound system requires interaction over a much longer period, $\gg 10^{-22}$ s.

When we are dealing with distinct resonances, we can estimate the lifetime from the resonance width, using the uncertainty relation (section 1.7.5). Since $\tau \approx (6.6 \times 10^{-16}/\Gamma)$ s if Γ is expressed in eV, we see that lifetimes from 10^{-14} s to 10^{-20} s and less are encountered.

2.18.4 Energy spectra

The two reaction modes, compound and direct, are not mutually exclusive. We can expect both to be present in a given reaction (and perhaps modes intermediate between these two extremes). They will, however, tend to vary in relative importance with the Q-value of the reaction. We can understand this in the following terms. In the compound picture, the available energy is distributed more or less uniformly over the compound nucleus. The energy spectrum of the evaporated particles will reflect a Maxwell-type of distribution of nucleon velocities within the nucleus. The probability of finding one particle carrying most of the energy is small, so the spectrum peaks for the emission of low-energy particles. Consider a particular reaction A(a, n)B; the spectrum of energies for the emitted neutrons might appear like the full curve in *Figure 2.32*, where $n(E)$ is the number emitted with energy between E and $E + dE$. The maximum energy E_m is determined by the bombarding energy E_a and the Q_0-value; in the CMS, $E_m = E_a + Q_0$. The dashed curve is the spectrum one would expect from simple evaporation. The position of the peak of this distribution is determined by the *temperature*; i.e. the degree of excitation of the compound nucleus. (We return to this in Chapter 4.) Because of the peaking of the evaporation spectrum for neutrons at low energy, if enough energy is available usually we will find two or more low-energy neutrons emitted instead of one high-energy neutron.

In practice, however, one finds an excess (over the evaporation predictions) of particles with the higher energies. These we can identify as arising from direct (or semi-direct) reactions which do not inhibit the emission of energetic particles. Rather, as already remarked, the fast direct collision tends not to excite the nucleus very much, hence the energy loss tends to be small.

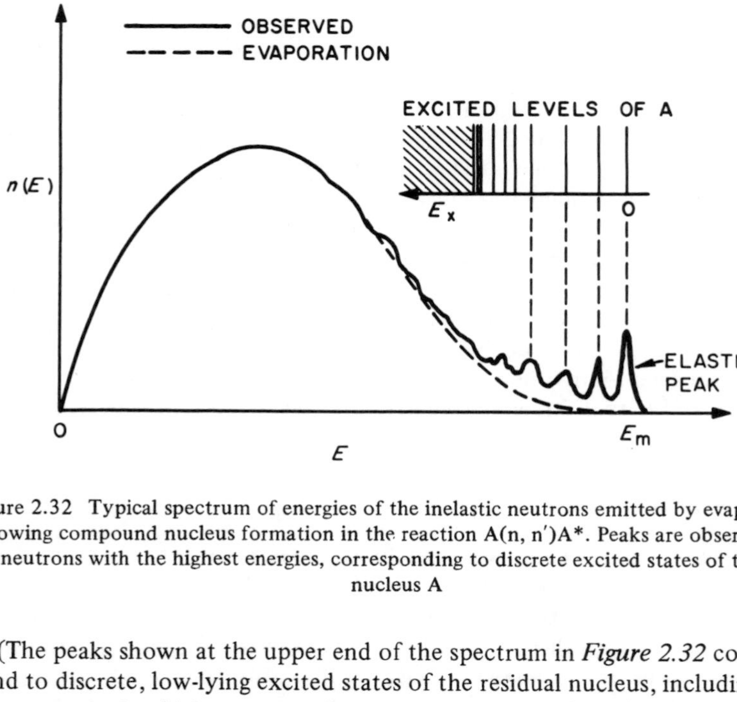

Figure 2.32 Typical spectrum of energies of the inelastic neutrons emitted by evaporation following compound nucleus formation in the reaction A(n, n')A*. Peaks are observed for the neutrons with the highest energies, corresponding to discrete excited states of the target nucleus A

(The peaks shown at the upper end of the spectrum in *Figure 2.32* correspond to discrete, low-lying excited states of the residual nucleus, including the uppermost peak which corresponds to its ground state. If the incident beam had a very precise energy and our detecting system was perfect, each of these would show up as a sharp line. In practice, the beam will have a finite energy spread, the target material will have a finite thickness and the detectors will have a finite energy resolution so the lines are broadened into peaks.)

2.18.5 Branching ratios

By this we mean the relative probabilities for a colliding pair A + a to result in the various exit channels B + b, etc. These relative probabilities are determined by *competition* when the reaction proceeds through a compound nucleus C. The hypothesis of the independence of formation and decay was expressed by equation 2.49. In this equation, the G^{C}_{Bb} represents the branching ratio for a particular B + b channel. Since the total probability for decay is unity, we must have

$$\sum_{bB} G^{C}_{Bb} = 1$$

where the sum runs over all open channels (including A + a for elastic scattering). Hence the value of G for a given channel depends strongly on how many other competing open channels there may be. For example, the elastic channel is always open, but if the bombarding energy is below the threshold for any other, non-elastic, channel, then we would have the maximum value of unity for G^{C}_{Aa}.

When the compound nucleus is highly excited and many decay channels are open, there are unlikely to be any special relationships between the compound system and a given final state. We then expect the branching probability G to depend smoothly only on simple properties like the excitation energy of the residual nucleus. This is the basis of the statistical model to be described in Chapter 4.

The factors governing the feeding of a particular channel by direct reaction are quite different. There is no formation and decay of an intermediate system so that competition between different modes of decay of the compound nucleus plays no role. The speed and simplicity of a direct reaction ensures that the target nucleus is only slightly rearranged to form the residual nucleus. In other words, the target and residual nuclei are very similar in structure; their wave functions have good overlap. As a corollary to this, if the structure of a particular residual nuclear state differs considerably from that of the target we will not expect it to be excited by a direct reaction mechanism. Hence there will be selectivity even among the low-lying excited states of the residual nucleus. Some will be fed strongly by the direct mode, some will only be fed weakly by the compound or some intermediate mode.

Although both compound and direct reaction components will generally contribute to a given reaction, some types of reactions will favour one over the other. For example, the spectrum shown in *Figure 2.32* is typical of those for the inelastic scattering of neutrons of, say, 10 or 20 MeV. Most of the reactions proceed through the compound mode, but with an important high-energy direct component. Some reactions, such as the (p, d) reaction with protons of a few tens of MeV, appear to proceed predominantly through the direct mode (which, in this case, is the simple pick-up of a neutron from the nuclear surface). The relative importance of direct and compound processes also depends upon the bombarding energy, as shown in *Figure 2.30*. The behaviour shown in *Figure 2.30* for the compound cross-section can be understood in terms of competition. As other channels become open, they compete for the flux from the compound nucleus decay so that the cross-section for this particular final state decreases. This is particularly marked for proton-induced reactions when the (p, n) threshold is reached; neutrons can escape much more readily than charged particles because they experience no Coulomb barrier (section 2.18.8) and tend to become the most important decay channel.

2.18.6 Importance of direct reactions

This tendency for the low-lying states of nuclei to be populated by direct reactions is one of the main reasons for the importance of these reactions. The nucleus represents a complicated many-body problem and in general we can only expect our theories of nuclear structure to give accurate and detailed explanations of the structure of these isolated low-energy quantum states. The direct reactions have a relatively simple nature, with no complications from an intermediate compound system, and are well suited to giving information about the relationship (overlap) between the ground state of the target nucleus and the ground or a particular excited state of the residual nucleus. For example, with a deuteron stripping (d, p) reaction we may learn to what degree a given residual nuclear state looks like the target ground state with the simple addition of one neutron in a particular shell model orbit. This gives us important information about the validity of the shell model and can provide tests of calculated shell model wave functions.

The dense, highly excited, states of nuclei generally have complicated structures and we can only expect to understand their properties in an average or statistical way. Hence, it is particularly appropriate that the compound reaction mechanism which tends to populate them is also one that can be described in similar terms.

2.18.7 Characteristic angular distributions

We have remarked that the differential cross-section for a reaction A(a, b)B depends upon the angle of emission of the residual particle b relative to the incident beam. The reaction products are not, in general, emitted isotropically, that is to say in all directions with equal probability. This might seem surprising for a compound nucleus reaction, where there is supposed to be independence of formation and decay, so that the reaction products have forgotten the direction of the incident beam. In fact, such reactions exhibit angular distributions which are *not* isotropic although in many cases they do not vary strongly with angle. The explanation is that we can never have *complete* independence of formation and decay because certain constants of the motion must be conserved and their initial values are determined by the conditions in the entrance channel. The particular conserved quantity of importance here is the angular momentum. The angular momentum **L** of the incident particle relative to the target nucleus is a vector which is always perpendicular to its direction of motion and hence has zero component in this direction. It is through this property that the incident direction influences the subsequent decay of the compound nucleus. If the intrinsic spins of the projectile and target are zero (or negligible compared to **L**), then **L** is also the spin of the compound nucleus, which is then oriented perpendicular to the direction of the incident beam. In the extreme classical limit of very large **L**, we can visualise the decay particles being flung preferentially from the equator of the rapidly rotating compound system. When we average over the orientation of this spinning system by averaging in azimuth around the incident direction, this 'water wheel' picture results in a preponderance of particles emitted equally in the forward and backward directions. Because the element of solid angle is $d\Omega = \sin\theta \, d\theta \, d\phi$, *uniform* emission from all parts of the equator gives a differential cross-section $d\sigma/d\Omega$ which is proportional to $1/\sin\theta$. However, unless the energy is high or the mass of the projectile is large (a heavy ion), the value of **L** will be rather small. In addition, the two particles often have intrinsic spins which will be randomly oriented and this will tend to reduce the degree of orientation of the spin of the compound nucleus. As a consequence, these compound reactions often do not show a marked anisotropy in the angular distribution of their products (*see Figure 2.33*). Exceptions occur when large **L** values are involved, as happens in heavy-ion collisions.

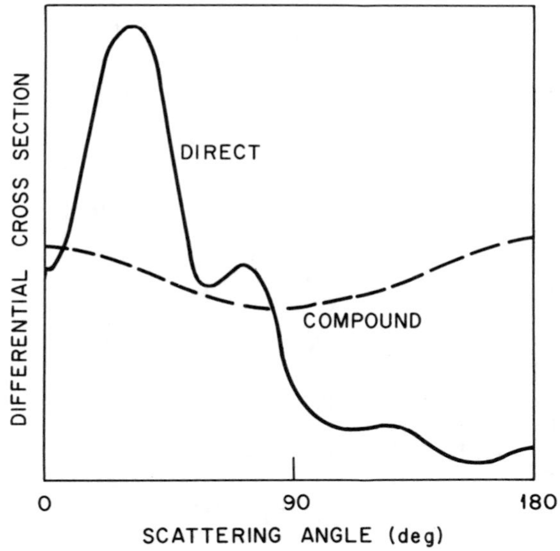

Figure 2.33 Typical angular distributions for direct and compound nucleus reactions induced by light ions with moderate bombarding energies (for example, some (d, p) reaction with 20-MeV deuterons)

Provided we are averaging over a sufficiently large number of compound nucleus states (either because our energy resolution is poor or because the states are dense and overlap strongly), the angular distribution will exhibit backward and forward symmetry. That is, the intensities at θ and $\pi-\theta$ will be equal. This is a characteristic of a compound nuclear reaction.

From our description of a direct reaction, the projectile suffers a glancing or peripheral collision with the target. It may lose some energy, or have one or a few nucleons transferred to or from it, but in any case, it will tend to continue moving in the forward direction. Hence the angular distribution of the emitted particle from a direct reaction will tend to be strongly peaked in the forward direction (*Figure 2.33*) and not exhibit a backward-forward symmetry. We shall see later that often we can learn from the form of these angular distributions important information about the angular momentum transferred between the particles during the collision, and in turn this can tell us about the spin and parity of the residual nuclear state.

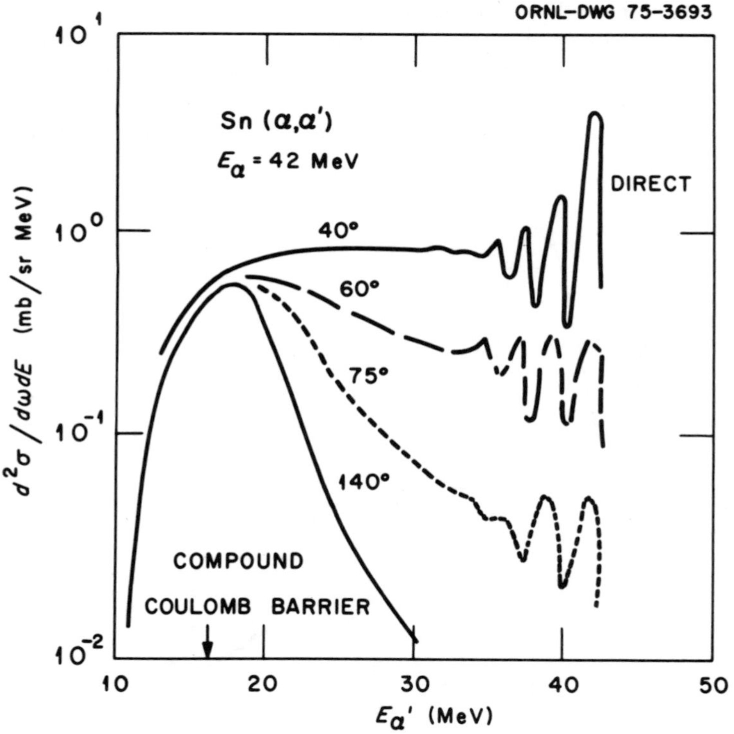

Figure 2.34 Spectra of the energies of α-particles inelastically scattered from Sn nuclei at various angles. α-particles with small energy losses (large $E_{\alpha'}$) show forward-peaked distributions characteristic of direct reactions. The low-energy α-particles are almost isotropic and more characteristic of evaporation from a compound nucleus. (After Chenevert *et al.*, 1971)

An example which illustrates several of the features just discussed is the Sn(α, α') reaction at 42 MeV. *Figure 2.34* shows the inelastic α-particle spectra as measured at several scattering angles. The most backward angle gives an evaporation type of spectrum which is typical of the compound nucleus process and in which the most probable energy is below 20 MeV (representing a large energy loss). As one comes to more forward angles the discrete low-excited states of the nucleus are more strongly excited as direct reactions become more important, illustrating both the feeding of low states (small energy loss) by the direct process and the forward-peaking associated with it.

2.18.8 Coulomb effects

Except for neutrons, all the particles involved in the usual nuclear reaction will be positively charged and hence repel each other. Once two nuclei a and A are close enough for the strong, attractive nuclear forces to act, this Coulomb repulsion is overwhelmed and is usually not very important (*see Figure 2.35*). At larger separations, however, it is far from being negligible and has an important influence on the probability of two nuclei coming into contact and undergoing a reaction. A proton approaching the lead nucleus, for example, will experience a repulsive Coulomb potential of about 13 MeV before the specifically nuclear forces begin to act. More highly charged projectiles experience correspondingly higher Coulomb barriers. Clearly the cross-section σ^C for forming a compound nucleus will be drastically reduced as the bombarding energy falls below this barrier. Direct reactions will be suppressed also, although, because they are peripheral, the two nuclei do not have to approach as closely

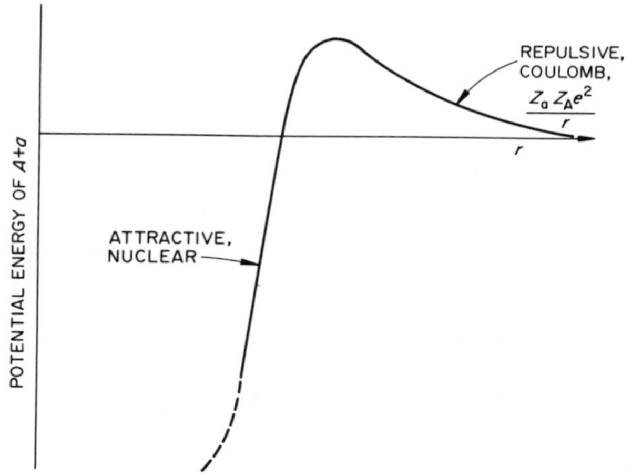

Figure 2.35 Schematic picture of the *Coulomb barrier* for two nuclei a + A. As the attractive nuclear potential begins to act, the total potential reaches a maximum then decreases to become negative (attractive). The repulsive part, due to the Coulomb potential, represents a barrier which tends to keep the nuclei apart

to experience a direct reaction as they do for complete coalescence into a compound system.

A charged particle within the nucleus, on the other hand, finds its escape is inhibited by this Coulomb barrier unless the energy available to it is comparable to or higher than the barrier (the classic example is the effect of barrier penetrability on α-decay; *see* Preston and Bhaduri, 1975). This means that the evaporation of low-energy charged particles from a compound nucleus will tend to be strongly inhibited relative to the emission of neutrons. It follows that the energy spectrum for a charged particle has its maximum shifted to a larger energy than that for neutron emission (*see Figure 2.36* for a comparison of the spectra of emitted neutrons and protons).

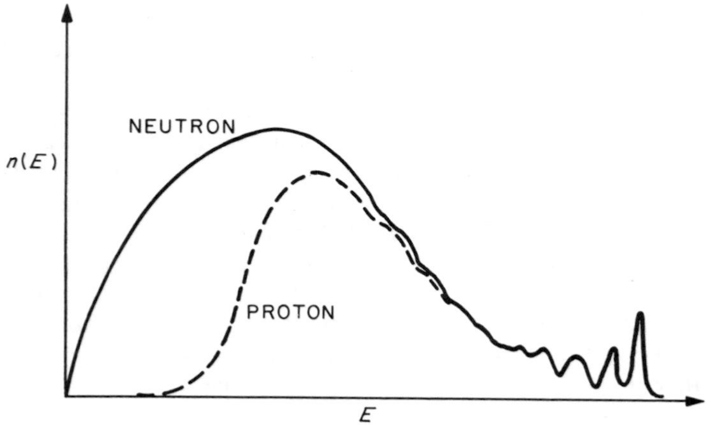

Figure 2.36 Comparison of the spectra of neutrons and protons evaporated from compound nuclei, leaving the same residual nucleus. The intensity of low-energy protons is depressed because of the Coulomb barrier they experience

The Coulomb barrier is very important for the collision of two heavy ions. For example, this barrier is about 400 MeV high for a collision between Xe and Pb nuclei. As a consequence, a high bombarding energy is required before these nuclei will fuse together or interact strongly.

2.18.9 Giant resonances and strength functions

As a general rule, the probability of exciting high-lying states in a nucleus varies smoothly with their excitation energy. At these excitations, individual states usually cannot be resolved and the spectrum of outgoing particles may look like those in *Figures 2.32, 2.34, 2.36*. Sometimes, however, a gross structure may be seen in the spectrum; this may be a concentration of excitation strength in a given region of excitation energy. An example was shown in *Figure 2.27*. Such a concentration is often referred to as a *giant resonance*. Although we associate definite quantum numbers, such as spin and parity, with this peak, we do not interpret it as the excitation of a single level of the nucleus, for such a resonance may have a width of one or more MeV in a region where the underlying fine-structure levels are known to be separated by only a few eV. When seen in inelastic scattering measurements it is natural to identify these resonances with some kind of collective motion (*see*, for example, Satchler, 1978). For example, a liquid drop will be able to undergo oscillations in shape and a quantum state of such oscillation would be strongly excited by inelastic scattering. However, we realise that this liquid-drop picture is only a rough approximation to the average behaviour of a nucleus; when we take account of the individual nucleon motions, we find that such an oscillatory state is only describing an average property of the very many actual states possible. Each of the actual states has a certain probability of being excited. This probability fluctuates in strength from level to level, but when studied as a function of excitation energy it will show on average a marked increase in the region of the giant resonance. This distribution of average probability can be called a *strength function*. The concept is not restricted to inelastic scattering; indeed it was first introduced to describe the average cross-section for formation of a compound nucleus by the capture of low-energy (*s*-wave) neutrons. This average cross-section shows a giant resonance behaviour as a function of the mass number A of the target nucleus (*see Figure 4.43*); it is large for nuclei with $A \sim 60$ and $A \sim 150$–190 but small in between for nuclei with $A \sim 100$. This behaviour is evidence for the partial transparency of nuclei to neutrons of low energy which then experience 'size-resonance' effects according to whether or not an integral number of their half-wavelengths will fit inside the nucleus. This can be described by the optical model (*see* Chapter 4 for more details).

2.18.10 Cross-section fluctuations

At low bombarding energies, the cross-section for a given reaction may show sharp resonances as the bombarding energy is varied (section 2.18.2 and *Figure 2.31*). As the energy is increased, these resonances broaden and become more dense until they overlap. However, careful measurements with good energy resolution may show that even then the cross-section does not vary smoothly with energy but rather exhibits rapid *fluctuations*. If the contributions from the many underlying levels are randomly distributed, these are *statistical fluctuations* and we may ask for the distribution of the values of the cross-section. Under simple assumptions, the distribution of amplitudes is *normal*; the probability of finding a cross-section value between σ and $\sigma + \mathrm{d}\sigma$ is then (Ericson, 1963)

$$P(\sigma)\mathrm{d}\sigma = (\mathrm{d}\sigma/\overline{\sigma}) \exp(-\sigma/\overline{\sigma})$$

where $\overline{\sigma}$ is the mean value. With this distribution, the root mean square deviation is equal to the average cross-section itself

$$\left(\overline{\sigma^2} - \overline{\sigma}^2\right)^{1/2} = \overline{\sigma}$$

Figure 2.37 shows the results of some measurements on the $^{12}C\,(^{16}O, \alpha)\,^{24}Mg$ reaction; *see* also *Figure 2.27*. A more detailed discussion of fluctuation phenomena is given in Chapter 4.

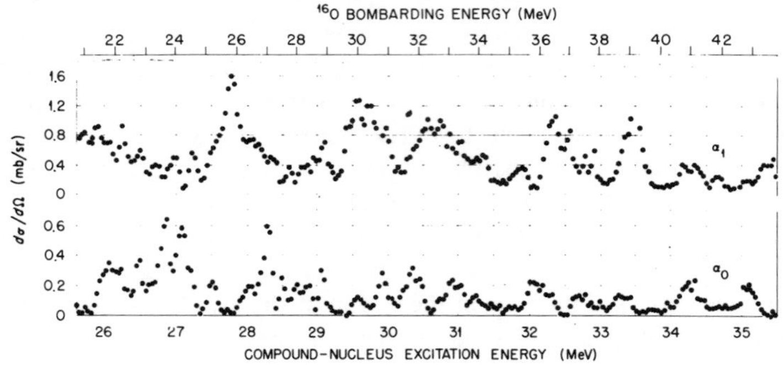

Figure 2.37 Excitation function (variation of the cross-section for a fixed angle as the energy is changed) for the reaction $^{12}C\,(^{16}O, \alpha)\,^{24}Mg$ at $\theta_L = 20°$ showing how the cross-section fluctuates rapidly as the bombarding energy is varied. (From Halbert *et al.*, 1967)

2.18.11 Strong and weak absorption: diffraction and the optical model

If a spherical object is black to radiation of wavelength $\lambda\!\!\!^{-}$ falling upon it, the cross-section for absorption is $\pi(R + \lambda\!\!\!^{-})^2$ where R is its radius; when $\lambda\!\!\!^{-}$ is small, this is just the geometric cross-section πR^2. The angular distribution of the scattered radiation shows a Fraunhofer diffraction pattern. Such patterns are seen in many nuclear reactions, for example in the elastic scattering of α-particles (*see* Chapter 4 for examples, especially *Figures 4.1 4.5, 4.7, 4.20*) and are said to be characteristic of ***strong absorption***. Classical diffraction theory predicts that these patterns are a universal function of $R\sin\theta/\lambda\!\!\!^{-} = kR\sin\theta$, where θ is the scattering angle, and this is found to be approximately true. However, the scattering of low- and medium-energy neutrons and protons does not show a sharp diffraction pattern; rather it implies that target nuclei are partially transparent to nucleons. For example, the total cross-section for neutrons scattering from nuclei (*Figure 2.6*) oscillates about the black nucleus value $2\pi(R + \lambda\!\!\!^{-})^2$ when the energy (or $\lambda\!\!\!^{-}$) or the target mass (or R) is varied. This can only mean that the neutron waves penetrate into the nuclear interior sufficiently for some resonance condition to be important. Thus the mean free path for neutrons within a nucleus must be comparable to the nuclear radius; this we call *weak absorption*. This observation of weak absorption phenomena led to the introduction of the 'cloudy crystal ball' model, now called the optical model (Chapter 4), which can explain these phenomena quantitatively.

2.18.12 Some characteristics of heavy-ion reactions

Reactions between two heavy nuclei exhibit an even richer variety of phenomena than those between a light projectile and a heavy target nucleus. (A 'heavy' ion is defined as a nucleus with mass $A > 4$.) However, a few simple characteristics can be remarked upon. Firstly, the large mass means a proportionately greater momentum for a given energy and consequently a greater amount of ***angular*** momentum* about the centre of a struck target nucleus (*see* equation 2.43). This facilitates the excitation of nuclear states with very high spin; for example, states with spin as large as $60\hbar$ have been excited in (HI, xn) reactions (this means a reaction in which a heavy ion (HI) projectile, e.g. ^{16}O, is absorbed by a target nucleus and some of the excitation energy is lost by the evaporation of a number x of neutrons). This large angular momentum and the associated kinetic energy is distributed over the many nucleons in the heavy projectile which makes the formation of a compound nucleus more probable than if it were concentrated upon one or just a few nucleons as in a high-energy light-ion reaction. We can then study, for example, whether there are limits to the amount of angular momentum that a compound nucleus can sustain so that it becomes unstable and flies apart under the influence of the centrifugal forces. Indeed, it is found that for angular momenta larger than a critical value, the compound nucleus rather quickly fissions into two large fragments.

Secondly, reactions between heavy ions are always characterised by strong absorption; that is to say, once they approach within the range of their strong interactions, some non-elastic event will occur. Further, heavy-ion reactions show much more strongly than reactions with light-ion projectiles classes of phenomena intermediate between the extreme peripheral (or direct) reactions occurring at grazing impact and the complete fusion (or compound nuclear) reactions resulting from more head-on collisions (*see*, for example, Schroder and Huizenga, 1984). One sees phenomena which appear to result from the two heavy nuclei sticking together for a short time, longer than one would associate with a direct reaction, but not long enough for the two systems to be regarded as having fused into a compound nucleus. There tends to be a large loss of kinetic energy in such collisions (hence they are sometimes called ***deep inelastic*** or ***strongly damped*** collisions), the energy reappearing as internal excitation ('heat') energy, but the system still retains 'memory' of how it was formed by showing, for example, a forward-peaked angular distribution. In addition, even when there is complete fusion, it is not always followed by evaporation, as is characteristic of light-ion reactions, but there is often a strong probability that the compound nucleus will fission into two large fragments.

The large mass of a heavy ion means the associated wavelength is shorter than

*For example, a Zn nucleus with a bombarding energy of 400 MeV (only 6.25 MeV per nucleon) carries an angular momentum of nearly $200\hbar$ when it makes a grazing collision with an Sn nucleus.

for a light ion with the same energy. In many cases the wavelength is short enough for it to be meaningful to use classical concepts to describe a collision, such as speaking of the ion following a classical trajectory. This has led to a resurgence of interest in semi-classical theories of scattering. Even if a full quantal treatment is needed to account for all the details of a reaction, often considerable physical insight can be gained from the semi-classical description.

Because the charge on a heavy-ion projectile may be quite large, high energies are required to overcome the Coulomb repulsion between projectile and target nuclei. As an extreme example, the Coulomb barrier to be surmounted before two uranium nuclei make contact is about 700 MeV. This requires a bombarding energy in the laboratory system of about 1400 MeV. However, the *average energy per nucleon* in the projectile is not high; only about 6 MeV per nucleon in this example and 5–50 MeV per nucleon are typical figures for many experiments, although some measurements are now being made at several hundred MeV per nucleon. The study of heavy-ion interactions is a rapidly expanding subject; accelerators are now available that can produce projectile ions with energies ranging from a few MeV per nucleon to as much as 200 GeV per nucleon, and which can accelerate ions as heavy as uranium.

It is useful to break up this enormous range of possibilities in a way which reflects the kind of physics which can be studied. At energies below the threshold for pion production (about 140 MeV in the CMS), we need only consider nuclei as made of individual nucleons. A characteristic length is ~ 1 fm (or the wavelength of the most energetic nucleon in a Fermi gas with the density of nuclear matter, $\lambdabar_F \approx 0.7$ fm). A characteristic energy is a few tens of MeV (or the maximum kinetic energy of a nucleon in the same Fermi gas, $E_F \approx 40$ MeV).

The velocity of sound in nuclear matter is estimated to be equal to that of a nucleon with a kinetic energy of about 20 MeV (or a speed of about one-fifth the velocity of light). This leads us to subdivide the 'low' energy region into subsonic and supersonic sections. In the subsonic region we expect nuclei to behave as though they were incompressible. However, when we have reactions at supersonic energies we may see compressibility playing a role; for example, we might generate shock waves in nuclei. The properties of nuclear matter at above-normal density have not yet been studied.

Above the pion-production threshold we can no longer expect to be able to ignore the substructure of nucleons. Nucleon isobars may be produced, for example. Eventually we reach energies (~ 1 GeV per nucleon) which are comparable to or greater than the energy of the rest mass of a nucleon; in this domain we cannot escape the consequences of relativity. It has been suggested that we might create a new abnormal or condensed state of nuclear matter in which the nucleons dissolve into a plasma of their constituent quarks and gluons. To induce this state, we need nuclear matter that is hotter and considerably more dense than normal. The possibility of producing this in the collision of ultrarelativistic heavy ions (~ 200 GeV per nucleon) with targets of heavy nuclei is presently being explored experimentally.

Because heavy-ion collisions involve the interactions between relatively large pieces of nuclear matter in which the kinetic energy is shared among all the nucleons, they are particularly suitable for the study of the *macroscopic* features of nuclear behaviour. These are most easily described in terms more appropriate to the motions of liquid droplets rather than the motions of one or a few nucleons. For example, we find the concepts of viscosity and friction being invoked. The deep inelastic scattering already mentioned is one of these features. Another possibility is the formation of nuclear 'molecules' (Vogt and McManus, 1960) in which two nuclei make sufficient contact to be attracted to one another, but in which they do not completely fuse to form the usual kind of compound nucleus. This structure could be recognised by the large probability that it would decay by breaking up into the same two nuclei by which it was formed; it is not a compound nucleus in the usual sense because it 'remembers' the entrance channel and the independence hypothesis is not obeyed.

It has been suggested that nuclei with masses $A \sim 310$ may be stable although they have not yet been found in nature. One of the attractions of heavy-ion physics has been the possibility of forming these *super-heavy* elements by fusing two heavy ions. At the other extreme there is the possibility of using the fusion of two very light nuclei as a power source (Appendix D).

A thorough treatment of heavy-ion reactions is to be found in Bass (1980) and Bromley (1984).

REFERENCES

Alder, K. and Winther, A. (1975). *Electromagnetic Excitation: Theory of Coulomb Excitation with Heavy Ions*. Amsterdam; North-Holland

Bass, R. (1980). *Nuclear Reactions with Heavy Ions*. Berlin; Springer-Verlag

Bellicard, J. B., Bounin, P., Frosch, R. F., Hofstadter, R., McCarthy, J. S., Clark, B. C., Herman, R. and Ravenhall, D. G. (1967). *Phys. Rev. Lett.* Vol. 19, 527

Biedenharn, L. C. and Brussard, P. J. (1965). *Coulomb Excitation*. Oxford; Clarendon Press

Blann, M. (1975). *Ann. Rev. Nucl. Sci.* Vol. 25, 123

Breit, G., Hull, M. H., Lassila, K. E. and Pyatt, K. D. (1960). *Phys. Rev.* Vol. 120, 2227

Bromley, D. A. (1974). In *Proc. Int. Conf. Nuclear Physics*. Eds J. de Boer and H. J. Mang, Amsterdam; North-Holland

Bromley, D. A. (1984). *Treatise on Heavy Ion Science*. New York; Plenum Press

Bromley, D. A., Keuhner, J. A. and Almquist, E. (1961). *Phys. Rev.* Vol. 123, 878

Cerny, J. ed. (1974). *Nuclear Spectroscopy and Reactions.* New York; Academic Press

Chenevert, G., Chant, N. S., Halpern, I., Glasshausser, C. and Hendrie, D. L. (1971). *Phys. Rev. Lett.* Vol. 27, 434

Christensen, P. R., Manko, V. I., Becchetti, F. D. and Nickles, R. J. (1973). *Nucl. Phys.* Vol. A207, 33

Daniels, J. M. (1965). *Oriented Nuclei*. New York; Academic Press

England, J. B. A. (1974). *Techniques in Nuclear Structure Physics*. New York; Halsted

Ericson, T. (1963). *Phys. Lett*. Vol. 4, 258

Ferguson, A. J. (1974). In *Nuclear Spectroscopy and Reactions*. Ed. J. Cerny. New York; Academic Press

Fowler, J. L., Minor, T. C., Martin, F. D., Montgomery, H. E. and Okun, L. M. (1969). *Am. J. Phys.* Vol. 37, 649

Frauenfelder, H. and Steffen, R. M. (1966). In *Alpha-, Beta- and Gamma-Ray Spectroscopy*. Ed. Siegbahn, K. Amsterdam; North-Holland

Friar, J. L. and Negele, J. W. (1973). *Nucl. Phys.* Vol. A212, 93

Geiger, H. and Marsden, E. (1913). *Phil. Mag.* Vol. 25, 604

Gill, R. D. (1975). *Gamma-Ray Angular Correlations*. New York; Academic Press

Goshal, S. N. (1950). *Phys. Rev.* Vol. 80, 939

Halbert, M. L., Durham, F. E. and van der Woude, A. (1967). *Phys. Rev.* Vol. 162, 899

Heisenberg, J., Hofstadter, R., McCarthy, J. S., Sick, I., Clark, B. C., Herman, R. and Ravenhall, D. G. (1969). *Phys. Rev. Lett.* Vol. 23, 1402

Heisenberg, J. H. and Sick, I. (1970). *Phys. Lett*. Vol. 32B, 249

Hess, W. N. (1958). *Rev. Mod. Phys.* Vol. 30, 368

Hodgson, P. E. (1987). *Reps. Prog. Phys*. Vol. 50, No. 9

Johnson, R. C. (1985). In *Clustering Aspects of Nuclear Structure*. Eds J. S. Lilley and M. A. Nagarajan. Dordrecht; Reidel

Kinney, W. E. and Perey, F. G. (1970). *Nucl. Sci. Eng*. Vol. 40, 396

Kocher, D. C. and Haeberli, W. (1972). *Nucl. Phys.* Vol. A196, 225

Kondo, M., Kobayashi, S., Tanifuji, M., Yamazaki, T., Kubo, K. I. and Onishi, N. (1986). *Polarization Phenomena in Nuclear Physics*, Suppl. to *J. Phys. Soc. Japan*. Vol. 55

Lee, L. L. and Schiffer, J. P. (1963). *Phys. Lett*. Vol. 4, 104

McGowan, F. K. and Stelson, P. H. (1974). In *Nuclear Spectroscopy and Reactions*. Ed. J. Cerny. New York; Academic Press

Mattauch, J. H. E., Thiele, W. and Wapstra, A. H. (1965). *Nucl. Phys.* Vol. 67, 1

Mott, N. F. and Massey, H. S. W. (1965). *The Theory of Atomic Collisions*, 3rd edn. Oxford; Oxford University Press

Preston, J. A. and Bhaduri, R. K. (1975). *Structure of the Nucleus*. Reading, Mass.; Addison-Wesley

Ramage, J. C., McKeown, J. and Ledingham, K. W. D. (1975). *Am. J. Phys.* Vol. 43, 51

Rand, R. E., Frosch, R. and Yearian, M. R. (1966). *Phys. Rev.* Vol. 144, 859

Rutherford, E. (1911). *Phil. Mag.* Vol. 21, 669

Satchler, G. R. (1978). In *Proc. Int. School of Physics 'Enrico Fermi', Course LXVIX*. Eds. A. Bohr and R. A. Broglia. Amersterdam: North-Holland

Schroder, W. U. and Huizenga, J. R. (1984). In *Treatise on Heavy Ion Science*,

Vol. 2. Ed. D. A. Bromley. New York; Plenum Press
Sinha, B. B. P., Peterson, G. A., Whitney, R. R., Sick, I. and McCarthy, J. S. (1973). *Phys. Rev.* Vol. C7, 1930
Uberall, H. (1971). *Electron Scattering from Complex Nuclei*. New York; Academic Press
Vogt, E. and McManus, H. (1960). *Phys. Rev. Lett.* Vol. 4, 518
von Witsch, W., Richter, A. and von Brentano, P. (1968). *Phys. Rev.* Vol. 169, 923

EXERCISES FOR CHAPTER 2

2.1 A particle of mass M_a with energy E is scattered elastically by a target with mass M_A. The target recoils at an angle θ_A (as measured in the laboratory). Derive an expression for the kinetic energy and the velocity of the recoiling target particle.

Chadwick estimated the mass of the neutron from measurements on the elastic scattering of neutrons by hydrogen and by nitrogen. The maximum velocity of the recoiling protons was observed to be 3.3×10^7 m s^{-1} while the maximum recoil velocity of the nitrogen ions was 4.7×10^6 m s^{-1}. Estimate the mass of the neutron. How does this compare with the present value? What was the energy of the incident neutrons?

2.2 Derive the expression equation B20 of Appendix B for x, the ratio of the speed of the centre of mass of a colliding system to the speed of the outgoing particle in the CMS.

Find the maximum allowed scattering angle θ_L in the LAB system for a reaction in which $x > 1$.

2.3 Using Appendix B, find the distribution (a) in energy and (b) in angle (in the LAB system) of the recoil protons from the s-wave scattering of 1-MeV neutrons by a small quantity of hydrogen.

2.4· What is the minimum bombarding energy of ^{12}C ions needed to initiate the reaction $^{12}C + ^{28}Si \rightarrow ^{16}O + ^{24}Mg$ if the Q-value is -2.82 MeV? If instead, ^{12}C were used as the target, what is the minimum energy of the ^{28}Si ions that would be required?

Suppose the $^{28}Si\,(^{12}C, ^{16}O)\,^{24}Mg$ reaction was studied with ^{12}C ions of 96 MeV. What bombarding energy of ^{16}O ions would give the inverse reaction at the same total energy?

2.5 If ^{16}O and ^{12}C at rest were combined to form ^{28}Si in its ground state there would be an energy release of $Q = 16.75$ MeV. Similarly, formation of ^{28}Si in its ground state through the capture of a proton by ^{27}Al would release $Q = 11.59$ MeV.

With what excitation energy would the compound nucleus be formed when ^{12}C is bombarded by ^{16}O ions of 50 MeV? What energy protons bombarding

^{27}Al would be required to form the same compound nucleus at the same excitation energy?

Estimate the highest angular momentum with which we could expect to form ^{28}Si in each reaction. (Assume that the proton is a point particle and that the other nuclei have radii of $1.3 \times A^{1/3}$ fm.)

2.6 A thin target of ^{24}Mg metal had a mass of 1 mg cm^{-2}. It was bombarded by a beam of α-particles with an intensity of 10^6 particles s^{-1}. The number of protons emitted from the target was observed to be 9 s^{-1}. Estimate the total cross-section for these Mg (α, p) Al reactions. (Assume that a ^{24}Mg atom has a mass of 24 amu.)

2.7 An α-particle approaches a ^{208}Pb nucleus. The potential energy due to their Coulomb repulsion is given by equation 2.19. Assume that the nuclear forces provide an additional potential energy of the form

$$V \exp\left\{-[(r - R_0)/a]\right\}$$

where $V = -135$ MeV, $R_0 = 1.5\ (208)^{1/3}$ fm and $a = 0.5$ fm. Calculate and compare these two potential energy terms when the centres of the two nuclei are separated by $r = 11$ fm. (The charge radius may be taken to be $R = 7.1$ fm.) What are the corresponding forces? At what distance r do the two forces just balance? Plot the total potential as a function of r for $r > R_0$.

Assuming that a nuclear potential of magnitude less than 0.1 MeV has negligible effect on the scattering, at what angle would you expect the scattering of α-particles with a bombarding energy of 25 MeV to show deviations from the Rutherford formula?

2.8 Verify equation 2.32 and deduce the next term in the expansion. Derive an exact expression for $F(q)$ when the nuclear charge is distributed uniformly over a sphere of radius R. Give explicit expressions for the coefficients $\langle r^n \rangle$ which enter the expansion equation 2.32.

Assume that the minima in the angular distribution shown in *Figure 2.15* of electrons scattered from ^{40}Ca correspond to the zeros of the expression you have derived for $F(q)$. Deduce the radius R of the charge distribution of ^{40}Ca. (Take the momentum transfer to be $q \approx k\theta$ where the scattering angle θ is measured in radians and the relativistic wave number is given by $k \approx E/\hbar c$. The bombarding energy E is 750 MeV.)

Why do the measured cross-sections not go to zero as this simple model predicts?

2.9 Why does the expansion of equation 2.37 involve only Legendre polynomials $P_\ell(\cos\theta)$ and not the more general spherical harmonics $Y_\ell^m(\theta, \phi)$ of Appendix A? (Hint: what choice of z-axis is implied by the form of equations 2.36 and 2.37?)

2.10 Explicitly derive equation 2.39 from the other equations given. Rewrite the expression for $f(\theta)$ in terms of the phase shifts δ_ℓ.

Derive an explicit expression for the differential cross-section in powers of $\cos\theta$, assuming that only *s*- and *p*-waves are scattered. Is the angular distribution necessarily symmetric about $\theta = \frac{1}{2}\pi$? It not, what is the origin of the non-symmetric terms?

Give an expression for the integrated cross-section as defined by equation 2.10 and discuss its form compared to that for the differential cross-section.

2.11 Neutrons are scattered from hydrogen. As the neutron-bombarding energy is increased from zero, at what energy do you expect the scattering to deviate from isotropy in the centre of mass system? (Assume that the range of the nuclear force is 1.4 fm.)

Compare your result with *Figure 2.22*. The experimental results shown there indicate anisotropy at 27 MeV. To what do you attribute this?

At higher energies, *Figure 2.22* shows the angular distribution is not symmetric about $\theta = 90°$. Discuss the origin of this asymmetry.

2.12 Protons are subjected to a magnetic field of 20 kG and cooled to a temperature of 0.1 K. Calculate the degree of polarisation of the proton spins. (The magnetic moment of the proton is 2.793 μ_N, where the nuclear magneton is $\mu_N = 0.50505 \times 10^{-23}$ erg G^{-1}.)

What temperature or magnetic field would be required to achieve a polarisation of 50%?

2.13 Two indistinguishable nuclei with zero spin, such as $^{12}C + {}^{12}C$, are elastically scattered through 90° in the CMS. Compare the symmetrised and the unsymmetrised expressions for the cross-section.

Deduce expressions for the corresponding cross-sections for the scattering of two indistinguishable nuclei with spin -$\frac{1}{2}$, such as $^{13}C + {}^{13}C$, when the scattering interaction is independent of the spin orientation. (Hint: it follows that the orientation of the spin of each nucleus, i.e. its magnetic quantum number, cannot be changed by the scattering. Therefore the scattering amplitudes for flipping this spin must vanish. Further, the amplitudes which do not flip the spin must be independent of the spin orientation and hence all be equal. Catalogue all possible pairs of initial and final spin orientation; each pair gives an incoherent contribution to the cross-section.)

2.14 The differential cross-section at 168° for the reaction $^{24}Mg\,(\alpha, p)\,^{27}Al$ was measured to be 5.9 ± 0.2 mb sr^{-1} for an α-bombarding energy of 13.6 MeV. The inverse reaction $^{27}Al\,(p, \alpha)\,^{24}Mg$ at the corresponding energy (*see Figure 2.25*) was found to have a cross-section of 2.0 ± 0.1 mb sr^{-1}. The *Q*-value of the (α, p) reaction is -1.59 MeV. Deduce the spin of the ground state of ^{27}Al.

2.15 Derive the classical expression discussed in section 2.18.7

$$\frac{d\sigma}{d\Omega} \propto \frac{1}{\sin\theta}$$

for the angular distribution of particles emitted from a compound nucleus with a very large angular momentum.

2.16 Dipole γ-radiation whose photons carry unit angular momentum with a z-component equal to M has an intensity whose angular distribution is $F_M(\theta)$, where

$$F_0(\theta) = \sin^2\theta, \quad F_{\pm 1}(\theta) = \tfrac{1}{2}(1 + \cos^2\theta)$$

and θ is the angle between the direction of the radiation and the z-axis. Compare this with the radiation pattern from a dipole antenna.

In the reaction ^{20}Ne (^{12}C, α) ^{28}Si, a state with spin-1 in ^{28}Si is excited which then decays by γ-emission to the spin-zero ground state of ^{28}Si. An angular correlation measurement is made by observing the γ-rays in coincidence with α-particles emitted at 0°. Deduce the angular distribution of these γ-rays with respect to the incident beam.

Suppose the state in ^{28}Si had been formed in such a way that all three magnetic substates were populated equally. What would the angular distribution of γ-rays then become?

2.17 A spherical heavy ion projectile, such as ^{40}Ar, is on a hyperbolic (Rutherford) orbit as it scatters from a target nucleus, such as ^{238}U, with an ellipsoidal shape. Such an ellipsoidal nucleus may rotate and possesses a rotational spectrum of excited states (*see* Chapter 1 and *Figure 1.4*). Discuss the electrical forces exerted on the target by the projectile and their effects. (*See*, for example, McGowan and Stelson, 1974.)

2.18 Two nuclei follow a Rutherford orbit under the influence of their mutual Coulomb repulsion. The distance d of their closest approach and their angular momentum L are related by equation 2.27.

Construct a table showing the maximum angular momentum that various representative pairs of projectile and target nuclei can bring into their compound nucleus at various bombarding energies. Assume that they will only fuse if they approach one another so that the distance between their centres is less than $d_{\text{fus}} = 1.1(A_1^{1/3} + A_2^{1/3}) + 2$ fm. In each case, calculate the scattering angles corresponding to these grazing orbits.

2.19 How should the discussion of section 2.10 be modified if the projectile and target have charges of opposite sign (such as for the scattering of an electron, a negative pion or an anti-proton by a nucleus)? How would *Figure 2.35* be changed for such a case? Would there be a 'Coulomb barrier'?

3

Elementary Scattering Theory

This chapter is concerned with developing the basic language to be used in describing nuclear reactions and with establishing some general properties that such descriptions must satisfy. We indicated briefly in section 2.15 how we obtain the solution of the Schrödinger equation for the scattering of a particle by a potential and how to use this wave function to obtain the scattering amplitude and cross-section. Here we show how the wave function and the Schrödinger equation need to be generalised to describe collisions between complex nuclei which allow inelastic scattering and rearrangement reactions. The construction of various models, by which we clothe the formal framework with physical ideas, is dealt with in a later chapter. Of course, we can do no more than summarise those features of scattering theory which are of particular importance to us here. More complete and detailed treatments can be found in quantum mechanics texts (e.g. Messiah, 1962; Merzbacher, 1961) and more advanced nuclear physics texts such as those listed in the references. Section 3.3 is somewhat more advanced than the rest of the material and could be omitted at first reading.

Everywhere in this chapter, except in the very first section, we shall be working in the centre-of-mass coordinate system (CMS) and shall restrict ourselves to non-relativistic situations. Further, we do not consider reactions in which there are more than two particles, or fragments, in the final state.

3.1 FORM OF THE WAVE FUNCTION

In a scattering experiment, the particle waves have the form shown in *Figure 3.1* where the lines plotted represent the wave crests and the arrows indicate the direction of motion of the wavefronts.*

*This account is oversimplified. Actual accelerators do not produce coherent wavetrains of arbitrary length but rather beams made up of wave packets. This is discussed in detail by Austern (1970), also by Messiah (1962, p. 372 *et seq.*) and Yoshida (1974).

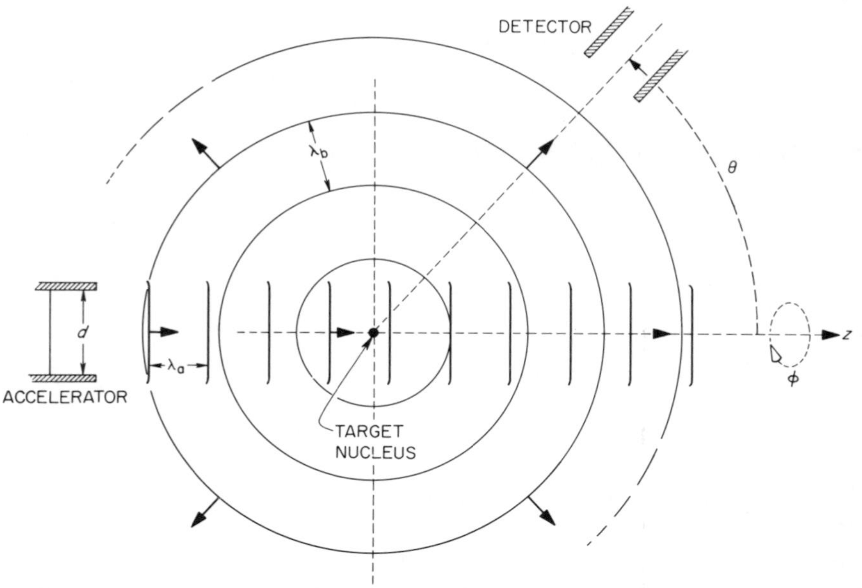

Figure 3.1 Schematic picture of the arrangement for a scattering experiment. The lines represent wavecrests, the straight lines for the incident beam and the circular ones for the waves scattered from the target

3.1.1 The incident wave

The length scales are very distorted in *Figure 3.1*. The wavelength λ_a ($\lesssim 10^{-12}$ cm) is *very* much smaller than the width d (~ 1 cm) of the beam. As a consequence, the wavefronts are essentially plane with very little blurring at the edges from diffraction. (If the wavelengths were truly as large as those shown in *Figure 3.1*, diffraction would cause the wavefronts of the beam to 'fan out' very quickly. One could reinterpret the figure as showing only every nth wavecrest, where n is a very large number, $n \gtrsim 10^{12}$.) Hence, except at the very edge of the beam, the wave describing it has the form

$$\chi_a = A_0 \exp i(k_a z - \omega_a t)$$

where we have taken the z-axis along the beam. Here, $k_a = 2\pi/\lambda_a = 1/\lambda\!\!\!^{-}_a$ and $\omega_a = 2\pi\nu_a = 2\pi E_a/h = E_a/\hbar$ where the frequency ν_a is related to the energy E_a of the beam particles by $E_a = h\nu_a$ and the wave number k_a is related to the momentum p_a by $p_a = \hbar k_a$. The amplitude A_0 is determined by the flux in the beam. If the flux is I_0 particles per unit area per unit time, then (remembering that the probability density for the particles is $|\chi_a|^2$)

$$A_0 = (I_0/v_a)^{1/2}$$

where v_a is the velocity of the particles. The spatial part of the wave function can be written

$$A_0 \exp(ik_a z) = A_0 \exp(ik_a r \cos\theta) = A_0 \exp(i\mathbf{k}_a \cdot \mathbf{r})$$

where the last form, involving the scalar product $\mathbf{k}_a \cdot \mathbf{r}$, is independent of our choice of z-axis.

Now suppose there are target particles A with momenta $\mathbf{p}_A = \hbar\mathbf{k}_A$ and energy $E_A = \hbar\omega_A$; in similar fashion they will be described by the wave function

$$\chi_A = N^{1/2} \exp i(\mathbf{k}_A \cdot \mathbf{r}_A - \omega_A t)$$

where the normalisation constant corresponds to N particles per unit volume. The wave function for the combined system is the product of these

$$\begin{aligned}\chi &= \chi_a \chi_A \\ &= N^{1/2} A_0 \exp[i(\mathbf{k}_a \cdot \mathbf{r}_a + \mathbf{k}_A \cdot \mathbf{r}_A)] \exp[i(\omega_a + \omega_A)t] \qquad (3.1)\\ &= N^{1/2} A_0 \exp[i(\mathbf{K} \cdot \mathbf{R})] \exp[i(\mathbf{k}_\alpha \cdot \mathbf{r}_\alpha)] \exp(i\omega t)\end{aligned}$$

where $\hbar\omega = E_A + E_a$ is the total energy, $\mathbf{K} = \mathbf{k}_a + \mathbf{k}_A$ is the total momentum* in units of $\hbar$, and

$$\mathbf{R} = \frac{m_a \mathbf{r}_a + m_A \mathbf{r}_A}{m_a + m_A}$$

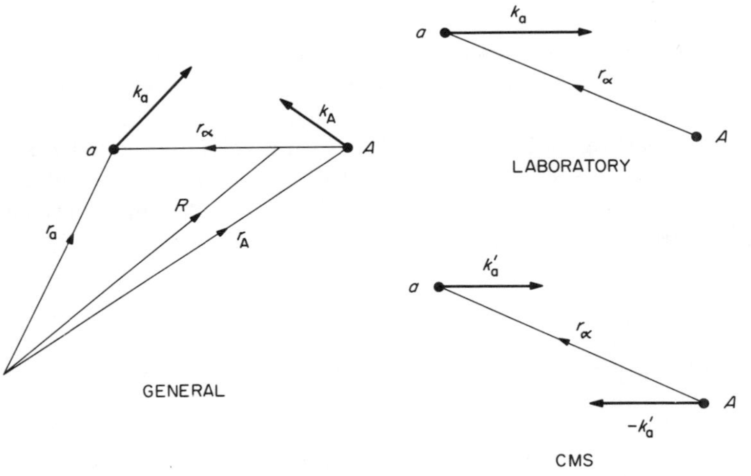

Figure 3.2 Coordinates and momenta for two colliding particles in general and also referred to coordinate frames fixed in the laboratory and moving with the centre of mass of the pair

*We shall often refer to the wave number k as the momentum; this is equivalent to using units such that $\hbar = 1$.

is the position of the centre of mass of the two particles (*see Figure 3.2*). Here m_i is the mass of particle i. Also $\mathbf{r}_\alpha = \mathbf{r}_a - \mathbf{r}_A$ is the separation of a and A, and

$$\mathbf{k}_\alpha = \frac{m_A \mathbf{k}_a - m_a \mathbf{k}_A}{m_a + m_A}$$

is called their relative momentum. Now in the absence of external fields the *total* energy is conserved, so the factor $\exp(i\omega t)$ does not change even after a collision and we do not need to write it explicitly. The total momentum **K** is also conserved and the position of the centre of mass continues in a uniform state of motion, so we do not need to carry the $\exp(i\mathbf{K}\cdot\mathbf{R})$ factor explicitly either. Consequently when the only forces acting are those between the two particles a and A, it is sufficient to use that part of the wave function 3.1 which describes their relative motion

$$N^{1/2} A_0 \exp i(\mathbf{k}_\alpha \cdot \mathbf{r}_\alpha) \tag{3.2}$$

Henceforth we shall treat the case $N = 1$, one target nucleus per unit volume.

3.1.2 Laboratory and centre-of-mass systems

In a coordinate system fixed in the laboratory, the target is at rest* so that $k_A = 0$, $\mathbf{K} = \mathbf{k}_a$ and

$$\mathbf{k}_\alpha = \frac{m_A}{m_a + m_A} \mathbf{k}_a \tag{3.3}$$

If we use a coordinate system moving so that the centre of mass is at rest (called the centre-of-mass system, or CMS), then $K = 0$. If we denote quantities measured in the CMS by primes, then $\mathbf{k}'_a = -\mathbf{k}'_A$ and we have the situation shown on the lower right of *Figure 3.2*. (Note that we do not need to have a and A making a head-on collision in the CMS; indeed in a plane-wave state we could not localise the particles in this way. We only require that they have equal and opposite momenta.) The CMS is moving in the laboratory with a velocity of

$$\mathbf{V}_{CM} = \mathbf{v}_a \frac{m_a}{m_a + m_A}$$

consequently the projectile has a velocity in the CMS of

$$\mathbf{v}'_a = \mathbf{v}_a - \mathbf{V}_{CM} = \frac{m_A}{m_a + m_A} \mathbf{v}_a$$

*The target nuclei will not be completely at rest because of thermal motion, but this may be neglected. Also, because of the uncertainty principle, they cannot have a precise momentum, such as zero, because we assume they are localised in the target material. strictly, they should be represented by a wave packet whose *average* momentum is zero. Fortunately, these added complications do not change any of the results obtained with our simplified picture.

while the target particle has a velocity $\mathbf{v}_A = -\mathbf{V}_{CM}$. The sum of their kinetic energies in the CMS is then

$$T_{CM} = \frac{1}{2} \frac{m_a m_A}{m_a + m_A} v_a^2$$

which is $m_A/(m_a + m_A)$ times the kinetic energy T_{LAB} of the projectile in the LAB system (the remainder being the kinetic energy due to the motion of the centre of mass itself). This is like the kinetic energy of a particle with velocity v_a and a *reduced mass*

$$\mu_\alpha = \frac{m_a m_A}{m_a + m_A}$$

The momentum $\mathbf{k}'_a$ of the projectile measured in the CMS is similarly reduced compared to that, $\mathbf{k}_a$, measured in a system at rest in the laboratory

$$\mathbf{k}'_a = \frac{m_A}{m_a + m_A} \mathbf{k}_a$$

so that the relative momentum $\mathbf{k}_\alpha$ is still given by the expression 3.3. We then have also that

$$T_{CM} = \frac{\hbar^2 k_\alpha^2}{2\mu_\alpha}$$

3.1.3 Internal states

Now the wave function 3.2 is still incomplete; it only describes the relative motion of the incident particle a and the target A. Certainly a complex nucleus has *internal* structure and even a single nucleon has spin. Hence, in general, both projectile and target have internal degrees of freedom which have to be described by wave functions ψ_a and ψ_A, say. These describe the orientation of the spin of the particle and its state of excitation, if any. (Normally the projectile and target will be in their ground states, but this need not be the case after the collision has taken place.) These internal wave functions will be normalised so that

$$\int |\psi_A(\tau_A)|^2 \, d\tau_A = 1, \quad \int |\psi_a(\tau_a)|^2 \, d\tau_a = 1$$

where τ_i represents the internal coordinates of the i nucleus. The wave function describing the state of the system before any collision occurs is then

$$A_0 \exp(i\mathbf{k}_\alpha \cdot \mathbf{r}_\alpha) \, \psi_a \, \psi_A \tag{3.4}$$

This is usually referred to as 'the incident wave'

3.1.4 The scattered waves

When the target nucleus is inserted into the beam and collisions occur, there will also be scattered waves radiating from the site of the collision for each open channel β corresponding to an outgoing particle b and residual nucleus B, so that the complete wave now has the form

$$\Psi = A_0 \exp(i\mathbf{k}_\alpha \cdot \mathbf{r}_\alpha)\, \psi_a\, \psi_A + \sum_\beta \Psi_{\text{scatt.}\,\beta} \tag{3.5}$$

The wavelength λ_β of these waves will be the same as, or different from, the initial λ_α according to whether the corresponding channel β represents elastic, or non-elastic, scattering. The elastically scattered wave with $\lambda_\beta = \lambda_\alpha$ will interfere with the incident plane wave in the region of space where they overlap; in the forward ($\theta = 0$) direction this interference, then destructive, gives rise to the attenuation of the incident wave as it passes the target (i.e. the shadow cast by the target). If the target is absorbent or non-transparent, i.e. if there is some non-elastic scattering, then the target casts a shadow. But in order to cast the shadow, there must be an elastically scattered wave $\Psi_{\text{scatt},\alpha}$ to interfere destructively with the incident wave 3.4 in the forward direction. Thus, because we define elastic scattering as the difference between the (a, A) part of the actual wave, equation 3.5, and the undisturbed wave, equation 3.4, before a collision occurs, it follows that non-elastic scattering must ***always*** be accompanied by elastic scattering. (We return to this point in section 3.4.10.)

At large distances from the target ($r_\beta \gg \lambda_\beta$, and beyond the influence of any interaction potential), the radial dependence of the outgoing spherical waves in channel β has the form $\exp(ik_\beta r_\beta)/r_\beta$, hence their intensity falls off as the inverse of the square of the distance r_β between the two residual particles. (The presence of *Coulomb* forces distorts the form of the spherical waves even at large distances—*see* below—but their intensity still obeys the inverse square law.) In addition the amplitude of these spherical waves will be modulated by a factor $f_\beta(\theta, \phi)$ which will in general vary with the scattering angles θ and ϕ, where ϕ is the angle in azimuth about the incident beam direction. (The amplitude f_β will also depend upon the bombarding energy.) The scattered wave in each channel β then has the form

$$\Psi_{\text{scatt},\,\beta}(r, \theta, \phi) \xrightarrow{\text{large } r} A_0\, f_\beta(\theta, \phi)\, \frac{\exp(ik_\beta r_\beta)}{r_\beta}\, \psi_b\, \psi_B \tag{3.6}$$

Here ψ_b, ψ_B represent the wave functions for the internal states of the nuclei b and B. We have also included the incident amplitude A_0 explicitly since clearly the amplitude in any final channel is proportional to that in the entrance channel. Then $f_\beta(\theta, \phi)$ is the scattering amplitude in the channel for an incident beam with 1 particle per unit volume ($A_0 = 1$) and a given bombarding energy.

If the various particles do not have spin, there is complete axial symmetry about the incident beam direction (which we have chosen to be the z-axis). It follows that f_β will not depend upon the azimuth angle ϕ, but only on the scattering angle θ. However, if any of the particles has spin, the specification of their states includes giving the orientation of the spin. The amplitude f then refers to transitions between states with definite spin orientations and it will only have axial symmetry if the interactions inducing the scattering are independent of these spins.

3.2 DIFFERENTIAL CROSS-SECTIONS

We may deduce an expression for the cross-section for the A(a, b)B reaction from the form 3.5 of the scattering wave function. (For an alternative derivation using the quantal expression for probability current, *see* Messiah, 1962.) Firstly we note that the detector is always at such a distance away from the target that it is appropriate to use the asymptotic form 3.6. Secondly, the wave function 3.5 corresponds to scattering from a single target nucleus per unit volume. Now the density of scattered particles b at the point (r, θ, ϕ) leaving the residual nucleus B in some particular state is the square modulus of the wave 3.5 integrated over the internal coordinates of the two nuclei; i.e. it is just* the square modulus of that part of Ψ_{scatt} which refers to the relative motion of b with respect to B; from equation 3.6 this is

$$\frac{1}{r^2}\,|A_0\, f_\beta(\theta, \phi)|^2$$

These particles are moving away from the residual nucleus with a (relative) velocity v_β, so the number emitted into an element $\mathrm{d}\Omega$ of solid angle per unit time is $v_\beta |A_0 f_b|^2 \mathrm{d}\Omega$. Dividing by the incident flux, $v_\alpha |A_0|^2$, we obtain the differential cross-section (*see* section 2.6) for the A(a, b)B reaction

$$\frac{\mathrm{d}\sigma_\beta}{\mathrm{d}\Omega} = \frac{v_\beta}{v_\alpha}\,|f_\beta(\theta, \phi)|^2 \tag{3.7}$$

For elastic scattering we have $v_\beta = v_\alpha$. When the scattering is non-elastic, the factor v_β/v_α enters because the cross-section concerns particle ***flux*** whereas the wave amplitude f describes particle *density*.†

If none of the particles has spin, then f_β and hence the cross-section is independent of the azimuth angle ϕ. With spins, f_β will depend on ϕ in general. In that case, however, the expression 3.7 for the cross-section is incomplete; it must be averaged over the initial spin orientations and summed over the final spin orientations. If neither target nor projectile were polarised (i.e. if their spins were not oriented preferentially in any way—*see* section 2.13) then on average

*For elastic scattering there is, in principle, interference between the incident wave and the scattered wave Ψ_{scatt}. However, in practice (*see Figure 3.1*) the detector is placed outside the region of the incident beam so that this interference vanishes.

†Some authors define the scattering amplitude to be $A_\beta(\theta) = (v_\beta/v_\alpha)^{1/2} f(\theta)$, so that $\mathrm{d}\sigma_\beta/\mathrm{d}\Omega = |A_\beta(\theta)|^2$.

the collision is clearly symmetric about the incident beam and so is the scattered flux. Hence the resulting cross-section is independent of ϕ. Azimuthal asymmetry can appear if the initial spins are *not* randomly oriented (polarised beam or polarised target or both) or if measurements are made on the orientation of the spins of one or both of the residual particles.

3.3 THE SCHRÖDINGER EQUATION

In order to calculate the scattering, we have to find a solution of the Schrödinger equation that has the form 3.5, 3.6. What is this equation? For a single particle of mass m (with no internal degrees of freedom) scattering from a fixed potential $V(\mathbf{r})$, it is

$$\left[-\frac{\hbar^2}{2m}\nabla^2 + V(\mathbf{r})\right]\chi(\mathbf{r}) = E\,\chi(\mathbf{r}) \tag{3.8}$$

where E is the energy of the particle. Outside the region where the potential V acts, E is just the kinetic energy. Sometimes equation 3.8 is abbreviated to the form

$$H\,\chi(\mathbf{r}) = E\,\chi(\mathbf{r}) \tag{3.9}$$

where H is the ***Hamiltonian*** for the system, i.e. the sum of the kinetic and potential energies (Messiah, 1962). Also, for convenience we often rewrite equation 3.8 as

$$[\nabla^2 - U(\mathbf{r}) + k^2]\,\chi(\mathbf{r}) = 0 \tag{3.10}$$

where the wave number $k = [2mE/\hbar^2]^{1/2}$ and $U = 2mV/\hbar^2$. When $V = 0$, the required solution is the plane wave $\chi = \exp(i\mathbf{k}\cdot\mathbf{r})$ which describes the undisturbed beam. Having $V \neq 0$ introduces elastically scattered waves. In section 2.15 we already discussed briefly the form of the solution to this equation that we require.

When two complex systems, such as nuclei a and A, are interacting, we may, as discussed above, interpret $\mathbf{r}$ as the distance $\mathbf{r}_\alpha$ between their centres of mass. If the interaction V cannot change their internal states, i.e. if only elastic scattering is allowed, the wave function for their relative motion satisfies the same equation 3.8 (except that m becomes the reduced mass of the pair, $m \to \mu_\alpha = m_a m_A/(m_a + m_A)$). In general, however, the interaction V is able to excite one or both of the nuclei, or even lead to a rearrangement of their constituent nucleons so that a different pair of nuclei, b + B, results. Then it is convenient to separate the Schrödinger equation into parts which describe the internal motions and the relative motions. For example, take the pair a + A. We may define ***internal*** Hamiltonians H_a, H_A (i.e. sums of terms for the kinetic and potential energies of the nucleons within a in the rest frame of a, and those within A in the rest frame of A, respectively). The internal wave functions ψ_a and ψ_A are solutions of the corresponding Schrödinger equations, written in the same abbreviated form as equation 3.9

$$H_a\psi_a = \epsilon_a\psi_a, \quad H_A\psi_A = \epsilon_A\psi_A \tag{3.11}$$

with eigenenergies ϵ_a and ϵ_A, respectively. These are then the energy levels of the separated, non-interacting, nuclei.

The *total* Hamiltonian is then obtained by adding to these internal Hamiltonians H_a and H_A the kinetic energy of relative motion and the interaction potential V_α between the two nuclei a and A

$$H = H_a + H_A - \frac{\hbar^2}{2\mu_\alpha}\nabla_\alpha^2 + V_\alpha \tag{3.12}$$

where ∇_α^2 acts upon $\mathbf{r}_\alpha$. The Schrödinger equation then becomes, formally

$$H\Psi = E\Psi \tag{3.13}$$

where E is the *total* energy of the system. We shall be looking for solutions of this equation which have the form of equation 3.5.

3.3.1 Coupled equations form of the Schrödinger equation

The wave functions ψ_a and ψ_A from equation 3.11 for the various internal states of the two nuclei a and A form complete sets, so we may expand the wave function Ψ, which describes the combined system, in terms of them

$$\Psi = \sum_{a'A'} \chi_{a'A'}(\mathbf{r}_\alpha)\,\psi_{a'}\psi_{A'} \tag{3.14}$$

where the (infinite) sum runs over all the possible internal states of a and A, here labelled by a′ and A′ respectively. If we insert this expansion into equation 3.13, using equations 3.11 and 3.12 we get

$$\sum_{a'A'}\left[(\epsilon_{a'} + \epsilon_{A'} - E) - \frac{\hbar^2}{2\mu_\alpha}\nabla_\alpha^2 + V_\alpha\right]\chi_{a'A'}(\mathbf{r}_\alpha)\,\psi_{a'}\psi_{A'} = 0 \tag{3.15}$$

If we now multiply from the left by $\psi_a^*\psi_A^*$ and integrate over all the *internal* coordinates τ_a and τ_A of the two nuclei, we can use the orthogonality property of the internal wave functions:

$$\int\psi_a^*(\tau_a)\,\psi_{a'}(\tau_a)\mathrm{d}\tau_a = \delta_{a,a'} \tag{3.16}$$

and similarly for the ψ_A functions. Here the Kronecker delta $\delta_{a,a'}$ means 1 if a = a′ and zero otherwise. When we do this, equation 3.15 becomes

$$[\nabla_\alpha^2 - U_{aA,aA}(\mathbf{r}_\alpha) + k_{aA}^2]\ \chi_{aA}(\mathbf{r}_\alpha) = \sum_{\substack{a' \neq a \\ A' \neq A}} \chi_{a'A'}(\mathbf{r}_\alpha)\ U_{aA,a'A'}(\mathbf{r}_\alpha) \qquad (3.17)$$

Here we have made the same transformation that led to equation 3.10; this simplifies the notation somewhat. In particular

$$k_{aA}^2 = 2\mu_\alpha (E - \epsilon_a - \epsilon_A)/\hbar^2 \qquad (3.18)$$

When the two nuclei are separated and non-interacting and in the internal states denoted by a and A, the quantity $(E - \epsilon_a - \epsilon_A)$ is the *kinetic* energy of their relative motion and k_{aA} is the corresponding wave number. Similar quantities $(E - \epsilon_{a'} - \epsilon_{A'})$ and $k_{a'A'}$ occur in the equations for every pair of possible states a′ and A′ of the two nuclei obtained by multiplying equation 3.15 by the other products $\psi_{a'}^* \psi_{A'}^*$ and integrating. We have also introduced into equation 3.17 the ***matrix elements*** of the interaction potential V_α:

$$U_{aA,a'A'}(\mathbf{r}_\alpha) = \frac{2\mu_\alpha}{\hbar^2} \iint \psi_a^*(\tau_a)\ \psi_A^*(\tau_A)\ V_\alpha\ \psi_{a'}(\tau_a)\ \psi_{A'}(\tau_A)\ d\tau_a d\tau_A$$

$$\equiv \frac{2\mu_\alpha}{\hbar^2} \langle aA|\ V_\alpha\ |a'A'\rangle \qquad (3.19)$$

(The second form is a very convenient abbreviation which makes use of Dirac's bracket notation (Messiah, 1962)). The integrations in equation 3.19 are only over the *internal* coordinates τ_a and τ_A, consequently the matrix element remains a function of the separation $\mathbf{r}_\alpha$ of the two nuclei. Each row and each column of the matrix is labelled by one of the double indices (a′, A′).

In equation 3.17 we put the *diagonal* element, $U_{aA,aA}$, on the left side and all the *off-diagonal* elements, with $a' \neq a$ and/or $A' \neq A$, on the right side. Then the left-hand side of the equation looks like that for the scattering of a particle by a fixed potential $U_{aA,aA}(\mathbf{r}_\alpha)$, as in equation 3.10. By itself, this would only describe elastic scattering. It is the coupling to other, excited, states with $a' \neq a$ and $A' \neq A$, which appears on the right-hand side of the equation, which enables us to describe inelastic scattering and its effect on the elastic scattering. Equation 3.17 actually is just one of an infinite set of similar *coupled equations*; the left side need not refer to the ground states and there is an equation 3.17 for every pair of states a′ and A′. Now $\chi_{a'A'}(\mathbf{r}_\alpha)$ describes the relative motion of the two nuclei when they are in the states a′ and A′; we must impose the boundary conditions that for large $\mathbf{r}_\alpha$ the $\chi_{a'A'}$ have the form 3.5 and 3.6, namely incoming plus outgoing waves in the entrance channel (when $a' = a$ and $A' = A$, representing the *ground* states) and outgoing waves in all the other channels which are energetically available or open.

It would seem that if we knew all the matrix elements $U_{aA,a'A'}$, we could then solve these coupled equations and obtain a complete description of the reaction. However, there are an infinite number of them. In practice we have to make an approximation; we truncate the infinite set of equations to a relatively few channels which we expect theoretically or know experimentally are strongly coupled and either neglect the rest or represent their effect by using a complex *optical potential* (*see* Chapter 4). It is at this point that *models* are introduced. (This truncation is sometimes called the strong coupling approximation or the coupled-channels method.)

A further inconvenience of the set of equations 3.17 is that while they are useful (and widely used) for describing the *inelastic* scattering of the two nuclei from, say, states a and A to states a′ and A′, they are not well suited for describing *rearrangement* collisions which result in two different nuclei b and B. In principle they include these events also (the ψ_a and ψ_A constitute complete sets of functions), but in practice a very large number of terms, involving highly excited states a′ and A′ (including unbound ones), would be required to represent a different partition b, B. Consequently, different methods are devised for treating rearrangement reactions. One such method, the Born approximation, is described later.

3.3.2 Integral form of the Schrödinger equation for scattering by a potential

The equations 3.8 or 3.10 for the scattering of a particle by a centre of force can be written

$$(\nabla^2 + k^2)\,\chi(\mathbf{r}) = U(\mathbf{r})\,\chi(\mathbf{r}) = F(\mathbf{r}) \tag{3.20}$$

(This same equation describes the relative motion of two structureless particles interacting via the potential $V(\mathbf{r})$.) When $V = 0$, this becomes the equation for a free particle of energy E

$$(\nabla^2 + k^2)\,\chi_0(\mathbf{r}) = 0 \tag{3.21}$$

which has as a solution the plane wave $\chi_0(\mathbf{r}) = \exp(i\mathbf{k}\cdot\mathbf{r})$. The presence of the potential V introduces scattered waves in addition to the 'incident' plane wave. We may think of the term $F(\mathbf{r}) = U(\mathbf{r})\,\chi(\mathbf{r})$ on the right side of equation 3.20 as a source term which generates these scattered waves. A solution of equation 3.20 then consists of waves generated by $F(\mathbf{r})$ plus any solution of the homogeneous equation 3.21. For the latter we choose the plane wave χ_0. The former is given by the particular integral corresponding to outgoing waves only (Messiah, 1962). Then we have

$$\chi(\mathbf{k}, \mathbf{r}) = \exp(i\mathbf{k}\cdot\mathbf{r}) - \frac{1}{4\pi}\int \frac{\exp(ik|\mathbf{r} - \mathbf{r}'|)}{|\mathbf{r} - \mathbf{r}'|}\, U(\mathbf{r}')\,\chi(\mathbf{k}, \mathbf{r}')\,d\mathbf{r}' \tag{3.22}$$

where we have indicated explicitly in χ the incident momentum $\mathbf{k}$. We may compare this to Poisson's equation for the potential $\chi(\mathbf{r})$ due to an electric charge distribution $\rho(\mathbf{r})$

$$\nabla^2\,\chi(\mathbf{r}) = -\,4\pi\,\rho(\mathbf{r})$$

The solution here is given by the integral

$$\chi(\mathbf{r}) = \int \frac{\rho(\mathbf{r}')\,\mathrm{d}\mathbf{r}}{|\mathbf{r} - \mathbf{r}'|}$$

We see that $F(\mathbf{r})$ plays the same role as the source term $4\pi\rho(\mathbf{r})$, while Poisson's equation corresponds to Schrödinger's equation for zero energy, $k = 0$.

The integral equation 3.22 is equivalent to the Schrödinger differential equation 3.20, but actually includes more information because it automatically incorporates the boundary conditions (plane wave plus outgoing spherical wave) that we require. To obtain the scattering amplitude we need $\chi(\mathbf{r})$ at large values of r. Now $V(\mathbf{r}')$ vanishes for large r' so that in the integrand in equation 3.22 only $r' \ll r$ contributes when r itself is large. Then for large r we can put $1/|\mathbf{r} - \mathbf{r}'| \approx 1/r$ and $k|\mathbf{r} - \mathbf{r}'| \approx kr - \mathbf{k}'\cdot\mathbf{r}'$, where $\mathbf{k}'$ is a vector with the magnitude of k but the direction of $\mathbf{r}$. So equation 3.22 becomes

$$\chi(\mathbf{k}, \mathbf{r}) \xrightarrow{\text{large } r} \exp(i\mathbf{k}\cdot\mathbf{r}) - \frac{\exp(ikr)}{4\pi r}\int \exp(-i\mathbf{k}'\cdot\mathbf{r}')\, U(r')\, \chi(\mathbf{k}, \mathbf{r}')\mathrm{d}\mathbf{r}' \qquad (3.23)$$

By comparison with equation 3.6, we can identify the scattering amplitude as the coefficient of the outgoing scattered wave exp $(ikr)/r$

$$f(\theta, \phi) = -\frac{1}{4\pi}\int \exp(-i\mathbf{k}'\cdot\mathbf{r}')\, U(\mathbf{r}')\, \chi(\mathbf{k}, \mathbf{r}')\,\mathrm{d}r' \qquad (3.24)$$

so that $\mathbf{k}'$ is the momentum of the outgoing particle, moving in the direction with polar angles (θ, ϕ).

3.3.3 The Born and the distorted-wave Born approximations

Despite the simple form of equation 3.24, it does not represent more than a formal solution of the scattering problem because the integrand still includes the unknown $\chi(\mathbf{r}')$ wave function. Nonetheless, it is a convenient starting point for approximations. For example, if the potential V is weak the scattered wave has a small amplitude and we may approximate $\chi(\mathbf{r}')$ in equation 3.24 by the first term of equation 3.22, the incident plane wave itself

$$f_{\mathrm{BA}}(\theta, \phi) = -\frac{1}{4\pi}\int \exp(-i\mathbf{k}'\cdot\mathbf{r}')\, U(\mathbf{r}')\, \exp(i\mathbf{k}\cdot\mathbf{r}')\mathrm{d}\mathbf{r}'$$

This is known as the *Born approximation*. Rearranged slightly this becomes

$$f_{\mathrm{BA}}(\theta, \phi) = -\frac{1}{4\pi}\int \exp(i\mathbf{q}\cdot\mathbf{r}')\, U(\mathbf{r}')\,\mathrm{d}\mathbf{r}' \qquad (3.25)$$

which we recognise as the Fourier transform of the potential evaluated at $\mathbf{q} = \mathbf{k} - \mathbf{k}'$. Here $\mathbf{q}$ is the change in momentum of the scattered particle. This momentum $\mathbf{q}$ is transferred to the target and reappears as the recoil of the target.

The amplitude f_{BA} does not depend upon ϕ, but depends upon θ because $q^2 = k^2 + k'^2 - 2kk' \cos\theta$.

If the potential $U(r)$ is spherically symmetric, that is it depends upon the magnitude of **r** but not its direction, the integral 3.25 reduces to

$$f_{BA}(\theta, \phi) = -\frac{1}{q}\int \sin(qr')U(r')r'\mathrm{d}r'$$

A more sophisticated version of this approximation is the *distorted-wave Born approximation* (DWBA) which is most widely used. Suppose the potential U can be written as the sum of two terms, $U = U_1 + U_2$, and suppose we know or can easily obtain the scattering solution for U_1

$$[\nabla^2 + k^2 - U_1(\mathbf{r})]\ \chi_1(\mathbf{k}, \mathbf{r}) = 0 \tag{3.26}$$

We distinguish two types of solution χ_1, one consisting of a plane wave plus an *outgoing* scattered wave denoted $\chi_1^{(+)}(\mathbf{k}, \mathbf{r})$ and one consisting of a plane wave plus an *ingoing* scattered wave denoted $\chi_1^{(-)}(\mathbf{k}, \mathbf{r})$. The latter is the time-inverse of the former and they are related by $\chi_1^{(-)}(\mathbf{k}, \mathbf{r}) = \chi_1^{(+)}(-\mathbf{k}, \mathbf{r})^*$. A generalisation of the derivation which leads to equation 3.23 (*see* Messiah, 1962) then yields a solution† for the full problem in terms of these $\chi_1^{(\pm)}$

$$\chi(\mathbf{k}, \mathbf{r}) \xrightarrow{\text{large } r} \chi_1^{(+)}(\mathbf{k}, \mathbf{r}) - \frac{\exp(ikr)}{4\pi r}\int \chi_1^{(-)}(\mathbf{k}', \mathbf{r}')^*\ U_2(r')\ \chi(\mathbf{k}, \mathbf{r}')\ \mathrm{d}\mathbf{r}' \tag{3.27}$$

Consequently the full scattering amplitude is the sum of that contributed by χ_1 (from potential U_1) and the second term of equation 3.27 (from potential U_2)

$$f(\theta, \phi) = f_1(\theta, \phi) - \frac{1}{4\pi}\int \chi_1^{(-)}(\mathbf{k}', \mathbf{r}')^*\ U_2(\mathbf{r}')\ \chi(\mathbf{k}, \mathbf{r}')\ \mathrm{d}\mathbf{r}' \tag{3.28}$$

This amplitude is identical to that given by the expression 3.24 provided $U = U_1 + U_2$. Again equation 3.28 is exact but not useful as it stands because it contains the exact solution χ. However, by analogy with the Born approximation, we may approximate χ by the solution χ_1 for the potential U_1, i.e. by the first term of equation 3.27. This will be a good approximation if U_2 is weak compared to U_1, and is called the *distorted-wave Born approximation*. It is 'Born' because it is first order in the potential U_2 but 'distorted-wave' because instead of using the plane waves as in equation 3.24 we use the distorted waves χ_1 which should be a better approximation to the exact solution. The corresponding amplitude is

$$f_{DWBA}(\theta, \phi) = f_1(\theta, \phi) - \frac{1}{4\pi}\int \chi_1^{(-)}(\mathbf{k}', \mathbf{r}')^*\ U_2(\mathbf{r}')\ \chi_1^{(+)}(\mathbf{k}, \mathbf{r}')\ \mathrm{d}\mathbf{r}' \tag{3.29}$$

This approximation can be generalised to inelastic and rearrangement collisions. In these cases, U_1 (and hence f_1) is chosen to describe the *elastic* scatter-

†Of course, the full solution we are seeking has the outgoing wave characteristic and should be denoted $\chi^{(+)}(\mathbf{k}, \mathbf{r})$, but we have omitted the superscript for simplicity.

ing (i.e. it is an optical potential, *see* Chapter 4), while U_2 is the interaction which induces the non-elastic transition. The validity of the DWBA then depends upon elastic scattering being the most important event which occurs when two nuclei collide so that other events can be treated as perturbations. We will not describe the derivation in detail, but the corresponding transition amplitude, say for a reaction A(a, b)B, has the form

$$f_{\text{DWBA}}(\theta, \phi) = -\frac{1}{4\pi}\int \chi_\beta^{(-)}(\mathbf{k}_\beta, \mathbf{r}_\beta)^* \langle \text{b}, \text{B} | U_2 | \text{a}, \text{A}\rangle \chi_\alpha^{(+)}(\mathbf{k}_\alpha, \mathbf{r}_\alpha)\, d\mathbf{r}_\alpha d\mathbf{r}_\beta \quad (3.30)$$

Here χ_1 has been generalised to χ_α and χ_β. The function χ_α describes the elastic scattering in the $\alpha = \text{a} + \text{A}$ entrance channel arising from an optical potential U_α, while χ_β describes the elastic scattering in the $\beta = \text{b} + \text{B}$ exit channel arising from an optical potential U_β. The potential U_2 which causes the non-elastic transition depends upon the type of reaction and the model chosen to describe it (*see* Chapter 4).

3.3.4 Integral equation for a general collision

The discussion of section 3.3.2 was for the simple case of a particle scattered by a centre of force, but a parallel approach can be made to the general scattering problem. A detailed discussion is beyond the scope of this book but the technique is widely used so that we shall briefly introduce the ideas and some of the language. The general Schrödinger equation may be written formally as in equation 3.13

$$H\Psi = E\Psi \quad \text{or} \quad (E - H)\Psi = 0$$

Following equation 3.12, the total Hamiltonian H may be split into a part H_0 describing the internal states of the two colliding nuclei, a + A, together with their relative kinetic energy

$$H_0 = H_\text{a} + H_\text{A} - \frac{\hbar^2}{2\mu_\alpha}\nabla_\alpha^2$$

plus the interaction V_α between a and A, so that $H = H_0 + V_\alpha$, or

$$(E - H_0)\Psi = V_\alpha\Psi \quad (3.31)$$

In the formal theory we may manipulate the various operators as though they were algebraic quantities (for example, introducing their inverses) provided we keep in mind that they may not commute, (ab $\neq$ ba in general), that inverse operators may introduce singularities that require special treatment, and that (as in the solution 3.22) the boundary conditions must be taken into account. For example, if we take the inverse operator $(E - H_0)^{-1}$, which is an *integral operator* because of the presence of ∇_α^2 in H_0, we must recognise that it is singular when acting upon an eigenstate of H_0 with the eigenvalue E. In scattering theory this is avoided by the trick of adding an infinitesimal imaginary quantity $i\epsilon$ with ϵ positive

$$\frac{1}{E - H_0} \rightarrow \underset{\epsilon\to 0}{\text{Lt}} \frac{1}{E - H_0 \pm i\epsilon}$$

It may be shown (Messiah, 1962) that the choice $+i\epsilon$ leads to a solution like equations 3.22 and 3.23 with an outgoing spherical wave, while the choice $-i\epsilon$ results in the time-reversed solution with an incoming spherical wave.

A formal 'solution' of the Schrödinger equation, equation 3.31, is simply

$$\Psi = \frac{1}{E - H_0} V_\alpha \Psi$$

or more correctly, if we demand an outgoing scattered wave

$$\Psi = \underset{\epsilon\to 0}{\text{Lt}} \frac{1}{E - H_0 + i\epsilon} V_\alpha \Psi$$

This is a particular integral solution. We may add any solution ϕ_0 of the homogeneous equation obtained by neglecting the right side of equation 3.31

$$(E - H_0)\phi_0 = 0$$

Then a general solution of equation 3.31 is the Lippman–Schwinger equation

$$\Psi = \phi_0 + \underset{\epsilon\to 0}{\text{Lt}} \frac{1}{E - H_0 + i\epsilon} V_\alpha \Psi \tag{3.32}$$

The addition of ϕ_0 allows us to fully satisfy the boundary conditions implied by the physical scattering problem in which we are interested. Not only do we require the outgoing scattered waves given by equation 3.32, but also that the incident beam is a plane wave as in equation 3.4. Hence we choose

$$\phi_0 = A_0 \exp(i\mathbf{k}_\alpha \cdot \mathbf{r}_\alpha)\, \psi_a\, \psi_A$$

where the constant A_0 is determined by the intensity of the incident beam.

The 'solution' 3.32, which is the generalisation of equation 3.22, has a deceptively simple form. Indeed equation 3.32 is an integral equation which is equivalent to the original Schrödinger equation, plus instructions how to satisfy the required boundary conditions. It cannot be used immediately because the wave function Ψ appears on the right side also, acted upon by the integral operator, just as in the simple, special case of equation 3.22. As there, however, the equation can be used as the starting point for approximations; e.g. if we replace Ψ on the right side by the unperturbed incident wave ϕ_0, we obtain the Born approximation

$$\Psi \approx \phi_0 + \underset{\epsilon\to 0}{\text{Lt}} \frac{1}{E - H_0 + i\epsilon} V_\alpha\, \phi_0$$

Successively iterating in this way generates the Born-series solution

$$\Psi = \phi_0 + \underset{\epsilon\to 0}{\text{Lt}} \frac{1}{E - H_0 + i\epsilon} V_\alpha \phi_0 + \underset{\epsilon\to 0}{\text{Lt}} \frac{1}{E - H_0 + i\epsilon} V_\alpha \frac{1}{E - H_0 + i\epsilon} V_\alpha \phi_0 + \ldots$$

The integral operator $\underset{\epsilon\to 0}{\text{Lt}} (E - H_0 - i\epsilon)^{-1}$ is often referred to as the Green operator; when it is given an explicit representation it becomes the *Green function* (or Green's function). For example, the factor in the integrand of equation 3.22 is the *Green function for a free particle*

$$G(\mathbf{r}, \mathbf{r}') = \frac{\exp(ik|\mathbf{r} - \mathbf{r}'|)}{|\mathbf{r} - \mathbf{r}'|}$$

For more details *see*, for example, Messiah (1962) and other texts on scattering theory given in the references.

3.4 PARTIAL WAVES

3.4.1 Significance of partial waves

The relative angular momentum $\ell\hbar$, of two colliding particles is quantised in units of $\hbar$. Three features induce us to make use of wave functions with definite ℓ, called partial waves. One is that since nuclear forces have a short range and nuclei have relatively sharp edges, only those particles in the beam with angular momentum $\ell\hbar$ relative to the target less than some maximum value will interact with the target. Another is that this maximum value is often quite small, so that only a few ℓ values need be considered. The third is that we are able to reduce the Schrödinger equation in three dimensions to a series of radial equations in one dimension, one for each ℓ value. In general these radial equations remain coupled, but (except for scatterings involving very short wavelengths) they are often more amenable to approximations than the original equation.

In addition, a particular reaction may be such as to select a particular angular momentum from the beam. An example could be the excitation of a compound nucleus resonance with a definite spin.

If two particles with relative momentum $\mathbf{p} = \hbar\mathbf{k}$ have an impact parameter b (*see Figure 2.3*), their relative angular momentum is $\ell\hbar = pb$, or $\ell = kb$. Suppose they do not interact unless they approach one another closer than some distance R (e.g. $R = R_1 + R_2$ in *Figure 2.3*). Then classically they can only interact if $\ell \leqslant kR$.

In quantum theory, the value of the impact parameter b has an uncertainty of the order of $\pm \frac{1}{2}\lambda\!\!\!^{-}$, while the magnitude of the angular momentum becomes $[\ell(\ell + 1)]^{1/2} \approx \ell + \frac{1}{2}$ instead of ℓ. The condition for interaction then becomes $\ell + \frac{1}{2} \leqslant k(R + \frac{1}{2}\lambda\!\!\!^{-}) = kR + \frac{1}{2}$, which is the same as we deduced from classical arguments.

If the scattering potential is central, the angular momentum is a constant of the motion. Then the wave function for a particular angular momentum ℓ (with z-component m) may be factored into radial and angular parts

$$\chi_{\ell m}(\mathbf{r}) = u_\ell(r)\, Y_\ell^m(\theta, \phi)$$

Correspondingly, the Schrödinger equation, equation 3.8, separates into radial and angular equations. The radial equation is most easily written in terms of $w_\ell(r) = r u_\ell(r)$

$$-\frac{\hbar^2}{2m}\frac{d^2 w_\ell}{dr^2} + \left[V(r) + \frac{\hbar^2}{2m}\frac{\ell(\ell+1)}{r^2}\right] w_\ell = E\, w_\ell \tag{3.33}$$

The term $\ell(\ell + 1)\hbar^2/2mr^2$ represents the rotational energy of a particle in an orbit with angular momentum $\ell\hbar$. This term looks like an additional repulsive (positive) potential; it is increasingly repulsive as one approaches the origin and the repulsion increases as ℓ increases. For this reason, it is referred to as the *centrifugal barrier* (or sometimes *centrifugal potential*, although it is not a true potential); its effect in the equation is to reduce the magnitude of the solutions w_ℓ at small r as ℓ increases, hence reflecting the classical notions described above. This is discussed explicitly in the next section for the case where $V = 0$ and the corresponding $u_\ell(r)^2$ are illustrated in *Figure 3.3*.

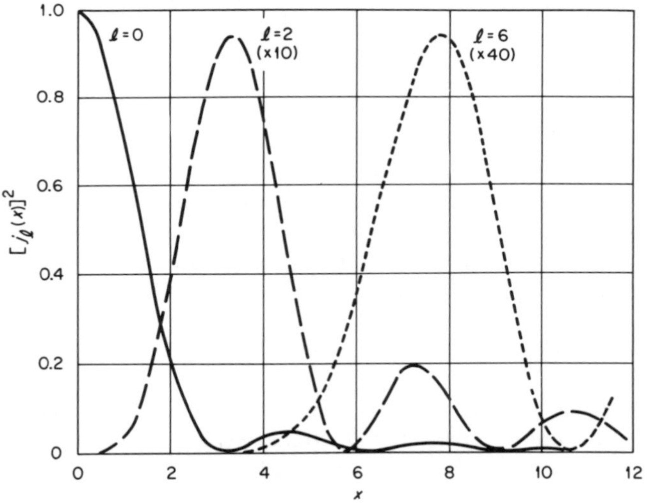

Figure 3.3 The squares of some typical spherical Bessel functions $j_\ell(x)$. (The curves for $\ell = 2$ and 6 have been multiplied by 10 and 40, respectively, to facilitate comparison with that for $\ell = 0$.) Explicit expressions for the first three functions are: $j_0(x) = (\sin x/x)$, $j_1(x) = (\sin x/x^2) - (\cos x/x)$ and $j_2(x) = [\{(3/x^2) - (1/x)\} \sin x] - \{(3/x^2) \cos x\}$

Now the wave number $k \approx 0.22(\mu E)^{1/2}$ fm^{-1}, where μ is the reduced mass of the pair in amu and E is the centre of mass energy in MeV. For nucleons ($\mu \approx 1$), k ranges from 7×10^{-3} fm^{-1} for $E = 1$ keV to 2 fm^{-1} for $E = 100$ MeV. Since typical R values run from a few fm to about 20 fm, it is clear that at the lower energy of 1 keV only s-waves ($\ell = 0$) are important and even with 100-MeV nucleons we need only 10 or 20 partial waves. By contrast, the description of the scattering of two molecules, whose size is measured in Ångstroms (1 Å = 10^5 fm), may require several thousand partial waves even if their energy is only about 1 keV.

The factor $\mu^{1/2}$ in the expression for k means that incident α-particles ($\mu \approx 4$) require about twice as many partial waves as nucleons of the same energy, and bromine ions will require nine times as many. Indeed the number of partial waves needed for a description of the scattering of a heavy ion is even larger because the interaction radius $R = R_1 + R_2$ for a given target is larger than for a nucleon, since the projectile itself has an appreciable size. However, the strong Coulomb repulsion between two heavy ions tends to keep them apart and reduce the importance of the larger angular momenta. For example, heavy ions with a mass of $A \sim 100$ and an energy of 50 MeV per nucleon colliding with a heavy target may require 2000 or more partial waves for a description of the collision. Even so, a precise calculation of the scattering may still be most conveniently made using partial waves if a high-speed computer is available, although one can often gain more physical insight into the scattering process by using simple approximations such as, for example, the diffraction and semi-classical models to be described later.

3.4.2 Partial wave expansions

It will be found in any text on quantum mechanics (*see* also Appendix A for some of their properties) that the eigenfunctions of orbital angular momentum are the spherical harmonics $Y_\ell^m(\theta, \phi)$ with eigenvalues of $\ell(\ell+1)$ for $\mathbf{L}^2$ and m for $\mathbf{L}_z$. Hence the partial wave technique is an expansion of the angular dependence of the wave function describing the relative motion of the scattering in a given channel in terms of these orthonormal functions Y_ℓ^m

$$\chi(r, \theta, \phi) = \sum_{\ell m} c_{\ell m} u_\ell(r) Y_\ell^m(\theta, \phi) \tag{3.34}$$

In general the weight coefficients $c_{\ell m}$ may depend upon both ℓ and the azimuthal quantum number m while the functions $u_\ell(r)$ depend only upon ℓ. To be physically acceptable, the $u_\ell(r)$ are those solutions of the radial equation 3.33 which are regular at the origin.

The prototype of equation 3.34 is the Rayleigh expansion of a plane wave, which is a solution of the Schrödinger equation when no potential is present, $V = 0$

$$\begin{aligned}\exp(i\mathbf{k}\cdot\mathbf{r}) = \exp(ikr\cos\theta) = \exp(ikz) &= \sum_{\ell=0}^{\infty} i^\ell\,[4\pi(2\ell+1)]^{1/2}\, j_\ell(kr)\, Y_\ell^0(\theta, \phi) \\ &= \sum_{\ell=0}^{\infty} i^\ell\,(2\ell+1)\, j_\ell(kr) P_\ell(\cos\theta)\end{aligned} \tag{3.35}$$

where we have chosen the z-axis along $\mathbf{k}$, the direction of motion. Because of the symmetry about this direction, only $m = 0$ appears and Y_ℓ^0 is actually independent of ϕ. The second, slightly simpler, form makes use of the relation

$$Y_\ell^0 \ (\theta, \phi) = [(2\ell + 1)/4\pi]^{1/2} \ P_\ell (\cos\theta)$$

where the P_ℓ are Legendre polynomials (Appendix A). When referred to arbitrarily oriented coordinate axes, equation 3.35 takes the more general form 3.34 which includes terms with $m \neq 0$

$$\exp(i\mathbf{k}\cdot\mathbf{r}) = 4\pi \sum_{\ell=0}^{\infty} i^\ell j_\ell(kr) \left[\sum_m Y_\ell^m (\theta, \phi) \ Y_\ell^m (\theta_k, \phi_k)^*\right]$$

Here (θ, ϕ) are the polar angles of the vector $\mathbf{r}$, and (θ_k, ϕ_k) are those of $\mathbf{k}$. (The appearance of the scalar product $\mathbf{k}\cdot\mathbf{r}$ shows that the plane wave is a scalar under rotations of the coordinate axes; this property is reflected on the right side by the scalar-invariant combination of the spherical harmonics (Brink and Satchler, 1968).

The radial coefficients in the expansion 3.35 are the spherical Bessel functions $j_\ell(kr)$ (Abramowitz and Stegun, 1970). A few examples are plotted in *Figure 3.3*. For small kr

$$j_\ell(kr) \approx (kr)^\ell / 1.3.5. \ \ldots \ (2\ell + 1) \tag{3.36}$$

while asymptotically they oscillate sinusoidally but with amplitudes decreasing as r^{-1}

$$j_\ell(kr) \rightarrow (kr)^{-1} \sin[kr - \tfrac{1}{2}\ell\pi] \tag{3.37}$$

The first (and largest) peak of j_ℓ occurs for a value of $kr \approx [\ell(\ell + 1)]^{1/2} \approx \ell + \frac{1}{2}$ when ℓ is large (and at values of kr somewhat larger than $\ell + \frac{1}{2}$ for small quantum numbers ℓ, as *Figure 3.3* shows). Thus j_ℓ peaks close to the radius of the classical orbit for a particle with angular momentum $\ell + \frac{1}{2}$.

If some interaction potential extends a distance R, we can see that only those partial waves with $\ell \lesssim kR$ will be affected by it. Equation 3.36 shows that the probability of finding particles with relative angular momenta greater than this close to the target decreases rapidly as ℓ increases.

3.4.3 Ingoing and outgoing waves

The asymptotic expression 3.37 can be rewritten

$$j_\ell(kr) \rightarrow \frac{i^{1-\ell}}{2kr} \left[(-)^\ell \exp(-ikr) - \exp(ikr)\right] \tag{3.38}$$

which we recognise as the sum of incoming and outgoing spherical waves with equal amplitudes. In other words, an infinite sum, equation 3.34 or 3.35, of incoming and outgoing *spherical* waves of equal amplitude and properly phased add up to a *plane* wave propagating uniformly along **k**.

Now equation 3.38 represents the ℓth part of the incident wave 3.4 before interaction with the target occurs. Physically it is clear that scattering from the target can only change the amplitude and phase of the *outgoing* part. Hence the total wave 3.5 must have the same incoming parts as the undisturbed incident wave. If we write the total wave 3.5 as

$$\Psi = \sum_{\mathrm{bB}} \chi_{\mathrm{bB}}(r, \theta, \phi)\, \psi_{\mathrm{b}} \psi_{\mathrm{B}}$$

as we did, for example, in equation 3.14, then asymptotically

$$\begin{aligned} \chi_{\mathrm{aA}}(r, \theta, \phi) &\rightarrow \exp(i\mathbf{k}_\alpha \cdot \mathbf{r}) + f_\alpha(\theta, \phi) \exp(ik_\alpha r)/r, && \text{elastic channel} \\ \chi_{\mathrm{bB}}(r, \theta, \phi) &\rightarrow f_\beta(\theta, \phi) \exp(ik_\beta r)/r, && \text{non-elastic channel} \end{aligned} \tag{3.39}$$

(For convenience we now put the normalising constant $A_0 = 1$, corresponding to one particle per unit volume in the incident beam.) Let us now make a partial wave expansion of the χ, as in equation 3.34.

For simplicity we will assume for the present that the various particles do not have any intrinsic spins so that the total angular momentum is just the orbital angular momentum ℓ. Then ℓ is conserved; a given ℓ in the entrance channel leads to the same ℓ in any exit channel. We also take $\mathbf{k}_\alpha$ as the z-axis so that $m = 0$ only. The ingoing parts of χ_{aA} must be the same asymptotically as in a plane wave with the same momentum k_α, as in equations 3.35 and 3.38; the non-elastic χ_{bB} have only outgoing parts. Hence the radial parts of the partial waves must have the asymptotic form

$$\begin{aligned} u_{\ell,\mathrm{aA}}(r) &\rightarrow \frac{y_\ell}{r} \left[(-)^\ell \exp(-ik_\alpha r) - \eta_{\ell,\alpha} \exp(ik_\alpha r)\right], && \text{elastic channel} \\ &\rightarrow -\frac{y_\ell}{r} \left(\frac{v_\alpha}{v_\beta}\right)^{1/2} \eta_{\ell,\beta} \exp(ik_\beta r) && \text{non-elastic channel} \end{aligned} \tag{3.40}$$

The initial factors on the right side of equations 3.40 are chosen to normalise the u_ℓ so that $\chi = \exp(i\mathbf{k}_\alpha \cdot \mathbf{r})$in the absence of any scattering; then

$$y_\ell = \frac{i\,[\pi(2\ell + 1)]^{1/2}}{k_\alpha}$$

The η_ℓ are known as the partial wave scattering amplitudes or scattering matrix elements and, of course, depend upon the bombarding energy. The factor $(v_\alpha/v_\beta)^{1/2}$, where v_β is the velocity in the exit channel, is inserted into the expression for the non-elastic wave because with this definition the properties of the η_ℓ are simplified, as we shall see below in equation 3.44.

Near the origin the u_ℓ are proportional to r^ℓ just as in the undisturbed plane wave. These two boundary conditions, at the origin and asymptotic, must be imposed upon the solutions of the Schrödinger equation in order that we may determine the η_ℓ coefficients.

3.4.4 Scattering matrix and phase shifts

When there is no scattering, the scattering amplitudes must be

$$\textit{no scattering: } \eta_{\ell,\beta} = \delta_{\beta,\alpha} = \begin{cases} 1 & \text{if b + B = a + A, elastic} \\ 0 & \text{if b + B} \neq \text{a + A, non-elastic} \end{cases} \tag{3.41}$$

(where the Kronecker delta $\delta_{\beta,\alpha} = 1$ if $\beta = \alpha$ and 0 otherwise. For this purpose, an excited nucleus is counted as different from an unexcited one: $A^* \neq A$, etc.). If there is only elastic scattering, there are no waves in the other channels; then the outgoing waves in the elastic channel cannot be changed in *amplitude* (because of conservation of the number of particles) but they can be changed in phase. Then $|\eta_{\ell,\alpha}| = 1$ or $\eta_{\ell,\alpha} = \exp 2i\delta_\ell$, say. Hence

$$\textit{elastic only: } \eta_{\ell,\beta} = \delta_{\beta,\alpha} \exp 2i\delta_\ell \tag{3.42}$$

The real angle δ_ℓ is called the *phase shift* of the ℓth partial wave because when 3.42 is inserted into 3.40 we see that asymptotically u_ℓ is then proportional to

$$(kr)^{-1} \sin [kr - \tfrac{1}{2}\ell\pi + \delta_\ell] \tag{3.43}$$

Comparison with equation 3.37 shows that it has the same radial form as the corresponding part of the undisturbed plane wave except that it is shifted in phase by δ_ℓ.

When non-elastic scattering also occurs, there will be a depletion of flux in the outgoing waves of the elastic channel so that† the amplitude $|\eta_{\ell,\alpha}| < 1$. This flux reappears in the non-elastic channels. If there are only simple two-body reactions of the type a + A → b + B, then for every α = (a, A) pair lost from the incident channel there must be a corresponding appearance of a β = (b, B) pair in one of the non-elastic channels. If none of these particles has spin, then conservation of angular momentum demands that the β pair has the same angular momentum ℓ of relative motion as the lost α pair. Since the amplitude of the outgoing wave is proportional to η_ℓ, this conservation of flux yields the relation

$$\sum_{\beta \neq \alpha} |\eta_{\ell,\beta}|^2 = 1 - |\eta_{\ell,\alpha}|^2$$

or (3.44)

$$\sum_{\beta} |\eta_{\ell,\beta}|^2 = 1$$

where now the sum extends over *all* channels, including the elastic channel with $\beta = \alpha$. This is the simplest example of the *unitarity* of the scattering matrix η_ℓ which follows from the conservation of the number of particles.

The set of amplitudes $\eta_{\ell,\beta}$ actually form a matrix. Each of them should also carry the label α because it is the amplitude for transition from the a + A channel to the b + B channel; i.e. it is the β, α element of a *unitary matrix* η_ℓ. We

†Even when $|\eta_{\ell,\alpha}| < 1$, it is still sometimes written in the form 3.42 but now with δ_ℓ complex. We shall not use these complex phase shifts here.

have omitted this extra label so far in order to simplify the notation. It must be clear now (and we shall show more explicitly in section 3.4.5) that once we know all the elements of this matrix for each ℓ value we have complete knowledge of the results of all the scattering measurements we can make on the a + A system.

The unitarity condition 3.44 has a very simple interpretation: when a projectile is incident upon a target, the probability that something will emerge is unity. There is another property which holds when (as we believe to be the case) the interactions are invariant under time-reversal. Then, as discussed in section 2.17, the amplitude for the **A**(a, b)**B** reaction must equal that for the inverse **B**(b, a)**A** reaction at the same total energy.* This has the consequence that the corresponding scattering matrix elements are also equal, which implies that the matrix is also symmetric.

A point concerning notation: when elastic scattering is being considered, it is common to use the notation η_ℓ for the scattering matrix elements, the subscript α being understood. For non-elastic collisions **A**(a, b)**B**, a more common notation is to use **S** for the scattering matrix, with elements

$$\eta_{\ell,\beta} \equiv S^{\ell}_{\beta,\alpha} \tag{3.45}$$

(A more complete specification of the labels will be discussed in section 3.8 when we consider the consequences of the particles having intrinsic spin.) The notation 3.45 exhibits explicitly the matrix nature of **S**. The unitarity property becomes

$$\sum_{\beta} | S^{\ell}_{\beta,\alpha} |^2 = 1 \tag{3.46}$$

while the symmetry property just discussed is

$$S^{\ell}_{\alpha,\beta} = S^{\ell}_{\beta,\alpha} \tag{3.47}$$

This latter has also been called the ***reciprocity theorem*** (Blatt and Weisskopf, 1952). The matrix character of **S** is discussed further in section 3.8.3.

3.4.5 Phase shifts for potential scattering

Elastic scattering from a simple central potential $U(r)$ provides a very physical interpretation of the phase shifts δ_ℓ appearing in equations 3.42 and 3.43. (The mathematical details are given in any text on quantum mechanics.) *Figure 3.4* illustrates this for $\ell = 0$ and a bombarding energy E. The uppermost section shows r times the undisturbed wave, sin kr, in the absence of any potential. The central section shows the change when a square, ***attractive*** ($V < 0$) potential well

*In practice one has to be careful about the meaning of 'inverse'. For example, if a particle has spin s and we specify its z-component, m_s say, then the time-inverse of this is a spin s with z-component $-m_s$, because an angular momentum vector changes sign under time-reversal, $t \to -t$; i.e. the direction of rotation is reversed.

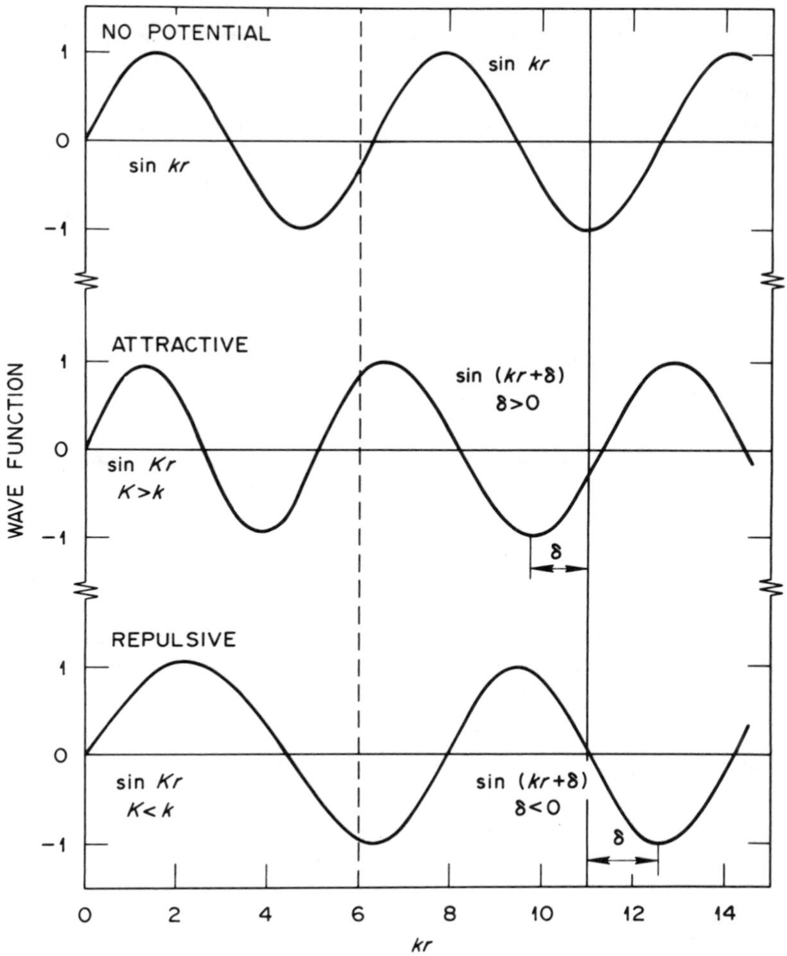

Figure 3.4 Scattering wave functions for $\ell = 0$ (S-wave) and (top) no potential, (middle) attractive square well potential and (bottom) repulsive square well potential. The strength of the repulsive potential is assumed to be less than the bombarding energy, $V < E$, so that there is positive kinetic energy in the interior. The curves are actually drawn for wells with a radius of $R = 6/k$ and strengths $|V| = \frac{1}{2}E$. Then $K = 1.225k$, $\delta = 1.258$ for the attractive potential and $K = 0.707k$, $\delta = -1.632$ for the repulsive potential. Note that the interior wave function amplitude is slightly *reduced* for the attractive potential (amplitude = 0.946) and *increased* for the repulsive potential (amplitude = 1.056). The reader may verify these numbers for himself

is introduced. In the interior the wave still behaves as sin Kr, but with a shorter wavelength, $K > k$, since the wavelength is determined by the *kinetic* energy which here is $E - V(r)$

$$K = [2\mu(E - V)/\hbar^2]^{1/2} \tag{3.48}$$

This interior wave is zero at the origin and must join continuously and smoothly onto the exterior wave. The latter has the same form as before (since $V = 0$ in the exterior) except that now it must be pulled in by an amount Δr in order to match onto the interior wave. It is clear that $\Delta r = -\delta_0/k$, where δ_0 is the phase shift, so that an attractive potential leads to a positive phase shift. In general, the need for smooth matching also results in a reduced amplitude for the interior wave. The lower part of *Figure 3.4* shows the opposite shift for a ***repulsive*** ($V > 0$) potential since in the region of this potential the kinetic energy and momentum are reduced and hence the wavelength is increased. Thus a repulsive potential induces a negative phase shift.

One limiting case of a repulsive potential is called a 'hard sphere'. This may be regarded as a potential which is infinitely repulsive for $r \leqslant a$ but zero for $r > a$. Hence the wave function must vanish at $r = a$. For example, for s-waves the wave function for $r \geqslant a$ is proportional to $\sin(kr + \delta_0)$; in order to vanish at $r = a$ we must have $\delta_0 = -ka$, the so-called hard sphere phase shift. Further, since the wave cannot penetrate the interior, there can be no absorption and the hard sphere behaves like a perfect reflector.

3.4.6 Partial wave expression for scattering amplitudes

By comparing equations 3.34–3.40 we can express the scattering amplitudes $f(\theta, \phi)$ in terms of the partial wave matrix elements η_ℓ. If we choose a z-axis along the incident beam, we find

$$f_\beta(\theta, \phi) = \frac{1}{2ik_\alpha}\left(\frac{v_\alpha}{v_\beta}\right)^{1/2} \sum_\ell (2\ell + 1)\,[\eta_{\ell,\beta} - \delta_{\alpha,\beta}]\, P_\ell(\cos\theta) \tag{3.49}$$

The same Kronecker delta appears here as in equation 3.41 and is unity for elastic scattering and zero otherwise. It appears for elastic scattering because the incident plane wave already has an outgoing part (*see* equation 3.38) and it is only the excess over this which represents true elastic scattering (*see* equation 3.5).

We should emphasise that equation 3.49 only holds if the various particles have no spin, or the interactions are independent of spin. The incident beam has zero component of angular momentum along the beam axis and this component is conserved in the absence of spin effects. As a consequence only Y_ℓ^m with $m = 0$ enter the expression 3.49 and, as discussed before, $f_{bB}(\theta, \phi)$ is then independent of ϕ. In the more general case with spin effects included, $m \neq 0$ will enter (section 3.8) and f will then depend on ϕ.

When we have elastic scattering only and write $\eta_{\ell,\alpha} = \exp 2i\delta_\ell$ as in equation 3.42, the expansion 3.49 can also be written as

$$f(\theta) = \frac{1}{k} \sum_\ell (2\ell + 1) \exp(i\delta_\ell) \sin\delta_\ell\, P_\ell(\cos\theta) \tag{3.50}$$

Now P_0 (cos θ) = 1, a constant, so it is clear from equation 3.49 or 3.50 that if only $\ell = 0$ contributes, then the amplitude f is independent of θ and the scattering is isotropic. It is necessary to have contributions from $\ell \geqslant 1$ for the differential cross-section to be anisotropic; i.e. for its angular distribution not to be constant.

3.4.7 Effects of Coulomb forces

The wave function described in section 3.1 is only fully appropriate for uncharged particles like neutrons. When there is a Coulomb interaction between the two residual nuclei b and B, we can no longer use equation 3.6 for the asymptotic form of the outgoing waves. Because of the long range of the r^{-1} dependence of the Coulomb potential, it distorts the wave even at large distances* (e.g. Messiah, 1962) so that the radial part of the asymptotic relative motion is described instead by

$$\frac{1}{r} \exp i(k_\beta r - n_\beta \ln 2k_\beta r) \tag{3.51}$$

with a phase shift (relative to an undisturbed wave) which depends logarithmically on the distance r. Here n_β is the Coulomb or Sommerfeld parameter

$$n_\beta = Z_b Z_B \, e^2/\hbar \, v_\beta \tag{3.52}$$

and $Z_b e$, $Z_B e$ are the charges on nuclei b and B, respectively. The presence of the Coulomb field also modifies the amplitude $f_\beta(\theta, \phi)$ relative to that which would be obtained in the absence of such a field. For example, a repulsive Coulomb barrier (*see Figure 2.35*) tends to shield the specifically nuclear interactions.

In addition, if there is a Coulomb interaction between the projectile a and target nucleus A, this alone gives rise to elastic scattering (Rutherford scattering if both are point charges or if the bombarding energy is below the Coulomb barrier—*see* section 2.10.3).

Now the wave equation for two point particles interacting via a Coulomb potential alone can be reduced to an equation of the Laguerre type whose solutions are well known (e.g. Messiah, 1962). Hence it is convenient to describe the scattering and reactions of charged particles starting from these Coulomb-distorted waves instead of the plane and spherical waves of equations 3.4 and 3.6. This means that for the asymptotic form we now have

*In most physical situations, these Coulomb fields are screened (e.g. by atomic electrons) at very large distances, but these distances are of atomic dimensions which are very much larger than nuclear dimensions and it is convenient to treat the Coulomb fields as if they were unscreened.

$$\chi_{aA}(r,\theta,\phi) \to \chi_{Coul}(r,\theta,\phi) + f'_\alpha(\theta,\phi)\frac{1}{r}\ \exp\left\{i(k_\alpha r - n_\alpha \ln 2k_\alpha r)\right\}$$
elastic channel (3.53)

$$\chi_{bB}(r,\theta,\phi) \to f'_\beta(\theta,\phi)\ \frac{1}{r}\exp\left\{i(k_\beta r - n_\beta \ln\ 2k_\beta r)\right\}\quad \text{non-elastic channel}$$

instead of equations 3.39 (compare with 3.27). Here χ_{Coul} is the wave for purely Coulomb (Rutherford) scattering of point charges; if the projectile or target were uncharged, we would have $\chi_{Coul} \to \exp(i\mathbf{k}_\alpha\cdot\mathbf{r})$. The f' amplitudes now describe the *additional* scattering due to the other interactions (plus any modifications of the Rutherford scattering that are due to the nuclear charges not being concentrated at a point but being spread over a finite region).

Since the Coulomb wave in equation 3.53 already includes the scattering due to point charges, the total scattering amplitude is now (compare with equation 3.28)

$$f_\beta(\theta,\phi) = f_C(\theta)\ \delta_{\alpha,\beta} + f'_\beta(\theta,\phi) \tag{3.54}$$

where $f_C(\theta)$ is the Rutherford scattering amplitude (e.g. Messiah, 1962) which, of course, only contributes to the elastic channel

$$f_C(\theta) = -\ \frac{n_\alpha}{2k_\alpha \sin^2(\frac{1}{2}\theta)}\ \exp[-in_\alpha \ln\ \sin^2(\tfrac{1}{2}\theta) + 2i\sigma_0] \tag{3.55}$$

Then $|f_C(\theta)|^2$ is the Rutherford cross-section, equation 2.16.

The partial wave expansion 3.34 is still valid, and the expansion of the $f'(\theta,\phi)$ corresponding to equation 3.49 is now

$$f'_\beta(\theta,\phi) = \frac{1}{2ik_\alpha}\left(\frac{v_\alpha}{v_\beta}\right)^{1/2}\ \sum_\ell\ (2\ell+1)\exp(2i\sigma_\ell)\ [\eta_{\ell,\beta} - \delta_{\alpha,\beta}]\ P_\ell(\cos\theta) \tag{3.56}$$

where we are still using the incident beam as z-axis. Also, σ_ℓ is called the Coulomb phase shift because the ℓth part of the Coulomb wave χ_{Coul} has the asymptotic form

$$(kr)^{-1}\ \sin[kr - n\ \ell n\ 2kr - \tfrac{1}{2}\ell\pi + \sigma_\ell] \tag{3.57}$$

which is just like the scattering wave 3.43 except for the additional logarithmic term. Explicitly (Messiah, 1962)

$$\begin{aligned}\sigma_\ell &= \arg\Gamma\ (\ell+1+i\,n)\\ &= 0\ \text{ if } n = 0\end{aligned} \tag{3.58}$$

where Γ is the gamma function (Abramowitz and Stegun, 1970).

The Coulomb effects show themselves explicitly through the appearance of this Coulomb phase in the amplitudes 3.56 but, as we have already remarked, the η_ℓ themselves are also modified by the presence of the Coulomb field compared to what would have been obtained if the particles had been uncharged.

3.4.8 *Partial wave expressions for cross-sections*

Equations 3.7 and 3.49 or 3.54 and 3.56 give the differential cross-section for the A(a, b)B reaction

$$\frac{d\sigma_\beta}{d\Omega} = \frac{v_\beta}{v_\alpha} \left| f_C(\theta)\, \delta_{\alpha,\beta} + f'_\beta\ (\theta\phi) \right|^2$$

charged particles:

$$= \left| f_C(\theta)\delta_{\alpha,\beta} + \frac{1}{2ik_\alpha} \sum_\ell (2\ell+1) \exp(2i\sigma_\ell)\,(\eta_{\ell,\beta} - \delta_{\alpha,\beta})\, P_\ell(\cos\theta) \right|^2 \tag{3.59a}$$

neutrons:

$$= \frac{1}{4k_\alpha^2} \left| \sum_\ell (2\ell+1)\,(\eta_{\ell,\beta} - \delta_{\alpha,\beta})\, P_\ell(\cos\theta) \right|^2 \tag{3.59b}$$

The first Kronecker delta in equation 3.59a appears because the Rutherford scattering only contributes to the elastic channel. In this channel, however, there is interference between the two amplitudes, Coulomb and nuclear.

The Rutherford amplitude 3.55 is inversely proportional to the bombarding energy E; it becomes less important as E increases and as the other amplitude f'_α begins to dominate. However, as equation 3.55 shows, f_C diverges for $\theta = 0$ so that it can never be neglected at the smallest angles.

3.4.9 *Integrated cross-sections*

In order to integrate the differential cross-section 3.59 over all angles we use the orthogonality properties of the Legendre polynomials (Appendix A)

$$\int P_\ell(\cos\theta)\, P_{\ell'}(\cos\theta) \sin\theta\, d\theta = \frac{2}{2\ell+1}\, \delta_{\ell,\ell'} \tag{3.60}$$

to get

$$\sigma_\beta = \int \left(\frac{d\sigma_\beta}{d\Omega}\right) d\Omega = \pi \lambda\!\!\!^{-}{}_\alpha^2 \sum_\ell (2\ell+1) \left| \delta_{\alpha,\beta} - \eta_{\ell,\beta} \right|^2 \tag{3.61}$$

(Remember that the integral over ϕ contributes a factor of 2π.) This expression can only be used for *elastic* scattering, $\beta = \alpha$, if either a or A is uncharged. For charged particles the integral over the Rutherford elastic scattering diverges. Equation 3.61 for $\beta = \alpha$ then represents the cross-section due to non-Coulombic interactions without the Rutherford scattering and without the interference with the Rutherford amplitude. Although of theoretical interest, this quantity is not directly observable (but *see* section 3.5).

We see that there is interference between contributions from different partial waves in the differential cross-sections (equations 3.59) but not in the integrated cross-section 3.61.

We define an *absorption cross-section* σ_{abs} (also called the *reaction cross-section*) as the sum of all the non-elastic integrated cross-sections

$$\sigma_{abs} = \sum_{\beta \neq \alpha} \sigma_\beta = \pi ƛ_\alpha^2 \sum_\ell (2\ell+1) \sum_{\beta\neq\alpha} |\eta_{\ell,\beta}|^2$$
$$= \pi ƛ_\alpha^2 \sum_\ell (2\ell+1)\,[1 - |\eta_{\ell,\alpha}|^2] \qquad (3.62)$$

where the last step follows from the unitary property 3.44. The physical interpretation of this is simple. In the absence of non-elastic events, the outgoing waves have unit intensity, $|\eta_{\ell,\alpha}|^2 = 1$, and $\sigma_{abs} = 0$. With non-elastic reactions occurring, these intensities are reduced below unity, and the amount of this reduction measures the absorption for that partial wave.

From equation 3.61, the integrated elastic cross-section for uncharged particles is

$$\sigma_{el} = \pi ƛ_\alpha^2 \sum_\ell (2\ell+1)|1 - \eta_{\ell,\alpha}|^2 \qquad (3.63)$$

3.4.10 Limits on partial cross-sections

Both σ_{el} and σ_{abs} are simple sums of contributions from each partial wave, $\sigma_{el,\ell}$ and $\sigma_{abs,\ell}$ say.The relationship between these contributions is shown in *Figure 3.5.* Since

$$0 \leqslant |\eta_{\ell,\alpha}| \leqslant 1$$

we have

$$\sigma_{e\ell,\ell} \leqslant 4\pi ƛ_\alpha^2\,(2\ell+1) \qquad (3.64)$$

The maximum occurs when $\eta_{\ell,\alpha} = -1$, and hence $\sigma_{abs,\ell} = 0$. The absorption is limited by

$$\sigma_{abs,\ell} \leqslant \pi ƛ_\alpha^2\,(2\ell+1) \qquad (3.65)$$

The maximum elastic cross-section is four times the maximum absorption allowed, but when absorption is at a maximum we have $\eta_{\ell,\alpha} = 0$ (corresponding to complete removal from the incident beam of particles with an angular momentum ℓ) and

$$\sigma_{abs,\ell} = \sigma_{el,\ell} = \pi ƛ_\alpha^2\,(2\ell+1) \qquad (3.66)$$

Only values of $\sigma_{el,\ell}$ and $\sigma_{abs,\ell}$ within the shaded area of *Figure 3.5* are allowed. (Remember that the expressions for elastic scattering only hold either for uncharged particles or for that part of the elastic cross-section obtained when the Rutherford amplitude is omitted.) The range of allowed values of σ_{el} for a given σ_{abs} can be understood by writing $\eta_{\ell,\alpha} = C_\ell \exp(i\gamma_\ell)$ in equations 3.62 and 3.63. We soon get the relation

$$\frac{\sigma_{el,\ell}}{\pi ƛ_\alpha^2\,(2\ell+1)} = 2\left\{1 - \left[1 - \frac{\sigma_{abs,\ell}}{\pi ƛ_\alpha^2\,(2\ell+1)}\right]^{1/2} \cos\gamma_\ell\right\} - \frac{\sigma_{abs,\ell}}{\pi ƛ_\alpha^2\,(2\ell+1)} \qquad (3.67)$$

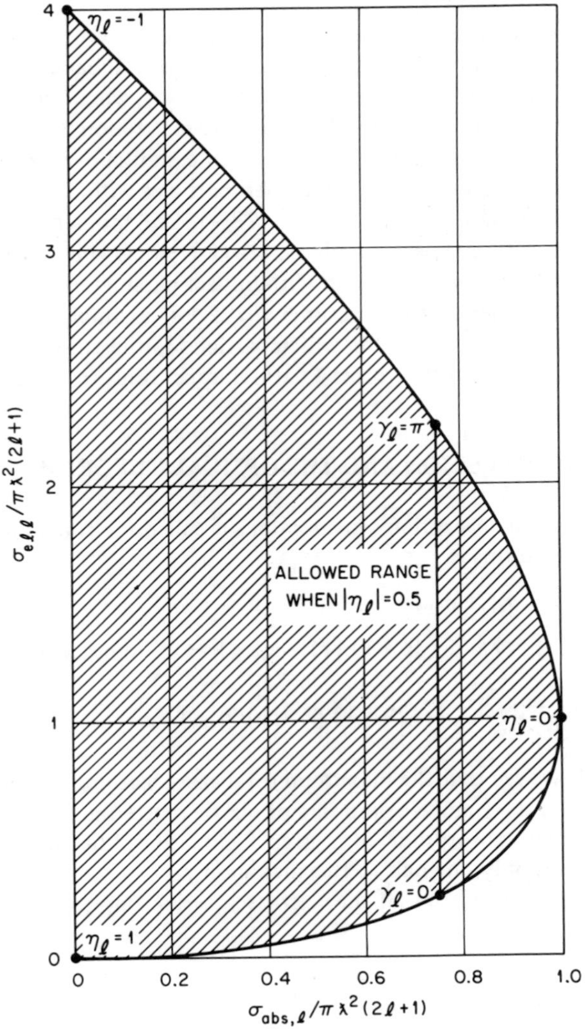

Figure 3.5 Limits on the elastic and absorption cross-sections for a given partial wave, illustrating equations 3.64–3.67. All allowed values fall in the hatched area. For example, when $|\eta_\ell| = \frac{1}{2}$, $\sigma_{abs}/\pi \lambda\!\!\!^{-2}(2\ell + 1) = \frac{3}{4}$ and $\frac{1}{4} \leqslant \sigma_{el}/\pi \lambda\!\!\!^{-2}(2\ell + 1) \leqslant \frac{9}{4}$

Thus, for a given $\sigma_{abs,\ell}$, the allowed values of $\sigma_{el,\ell}$ are limited by $-1 \leqslant \cos \gamma_\ell \leqslant 1$. The upper part of the curve in *Figure 3.5* corresponds to $\cos \gamma_\ell = -1$, the lower part to $\cos \gamma_\ell = +1$.

We see from these expressions that it is possible to have elastic scattering without absorption ($\sigma_{abs,\ell} = 0$ when $C_\ell = 1$, i.e. when $\eta_{\ell,\alpha}$ is unimodular) but we can never have absorption without some attendant elastic scattering. We already met this result in section 3.1.4 as the consequence of needing a scattered wave to produce the shadow cast by the absorbing target. It can also be derived explicitly by matching waves at the surface of the target, when the presence of absorption always introduces some reflection also—*see* section 3.6.1.

The factor $\pi \lambda\!\!\!^{-2} (2\ell + 1)$ appearing in the expression above has a simple geometrical interpretation. As discussed earlier, particles with quantum number ℓ have angular momentum $[\ell(\ell + 1)]^{1/2}\hbar \approx (\ell + \frac{1}{2})\hbar$ and momentum $p = \hbar/\lambda\!\!\!^{-}$. Classically their impact parameter would be $b = (\ell + \frac{1}{2})\lambda\!\!\!^{-}$. Owing to quantal uncertainties, these particles will be found with impact parameters mostly in the range $b \pm \frac{1}{2}\lambda\!\!\!^{-}$. We then see that the factor $\pi \lambda\!\!\!^{-2} (2\ell + 1)$ is simply the area of a ring with radius b and width $\lambda\!\!\!^{-}$ within which particles in the ℓth partial wave are most likely to be found and this forms a natural upper limit to the cross-section for this partial wave.

3.5 TOTAL CROSS-SECTION AND THE OPTICAL THEOREM

The sum of elastic and absorption cross-sections is called the total cross-section; if a or A is uncharged this is

$$\begin{aligned} \sigma_{\text{tot}} &= \sigma_{\text{el}} + \sigma_{\text{abs}} \\ &= \pi \lambda\!\!\!^{-2}_{\alpha} \sum_{\ell} (2\ell + 1) [|1 - \eta_{\ell,\alpha}|^2 + 1 - |\eta_{\ell,\alpha}|^2] \\ &= 2\pi \lambda\!\!\!^{-2}_{\alpha} \sum_{\ell} (2\ell + 1) [1 - Re\, \eta_{\ell,\alpha}] \end{aligned} \tag{3.68}$$

Since $|\eta_{\ell,\alpha}| \leqslant 1$, we see that the maximum total cross-section for a given ℓ is obtained when $\eta_{\ell,\alpha} = -1$ and is composed of elastic scattering only (*see* equation 3.62). 'Complete absorption' of a given ℓ, equation 3.66, which occurs when $\eta_{\ell,\alpha} = 0$, gives only one-half of this maximum.

From equation 3.49 we see that the elastic scattering amplitude at $\theta = 0°$ is just

$$f_\alpha (\theta = 0) = \frac{1}{2ik_\alpha} \sum_{\ell} (2\ell + 1) [\eta_{\ell,\alpha} - 1]$$

since $P_\ell (\cos \theta = 1) = 1$. Comparison with equation 3.68 gives *the optical theorem* which relates the total cross-section to the imaginary part of the forward ($\theta = 0$) elastic scattering amplitude

$$\sigma_{\text{tot}} = 4\pi \lambda\!\!\!^{-}_{\alpha}\, Im\, f_\alpha (\theta = 0) \tag{3.69}$$

Since the elastic differential cross-section, equation 3.7, is just $d\sigma_{\text{el}}/d\Omega = |f_\alpha(\theta, \phi)|^2$, the optical theorem 3.69 gives rise to *Wick's inequality*

$$\sigma_{\text{tot}} \leqslant 4\pi \lambda\!\!\!^{-}_{\alpha} [d\sigma_{\text{el}}(\theta = 0°)/d\Omega]^{1/2} \tag{3.70}$$

Of course, as we have repeatedly stressed, σ_{el} and $f_\alpha(0)$ are both infinite for charged particles, so that σ_{tot} is infinite also. Hence the optical theorem 3.69

and Wick's inequality 3.70 are of practical interest primarily for neutron scattering. Nevertheless, a theorem equivalent to the optical theorem 3.69 can be stated for charged-particle scattering (Cooper and Johnson, 1976; Schwarzschild *et al.*, 1976):

$$[\sigma_{el} - \sigma_R] + \sigma_{abs} = 4\pi ƛ_\alpha \, Im\, f'_\alpha\,(\theta = 0) \tag{3.71}$$

Here f'_α is the additional scattering amplitude arising from the specifically nuclear forces (and to the deviation of the nuclear charge distribution from a point charge) as defined by equation 3.54. Also, while σ_{el} is the full integrated elastic cross-section, σ_R is the integral of just the Rutherford cross-section, equation 2.16. Both of these are infinite, but their difference can be shown to remain finite and to be given by equation 3.71.

If $4\pi ƛ_\alpha \, Im\, f'_\alpha\,(\theta) \ll \sigma_{abs}$, as may happen when there is strong absorption (section 2.18.11) such as in heavy-ion scattering, then

$$\sigma_{abs} = [\sigma_R - \sigma_{el}] = \int \left[\frac{d\sigma_R}{d\Omega} - \frac{d\sigma_{el}}{d\Omega}\right] d\Omega$$

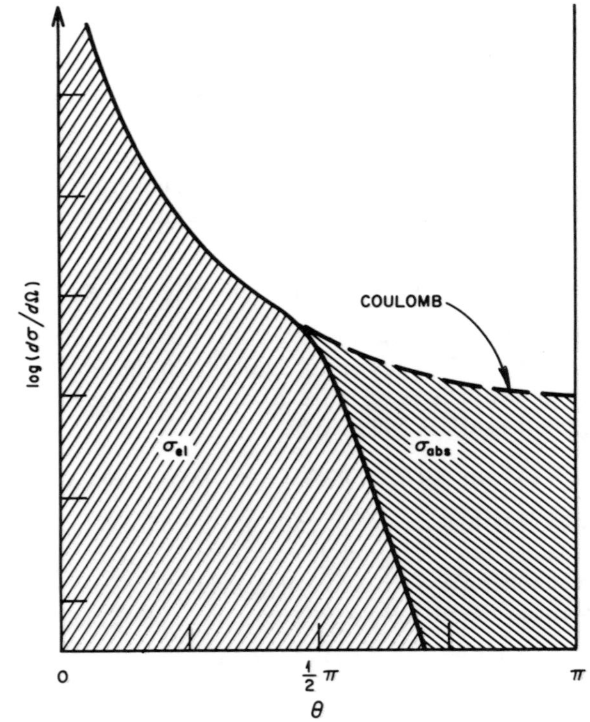

Figure 3.6 Illustrating the optical theorem, equation 3.71, for charged particles when there is strong absorption. In the classical limit when $f'_\alpha = 0$, the reduction of the elastic differential cross-section below the Rutherford (or Coulomb) value is just equal to the absorption cross-section

There is a simple classical picture for this relation. The differential cross-sections have the form shown in *Figure 3.6*; the actual elastic cross-section suddenly drops below the Rutherford or Coulomb value at an angle for which the corresponding Rutherford orbit allows the two particles to make contact (section 2.10.3) and interact strongly. There is absorption of flux from the elastic channel along orbits corresponding to angles larger than this (that is, those orbits which allow the two particles to approach more closely). Then the difference between the actual elastic cross-section σ_{el} and the Rutherford value σ_R is just the absorption cross-section σ_{abs}. However, in general, the wave nature of the scattering leads to $Im\, f'_\alpha(0) \neq 0$ and this simple identification does not hold exactly.

3.6 PENETRATION AND REFLECTION AT POTENTIAL BARRIERS

3.6.1 Reflection by an absorptive region

The wave-mechanical penetration of potential barriers, by which particles may penetrate into regions which would be forbidden classically, plays an important role in many aspects of nuclear physics. Even when a particle has sufficient kinetic energy to surmount a potential barrier, wave mechanics predict some reflection at the barrier which is not present classically. This penetration and reflection is treated in all standard texts on quantum mechanics (e.g. Messiah, 1962) so we will confine ourselves to a few illustrative examples.

Consider first an s-wave ($\ell = 0$) of energy E incident upon a strongly absorbing region of radius R. Outside, for $r > R$, the wave number is $k = (2mE)^{1/2}/\hbar$ and the general solution $u(r)$ is an ingoing plus an outgoing wave; putting $u(r) = w(r)/r$ as in equation 3.33

$$w_> = A\ [\exp(-ikr) - \eta_0 \exp(ikr)]\,, \quad r > R \tag{3.72}$$

Complete absorption within the sphere means that there are only *ingoing* waves just inside the surface. Further, absorption implies that the amplitude of these waves decreases exponentially as we move in from the surface $r = R$. The form of the wave just inside the surface will then be

$$w_< = B_0 \exp\ \{-(R-r)/2\Lambda\}\ \exp(-iK_0 r), \quad r < R \tag{3.73}$$

Here Λ is the decay length or mean free path; at a distance Λ inside the surface the intensity has fallen by e^{-1}. Strong absorption means $R \gg \Lambda$. Further, we may have a wave number $K_0 \neq k$ because, in general, the particle will encounter a potential as it crosses the surface $r = R$. This interior wave may be rewritten

$$w_< = B \exp(-iKr)$$

where

$$B = B_0 \exp(-R/2\Lambda), \quad K = K_0 + (i/2\Lambda) \tag{3.74}$$

(This shows, incidentally, that the damping of a wave arising from absorption can be represented by a complex wave number which can then be thought of as arising from motion under the influence of a complex potential—*see* Chapter 4.)

The amplitude A of the incident wave (equation 3.72) will be determined by the incident flux as in section 3.1.1. The reflected amplitude $\eta_0 A$ and the transmitted amplitude B may be related to A by matching the values and first derivatives of $w_>$ and $w_<$ at $r = R$. We soon find that

$$\eta_0 = \left(\frac{K - k}{K + k}\right) \exp(-2ikR)$$
$$B_0 = A\left(\frac{2k}{k + K}\right) \exp(i(K_0 - k)R) \tag{3.75}$$

A transmission or penetration coefficient T_0 can be defined as the fraction of the incoming flux which crosses the surface $r = R$. Since the flux is the wave intensity times the velocity, we have

$$T_0 = \frac{K_0}{k}\left|\frac{B_0}{A}\right|^2 = \frac{4kK_0}{|K + k|^2} \tag{3.76}$$

This can also be written from equation 3.75 as

$$T_0 = 1 - |\eta_0|^2 \tag{3.77}$$

From equation 3.62 we see that the corresponding absorption cross-section is just

$$\sigma_{abs} = (\pi/k^2)T_0 \tag{3.78}$$

In order to have $T_0 \neq 0$, we must have $|\eta_0| < 1$; that is we cannot have absorption without also having elastic scattering, as we already deduced in section 3.4.10. We see here that this arises because of the matching conditions; we cannot have wave transmission into an absorptive region without also having reflection at the boundary. Further we see that, since by hypothesis $K \neq k$, we cannot have $\eta_0 = 0$ or complete transmission, $T_0 = 1$. Even if $K_0 = k$, we still require $\Lambda^{-1} \neq 0$ and this implies that $T_0 = [1 + (1/4\,\Lambda k)^2]^{-1} < 1$. In practice, however, the absorption does not, as we have assumed, set in abruptly at some radius $r = R$ but increases slowly over a range of radii; this ameliorates the reflection and allows us to attain $\eta_0 \approx 0$ (*see* Chapter 4).

3.6.2 Coulomb barriers

The repulsive Coulomb potential between two particles with like charges provides a potential barrier, the Coulomb barrier (section 2.10.3 and *Figure 2.35*) which hinders the close approach of the particles. The penetration through this barrier is of importance for understanding the reactions between charged particles as well as for understanding α-decay (Preston and Bhaduri, 1975; Preston, 1962). In simplified form, we have a potential $(Z_a Z_A e^2/r)$ between the two particles for $r > R$; when $r = R$ the particles touch and the attractive nuclear forces come into play (*Figure 3.7*). The penetration factor T is given by the intensity of the wave

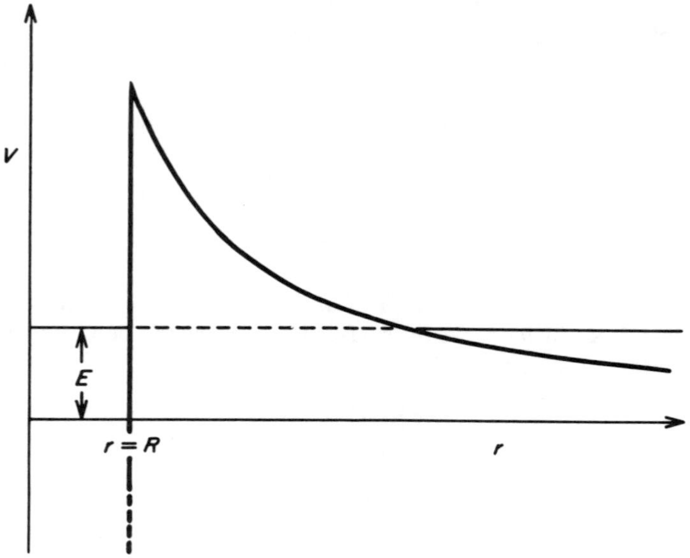

Figure 3.7 Simplified form of the Coulomb barrier between two charged particles. The attractive nuclear force is assumed to act abruptly when the two are separated by $r = R$ or less

function at $r = R$ relative to its intensity asymptotically. It may be written $T = e^{-G}$, where G is called the Gamow factor. For energies which are much lower than the top of the Coulomb barrier, it may be shown that

$$G \approx 2\pi n - \left(\frac{8 Z_a Z_A e^2 m R}{\hbar^2}\right)^{1/2}$$

where n is the Sommerfeld parameter of equation 2.21 which we may rewrite as $n = Z_a Z_A e^2 (m/2E\hbar^2)^{1/2}$. The second term may be neglected as E goes to zero so that G and T become independent of R

$$T \approx e^{-2\pi n} \tag{3.79}$$

Although this gives an indication of the strong dependence of the transmission through the Coulomb barrier upon the energy of the particles, it is only valid at very low energies. More complicated expressions are available for higher energies (Rasmussen, 1965).

3.6.3 Transmission across a rounded barrier

Another situation of interest for the description of nuclear reactions concerns the probability of transmission for a particle whose energy is close to the top of a barrier like the Coulomb barrier (*see*, for example, Wong, 1973). In practice

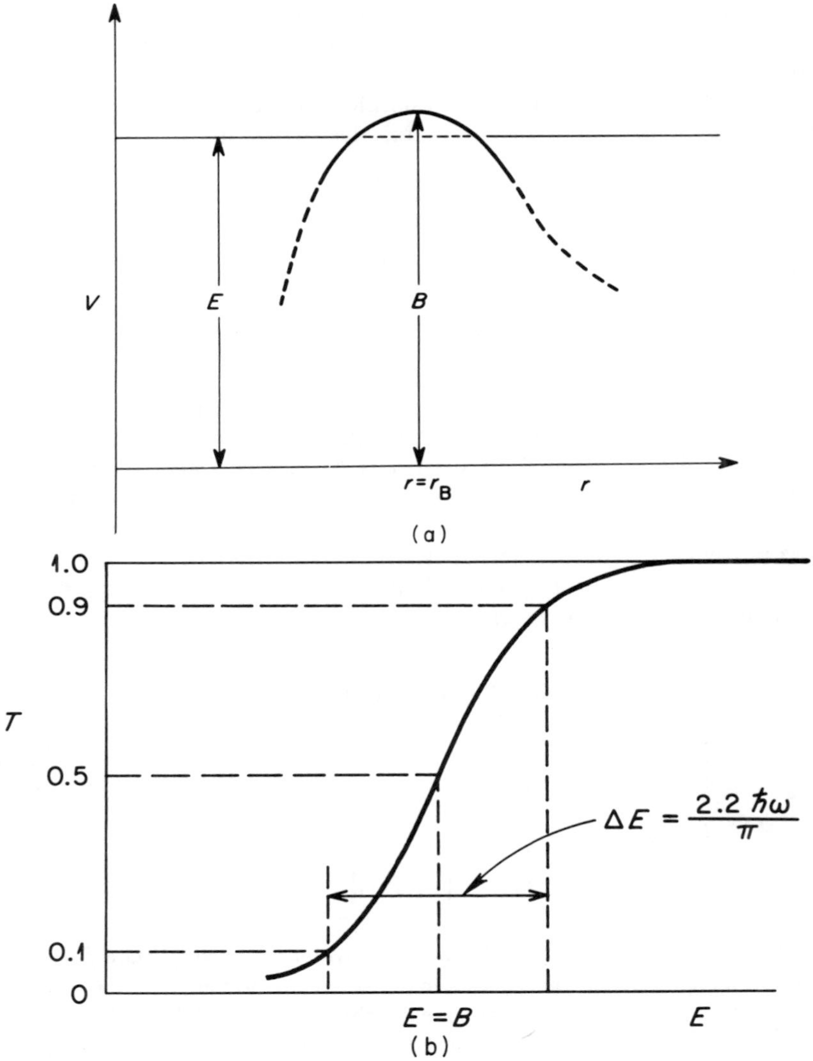

Figure 3.8 (a) An inverted parabolic or 'oscillator' potential barrier as given by equation 3.80; (b) transmission probability through this barrier for particles with energies E close to the top of the barrier, $E \approx B$, as given by equation 3.81

such a barrier does not have a sharp peak like that shown in *Figure 3.7* but is rounded by the addition of the nuclear forces as shown in *Figure 2.35*. Near the top we may approximate it by the parabolic form (*see Figure 3.8a*)

$$V(r) = B - b(r - r_B)^2 \tag{3.80}$$

(If b appeared with the opposite sign, this would be the potential energy of a simple harmonic oscillator with a natural frequency $\omega = (2b/m)^{1/2}$; for this reason it is sometimes referred to as an 'inverted oscillator' barrier.) It may be shown (Ford *et al.*, 1959) that the transmission coefficient for a particle of mass m and energy E incident on such a potential barrier is just

$$T = \frac{1}{1 + \exp(-2\pi x)} \tag{3.81}$$

where

$$x = \frac{E - B}{\hbar\omega}, \qquad \omega = \left(\frac{2b}{m}\right)^{1/2}$$

T is plotted as a function of energy in *Figure 3.8b*. The interesting feature is that T gradually rises from zero to unity as the energy E surmounts the barrier height B, being only $T = 1/2$ when $E = B$. Classically T would be a step function, $T = 0$ for $E < B$ and $T = 1$ for $E > B$. The rapidity with which the wave-mechanical T rises is determined by the curvature and hence the thickness of the barrier; T increases from 0.1 to 0.9 as E increases from $B - 1.1\hbar\omega/\pi$ to $B + 1.1\,\hbar\omega/\pi$. We have $T \approx 1/2$ for E slightly above B rather than the classical value $T = 1$ as another consequence of wave reflection at the barrier.

3.7 BEHAVIOUR OF CROSS-SECTIONS NEAR THRESHOLD

The general features of the cross-section for a particular reaction as a function of energy (i.e. the excitation function) near the threshold can be deduced from the expression 3.7

$$\frac{d\sigma_\beta}{d\Omega} = \left(\frac{v_\beta}{v_\alpha}\right) |f_\beta(\theta)|^2$$

if we make simple assumptions about the behaviour of the amplitude f_β of the scattered wave. In particular we assume that there are no narrow resonances close to the threshold so that f_β varies smoothly with energy. If the low-energy particle is a neutron, it can be shown that f_β is approximately constant (Blatt and Weisskopf, 1952; Lane and Thomas, 1958; Mott and Massey, 1965). If it is a charged particle, $|f_\beta|^2$ will be proportional to the penetrability through the Coulomb barrier. As discussed in the preceding section, near threshold this penetrability is approximately proportional to

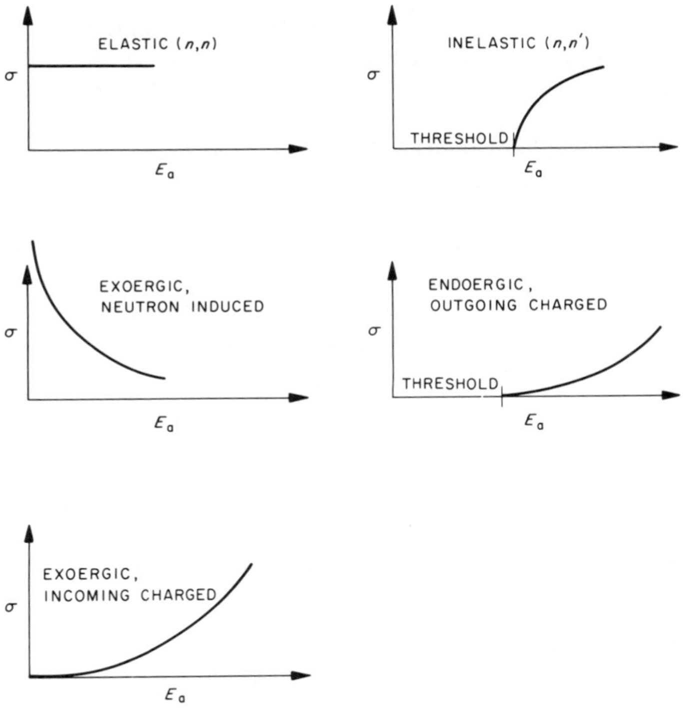

Figure 3.9 Schematic behaviour of various cross-sections near threshold

$$\exp(-2\pi n) = \exp(-C/E^{1/2})$$

where E is the energy of the low-energy charged particle and C is a constant.

The consequences are summarised schematically in *Figure 3.9*; they are as follows.

(i) *Elastic neutron scattering*: the threshold is at zero energy and $v_\beta = v_\alpha$. The cross-section will be approximately constant at the lowest energies.

(ii) *Inelastic neutron scattering*: close to threshold the incident velocity v_α varies slowly, while v_β is proportional to $E_\beta^{1/2}$ where E_β is the energy of the *outgoing* neutron. The cross-section varies like $E_\beta^{1/2}$.

(iii) *Exoergic neutron-induced reaction*: e.g. many (n, α) reactions. Both the velocity and the Coulomb penetrability of the outgoing fast particle are varying slowly near zero bombarding energy so the cross-section is proportional to v_α^{-1} or $E_\alpha^{-1/2}$ for the *incident* neutron. This is known as 'the $1/v$ law'.

(iv) *Endoergic reaction, outgoing neutron*: this is like inelastic neutron scattering, the behaviour near threshold being determined largely by v_β or $E_\beta^{1/2}$ for the *outgoing* neutron.

(v) *Endoergic reaction, outgoing particle charged*: the behaviour near threshold is mostly determined by the Coulomb penetrability of the slow *outgoing* particle and so increases like $\exp(-2\pi n_\beta) = \exp(-C_\beta/E_\beta^{1/2})$.

(vi) *Exoergic reaction induced by charged particle*: now the cross-section is most strongly affected by the Coulomb penetrability of the slow *incident* particle. It is proportional to $E_\alpha^{-1}\exp(-2\pi n_\alpha) = E_\alpha^{-1}\exp(-C_\alpha/E_\alpha^{1/2})$, where the additional E_α^{-1} term comes from the $\pi \lambda_\alpha^2$ factor in the expression for the cross-section.

3.8 COLLISIONS WITH SPIN: GENERAL THEORY

So far we have ignored the possibility that the colliding nuclei or their products could have intrinsic spins. In the majority of cases one or more of the particles *will* have non-zero spin. The general theory of collisions including the spins of the particles is beyond the scope of this book (*see*, for example, Preston, 1962; Blatt and Weisskopf, 1952; Lane and Thomas, 1958; Mott and Massey, 1965), but we can indicate here the approach and the structure of the theory.

3.8.1 Spins and channel spin

A projectile with spin i may be found in any one of $(2i + 1)$ different quantum states corresponding to the various orientations of its spin; i.e. its z-component may be $m_i = i, i - 1, \ldots -i$. An unpolarised beam of such particles consists of an *incoherent*, random mixture of these states.* The probability of any given value of m_i is then $(2i + 1)^{-1}$. Similarly, if the target has spin I and is also unpolarised, the probability of finding a particular z-component M_I is $(2I + 1)^{-1}$. Consequently, any cross-section for the collision of these two is the sum of cross-sections for each pair of initial values m_i, M_I weighted by $(2I + 1)^{-1}(2i + 1)^{-1}$.

It is often inconvenient to work with the z-components m_i and M_I because their values depend upon the choice of z-axis. The outcome of a collision cannot depend upon the coordinate axes chosen to describe it, so it is useful to choose a representation which does not depend upon these axes either. One possible step in this direction is to define the vector sum of the spins i and I, called the *channel spin*

$$\mathbf{S} = \mathbf{I} + \mathbf{i} \tag{3.82}$$

The magnitude of $\mathbf{S}$ covers the range

$$|I - i| \leqslant S \leqslant I + i \tag{3.83}$$

*If the beam is *polarised*, the particles are not uniformly distributed among the different m_i states; indeed, m_i may not be a good quantum number so that the beam is a *coherent* mixture of m_i states.

in integer steps. We now have but one z-component, M_S, to be concerned with. There are $2S + 1$ values of M_S for a given S; the *total* number of independent states has not changed and is $(2i + 1)(2I + 1)$, so the initial probability (or statistical weight) of finding a given S when the two particles are randomly oriented (unpolarised) must be

$$g(S) = \frac{2S + 1}{(2i + 1)(2I + 1)} \tag{3.84}$$

If the collision transition amplitude is independent of the spins, nothing is changed because $\sum_S g(S) = 1$. Such is the case for scattering by a simple central potential. However, as an example, we may see the significance of the channel spin in one special situation. Suppose we have a collision (such as with slow neutrons) in which only S-waves, $\ell = 0$, contribute. Then S is the total angular momentum of the system and must be conserved. Further, if the collision is unpolarised, the cross-section cannot depend upon the component M_S (i.e. it cannot depend upon the choice of z-axis). The cross-section must then have the form

$$\sigma = \sum_S g(S)\,\sigma(S) \tag{3.85}$$

Suppose now that this collision excites a compound nucleus resonance. This compound state itself will have a definite spin J, and in our example we must have $J = S$, so that only the amplitude for one of the allowed S values will resonate.

We now consider more general cases where the relative or orbital angular momentum ℓ is not zero. The total angular momentum J of the system is now the vector sum of ℓ and S

$$\mathbf{J} = \mathbf{l} + \mathbf{S}, \quad |\ell - S| \leqslant J \leqslant \ell + S \tag{3.86}$$

J is a valuable quantum number because the total angular momentum of the system and its z-component M_J are conserved; after the collision it may be composed of different values of ℓ' and S' but its magnitude and orientation are unchanged. Further, if the projectile and target are unpolarised and we do not measure the polarisation of the reaction products, the cross-sections will be independent of the orientation of J. The probability amplitude for finding a given J and M_J in a beam of particles is obtained from the algebra of angular momentum addition (*see* Appendix A; Brink and Satchler, 1968); it is (*see* equation A.54) the product of Clebsch–Gordan coefficients corresponding to equations 3.82 and 3.86.

The statistical weight for a given value of J follows from arguments parallel to those which led to equation 3.84. The probability of finding a resultant J when the angular momenta S and ℓ are randomly oriented relative to one another is (following equation 3.84)

$$\frac{2J+1}{(2S+1)(2\ell+1)}$$

Multiplying this by the probability $g(S)$ of finding S in an unpolarised beam of nuclei a + A gives the statistical spin factor

$$\frac{2J+1}{(2I_A+1)(2i_a+1)(2\ell+1)} \tag{3.87}$$

This factor should be included in the expressions for cross-sections derived earlier when the nuclei have spin. These expressions may be regarded as applying to a particular pair of values S and J and may have different values for different S and J. Then the cross-sections have the form

$$\sigma = \sum_{\ell SJ} \frac{2J+1}{(2I_A+1)(2i_a+1)(2\ell+1)}\, \sigma(\ell SJ) \tag{3.88}$$

For example, from equation 3.62 the generalised absorption cross-section would be

$$\sigma_{\text{abs}}(\ell SJ) = \pi \lambda\!\!\!^{-}{}_{\alpha}^{2}\, (2\ell+1)\, [1 - |\eta_{\ell SJ,\alpha}|^2]$$

where the scattering matrix element η may now depend upon S and J as well as ℓ. If perchance η is independent of S and J, the weight factor 3.87 may be summed over these quantities; the result is unity. In that case, the expressions derived before are unchanged by the presence of spins.

The formulae for the cross-sections always contain the term $\pi\lambda\!\!\!^{-}{}^2(2\ell+1)$. As explained in section 3.4.10, this is the probability of finding particles with relative angular momentum ℓ in the plane wave which represents the incident beam. This $(2\ell+1)$ cancels the one in equation 3.87 and the remainder is often written as

$$g(J) = \frac{2J+1}{(2I_A+1)(2i_a+1)}$$

and called the statistical factor for the total angular momentum J. Note, however, that $\Sigma_J g(J) \neq 1$ unless $\ell = 0$, although $\Sigma_{SJ} g(J) = (2\ell+1)$. When $\ell = 0$, we have $J = S$ and we retrieve equation 3.84.

The coupling scheme represented by equations 3.82 and 3.86 is not unique. Sometimes it is more convenient to couple the three vectors **i**, **I** and **l** to a resultant **J** in a different order. For example, suppose we are scattering nucleons. The success of the shell model of nuclear structure, in which a spin-orbit coupling interaction plays an important role, suggests that j, the orbital plus spin angular momentum, could be a useful quantum number for nucleons. Indeed, this is so for the optical model of nucleon scattering (*see* Chapter 4). We then couple the vectors in the following order

$$\mathbf{j} = \mathbf{l} + \mathbf{i}, \quad \mathbf{J} = \mathbf{I} + \mathbf{j} \tag{3.89}$$

The two schemes are not independent; they are related by a unitary transformation (Appendix A; Brink and Satchler, 1968). A state with definite quantum numbers ℓ, j and J can be expanded as a sum over states with various S values but the same ℓ and J, and vice-versa.

3.8.2 Collision channels with spins

We introduced the idea of a collision channel as consisting of two nuclei in particular states of excitation (including their ground states) and with a given energy of relative motion. Next we extended this notion to partial channels where the pair of nuclei had a particular value of the relative angular momentum ℓ (a partial wave). The outline of the previous section shows how we can generalise this to include spins by introducing two other quantum numbers, the total angular momentum J and, for example, the channel spin S.

So now the complete specification of a partial channel is two nuclei, a and A say, in definite internal states, with a given total energy E, and in a definite angular momentum state specified by, for example, ℓ, S and J. Let us denote such a channel by a Greek letter, α say. Then α stands for the set of numbers

$$\alpha = \{E, \mathrm{a}, \mathrm{A}, i_\mathrm{a}, I_\mathrm{A}, \ell_\alpha, S_\alpha, J\} \tag{3.90}$$

and the corresponding states or generalised partial waves form an orthogonal set which may be used to describe any reaction. (We have omitted the subscript α from E and J because these are always conserved quantities. The collision can only cause transitions between channels α and β, say, which have the same values of E and J.)

We see now that the expression 3.42 for elastic scattering needs to be generalised. Elastic scattering may include transitions between channels α and α' in which the internal states of a and A and their relative kinetic energy do not change but the final quantum numbers, $\ell_{\alpha'}$ and $S_{\alpha'}$ may differ from the initial ℓ_α and S_α. The interpretation of this is simple. When the particles have spin, the orientation of their spins and the orientation of the plane of their orbital motion may be changed without changing their internal states. The scattering is 'elastic' as long as the internal states do not change.

The optical theorem 3.69 also needs generalising. The forward-scattering elastic amplitude appearing there is now an average of the various partial amplitudes in which channel spin does not change, $S_{\alpha'} = S_\alpha$ (Lane and Thomas, 1958). Amplitudes corresponding to changes in S do not contribute to the optical theorem (but may contribute to the elastic cross-section). Wick's inequality remains unchanged.

3.8.3 The scattering wave function and the scattering matrix

The basic approach to the general theory of collisions is simply a generalisation of what we have already discussed, especially equations 3.39 and 3.40. The total wave function Ψ is built up from pieces in each channel $\alpha, \beta, \ldots$ defined as in the previous section; outside the region of interaction we may write

$$\Psi = \sum_{\alpha} C_\alpha \, [I_\alpha - \sum_{\beta} S_{\beta\alpha} \, O_\beta] \tag{3.91}$$

The I_α, O_α are incoming and outgoing waves, respectively, in the partial channel α; that is, asymptotically their radial parts behave like the corresponding parts of equation 3.40 (or the equivalent Coulomb waves discussed in section 3.4.7). $S_{\beta\alpha}$ is an element of **S**, the generalisation of the scattering matrix introduced in section 3.4.4, where now the labels α, β refer to the more complete sets 3.90. (In the description of nuclear reactions, the *collision matrix* **U** is often used instead of **S** (Lane and Thomas, 1958). **U** is the same as **S** for uncharged particles and differs only by a phase otherwise.) In the absence of any scattering, **S** becomes the unit matrix $S_{\beta\alpha} = \delta_{\beta\alpha}$. In general $S_{\alpha\beta}$ is a complex number and $|S_{\beta\alpha}| \leqslant 1$. The properties of S already discussed in section 3.4.4 still hold, for example

$$\mathbf{S}\mathbf{S}^\dagger = \mathbf{1}, \quad \text{or} \quad \sum_{\gamma} S_{\alpha\gamma} S^*_{\beta\gamma} = \delta_{\alpha,\beta}: \textit{unitarity} \tag{3.92}$$

In particular

$$\sum_{\gamma} |S_{\alpha\gamma}|^2 = 1: \textit{conservation of probability}$$

In addition

$$\mathbf{S}^* = \mathbf{S}^\dagger, \quad \text{or} \quad S_{\alpha\beta} = S_{\beta\alpha}: \textit{symmetry or reciprocity} \tag{3.93}$$

The physical significance of the form 3.91 should be clear by now. Only the outgoing waves are modified by the collision and the information about the collision is then contained in the matrix elements $S_{\beta\alpha}$. The presence of off-diagonal elements, $\beta \neq \alpha$, implies that an incoming wave in channel α gives rise to an outgoing wave in channel β. These events include non-elastic collisions although, as we have just seen, elastic scattering may also involve transitions between partial channels when the particles have spins. The coefficients C_α must then be chosen to satisfy the initial conditions. Normally these correspond to the physical situation we have already described, namely a plane wave of particles a incident on a target such as in equation 3.4 and *Figure 3.1*.

3.8.4 Cross-sections and inverse reactions

Arguments parallel to those used earlier result in expressions for the integrated cross-sections. The partial cross-section for transitions between two particular channels is

$$\sigma_{\alpha\beta} = \pi \lambda\!\!\!^{-}{}_\alpha^2 \, (2\ell_\alpha + 1) \, |\delta_{\alpha\beta} - S_{\alpha\beta}|^2 \tag{3.94}$$

while the complete cross-section for the A(a, b)B reaction may be written

$$\sigma_{\mathrm{aA,bB}} = \sum_{\beta} \langle \sigma_{\alpha\beta} \rangle_{\alpha} \tag{3.95}$$

where $\sum_{\beta}\langle\,\rangle_{\alpha}$ stands for an *average* over all of the contributing *incident* a + A partial waves and a *sum* over the *final* b + B partial waves. When taking the average, the appropriate statistical weights $g(\alpha)$, such as in equations 3.84 and 3.87, must be included.

These expressions provide another proof of the reciprocity relation 2.48 for inverse reactions. Using the statistical weight 3.87 and the symmetry 3.93 in the cross-section equations 3.94 and 3.95, we have

$$\frac{\sigma_{\mathrm{aA,bB}}}{\sigma_{\mathrm{bB,aA}}} = \frac{\lambda\!\!\!^{-2}_{\alpha}\,(2I_{\mathrm{B}} + 1)\,(2i_{\mathrm{b}} + 1)}{\lambda\!\!\!^{-2}_{\beta}\,(2I_{\mathrm{A}} + 1)\,(2i_{\mathrm{a}} + 1)} \tag{3.96}$$

The expressions for the differential cross-sections are more complicated (although having the same general structure as in the cases without spin) and we shall not describe them here (*see*, for example, Preston, 1962; Blatt and Weisskopf, 1952; Lane and Thomas, 1958). No longer are the contributions from different partial channels incoherent as in equation 3.95 and the interference between them must be included.

3.9 *R*-MATRIX AND BOUNDARY-MATCHING THEORIES

An elaborate and completely general mathematical apparatus for handling nuclear collision problems has been constructed in a number of forms which may be included under the heading of *R*-matrix or boundary-matching (*see*, for example, Preston, 1962; Blatt and Weisskopf, 1952; Lane and Thomas, 1958; Mott and Massey, 1965).

The general problem of nuclear reactions is to relate the values of the scattering or collision matrix elements (which in principle can be obtained from measurements) to the dynamics of nuclear structure. The *R*-matrix or boundary-matching theories do not solve this problem. Rather, they provide a mathematical reformulation of the problem which on the one hand makes explicit some of the phenomena (such as resonances) that one may expect to observe and on the other hand expresses the measurable quantities in terms of parameters whose physical significance is often more easily understood.

These approaches are useful because of the short range of nuclear forces. Two nuclei may be assumed not to be interacting (except for Coulomb forces) if they are separated by a distance greater than $r = a$, where a is not much larger than the nuclear radius. We know the form of the wave function in the exterior region $r > a$; it is comprised of spherical incoming and outgoing waves with amplitudes given by the scattering matrix elements as in equations 3.40 and 3.91. In the interior region, $r < a$, we do not know the wave function *a priori*; all we know is that its magnitude and derivative must match smoothly onto the exterior waves across the boundary $r = a$. So we define a complete set of states X_{λ} in the region

$r \leqslant a$ as solutions of the wave equation in that region with energy eigenvalues E_λ, but satisfying certain boundary conditions at $r = a$. The actual wave function Ψ in the interior is then expanded in terms of this complete set of 'compound system' states X_λ

$$\Psi = \sum_\lambda c_\lambda X_\lambda \tag{3.97}$$

The unknown expansion coefficients c_λ are related to the observable scattering matrix elements through the continuity conditions at $r = a$. The hope is that these states X_λ may correspond closely to the actual states of the compound nucleus which satisfy the same wave equation but do not have to satisfy the artificial boundary conditions at $r = a$.

As an example of the way this representation can be used, consider what happens when the E_λ are well separated and the bombarding energy E is very close to just one of them. Then the true wave function Ψ will be very similar to the corresponding X_λ; that is, just one compound state is excited strongly and we have a resonance at $E = E_\lambda$ (*see* lower part of spectrum in *Figure 2.31*). On the other hand, when the energy eigenvalues E_λ are very dense near the energy E, the reaction will excite many such compound states (*see* upper part of spectrum in *Figure 2.31*) and we are led to the statistical model which treats the corresponding expansion coefficients c_λ as random variables (*see* Chapter 4).

Although the mathematical formalism is perfectly general, it is clear that it is particularly suited to describing reactions of the compound nucleus type. Direct reactions do not involve just a single compound state nor a random mixture. Rather there is a correlation between the contributions from various X_λ. Boundary-matching theories often provide an insight into how this correlation comes about, but perturbation theory usually provides a more convenient language for describing direct reactions.

3.10 CLASSICAL AND SEMI-CLASSICAL DESCRIPTIONS OF SCATTERING

The earliest theory of scattering by nuclei (that of Rutherford) was a classical one (*see* section 2.10). Even today it is sometimes profitable to think of a nuclear collision in terms of classical particles moving along localised orbits or trajectories. This has long been true for the scattering of atoms, and classical ideas have been the subject of a resurgence of interest among nuclear physicists because of the considerable expansion in recent years of studies of reactions between heavy ions. This interest is appropriate because the heavy ions used have very short wavelengths and often large amounts of angular momentum are involved in a collision. However, as we shall see, it is often important not to forget that specific quantum (or wave) effects such as interference, diffraction and tunnelling may be important in the cases of physical interest. When the theory is modified to include such effects approximately, we have a *semi-classical* description (see Brink, 1985; Broglia and Winther, 1981).

Usually the wavelengths involved in light-ion reactions are long and wave effects dominate; however, at high enough energies it becomes appropriate here also to think of localised trajectories and to use a semi-classical approach.

Two things make the classical or semi-classical approaches useful: (i) the correspondence principle ensures that quantum mechanics reduces to classical mechanics in the limit that $\hbar \to 0$ (i.e. the limit in which the action integrals involved are very large compared to $\hbar$), and (ii) the classical and quantal cross-sections are the same for the scattering of two (non-identical) particles interacting via a force obeying an inverse square law like the Coulomb force.

3.10.1 Classical elastic scattering of particles

The classical theory of the scattering of two particles with reduced mass μ under the influence of a mutual interaction potential $V(r)$ is described in standard texts (e.g. Goldstein, 1950; Newton, 1966). If we take the centre of mass as origin,

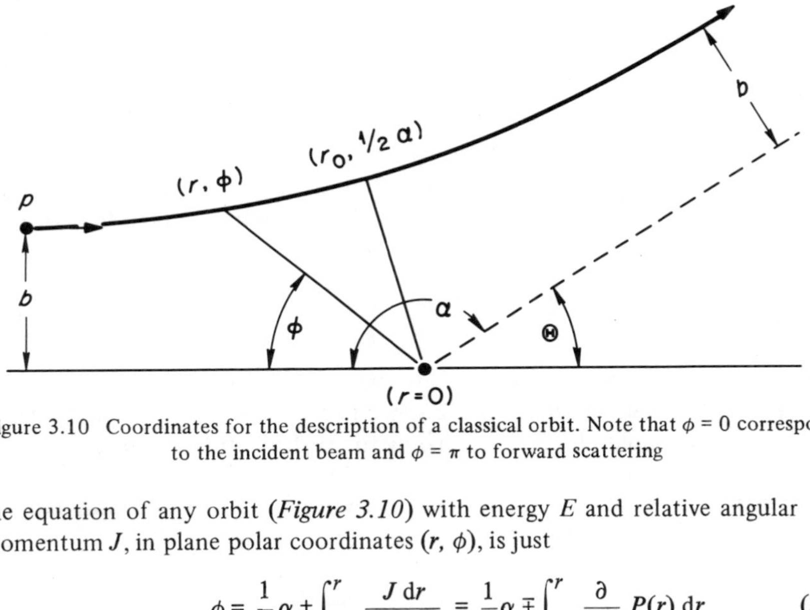

Figure 3.10 Coordinates for the description of a classical orbit. Note that $\phi = 0$ corresponds to the incident beam and $\phi = \pi$ to forward scattering

the equation of any orbit (*Figure 3.10*) with energy E and relative angular momentum J, in plane polar coordinates (r, ϕ), is just

$$\phi = \frac{1}{2}\alpha \pm \int_{r_0}^{r} \frac{J\,\mathrm{d}r}{r^2 P(r)} = \frac{1}{2}\alpha \mp \int_{r_0}^{r} \frac{\partial}{\partial J} P(r)\,\mathrm{d}r \tag{3.98}$$

where the upper sign applies to the receding $(\phi \geqslant \frac{1}{2}\alpha)$, and the lower sign to the approaching $(\phi \leqslant \frac{1}{2}\alpha)$ segment of the orbit. Also $P(r)$ is the local momentum at the point r on the orbit, and

$$P(r) = [2\mu(E - V) - J^2/r^2]^{1/2}$$

At large r, where V goes to zero and r^{-2} becomes negligible, $P(r)$ becomes the ordinary momentum p.

3.10.1.1 Deflection function, orbits and cross-sections

If $P(r) = 0$ at some $r = r_0$ (the turning point or distance of closest approach) and if α is the angle between the asymptotes of the orbit, then $\frac{1}{2}\alpha = \phi(r = \infty) - \phi(r = r_0)$. Further, the angle through which the orbit has been deflected is $\Theta = \pi - \alpha$, so we have for the orbit with angular momentum J

$$\Theta(J) = \pi - 2\int_{r_0}^{\infty} \frac{J\,\mathrm{d}r}{r^2 P(r)} \tag{3.99}$$

Now $\Theta(J)$ is called the *classical deflection function*.* From it we may obtain the elastic scattering cross-section using a derivation similar to that in section 2.10.1 for the Rutherford cross-section; we get

$$\frac{\mathrm{d}\sigma}{\mathrm{d}\Omega} = \frac{J}{p^2 \sin\theta\,|\mathrm{d}\Theta/\mathrm{d}J|_\theta} = \frac{b}{\sin\theta\,|\mathrm{d}\Theta/\mathrm{d}b|_\theta}$$

or (3.100)

$$\frac{\mathrm{d}\sigma}{\mathrm{d}\theta} = \frac{2\pi b}{|\mathrm{d}\Theta/\mathrm{d}b|_\theta}$$

where $|\mathrm{d}\Theta/\mathrm{d}J|$, etc., is evaluated for $|\Theta| = \theta$ or $|\theta \pm 2n\pi|$ with n integer. Also we have used $J = pb$ and $\mathrm{d}\Omega = 2\pi \sin\theta\,\mathrm{d}\theta$. If more than one J or b value results in the same θ value, each contributes a term like equation 3.100 to the cross-section; e.g.

$$\frac{\mathrm{d}\sigma}{\mathrm{d}\theta} = \sum_i \frac{\mathrm{d}\sigma_i}{\mathrm{d}\theta} = \sum_i \frac{2\pi\, b_i}{|\mathrm{d}\Theta/\mathrm{d}b|_i} \tag{3.101}$$

(This last result is in contrast to the scattering of waves where there would be *interference* between different contributions to the scattering; we would have to add the various amplitudes *before* squaring to get the scattered intensity.)

For Rutherford scattering by a purely Coulomb potential, $V \sim r^{-1}$, we see from Chapter 2, equation 2.15, where Θ was written as θ, that the Coulomb deflection function is simply

*Sometimes Θ is expressed as a function of the asymptotic impact parameter b (*see Figures 2.8, 3.10*) where $J = pb$, instead of J itself. Further, it is common practice to use the capitalised symbol Θ for the deflection angle in this classical context instead of the θ that we have used previously. We reserve θ for the *observed* scattering angle, that is the angle between the direction of the final motion of the particle and its incident direction; hence $0 \leqslant \theta \leqslant \pi$. The deflection angle Θ may be negative and may exceed π in magnitude if the potential V has an attractive part. Note that deflection angles of Θ and $\Theta - 2n\pi$, where n is integer, lead to the same scattering angle θ; the particle may go around the scattering centre n times before escaping. If we cannot follow the orbit in detail but can only observe the direction of emission of the particles (which is the case in nuclear-reaction experiments), we cannot distinguish between Θ and $\Theta - 2n\pi$, nor between Θ and $-\Theta$. Hence, in general, Θ and θ are related by $\Theta = \pm\,\theta - 2n\pi$.

$$\Theta(J) = 2\cot^{-1}\left(\frac{2b}{d_0}\right) = 2\cot^{-1}\left(\frac{2J}{pd_0}\right)$$
$$= \pi - 2\tan^{-1}\left(\frac{2J}{pd_0}\right) \tag{3.102}$$

Hence the Rutherford or Coulomb Θ decreases monotonically from π for $J = 0$ to 0 for $J = \infty$ so that $\theta = \Theta$ always; $(\sin\theta)^{-1}$ is infinite at both limits but only for $J = \infty$ does the cross-section 3.100 diverge.

When we add an attractive nuclear field to a repulsive Coulomb one, the form of Θ is modified from that of equation 3.102. *Figure 3.11* illustrates some of the classical trajectories in this case, when the energy is greater than the Coulomb barrier. At large impact parameters b (i.e. large J), the scattering is Coulombic. At a closer approach, the attractive nuclear field begins to act and pulls the orbit forward away from its Coulombic path (*see* dashed curve on right side of *Figure 3.11*). The orbits for which this begins to occur are often called *grazing* or *skimming* orbits. As b or J is reduced, the turning point for the corresponding orbit may reach a radius where the attractive nuclear and repulsive Coulomb and centrifugal forces just balance, or $(\mathrm{d}S(r)/\mathrm{d}r) = 0$ at $r = r_0$ where

$$S(r) = V_N(r) + V_C(r) + \frac{J^2}{2\mu r^2}$$

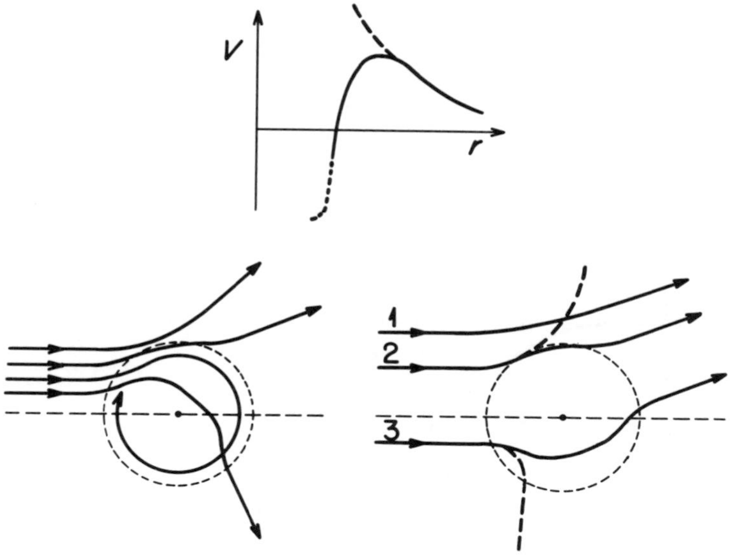

Figure 3.11 Some classical trajectories for scattering by the potential shown above. On the left Rutherford, grazing, orbiting and plunging orbits are shown; on the right are shown three different trajectories which result in the same scattering angle. Above is shown the repulsive Coulomb plus attractive nuclear potential, $V(r)$

In that case there is neither radial velocity nor acceleration at $r = r_0$ so that the system continues *orbiting* or *spiralling* at that radius (*see*, for example, Hirschfelder *et al.*, 1954). This can only occur for energies E below some critical value E_{crit} for which there is a J value such that $E = S$ and $dS/dr = 0$ can be satisfied simultaneously. (*See Figure 4.21* for some typical $S(r)$ curves.) When $E > E_{crit}$, orbiting can no longer occur. Clearly E_{crit} is the value of S at the point where $dS/dr = 0$ in the S curve for which the 'pocket' (*see Figure 4.21*) just disappears as J is increased. It follows that the value of E_{crit} depends upon the nuclear potential V_N; a deeper V_N will mean a larger E_{crit} because pockets will appear in the S curves for larger J values. This fact may be used as a tool to deduce information on the depth of V_N from scattering data.

For smaller impact parameters the nuclear attraction may overwhelm the Coulomb and centrifugal repulsion and the orbit *plunges* into the potential well to emerge on the other side with a negative deflection angle Θ. The right side of *Figure 3.11* also indicates how these characteristics allow three or more different orbits (that is, orbits corresponding to different angular momenta J_i or impact parameters b_i) to emerge with the same scattering angle θ. The corresponding cross-section is then a sum of terms as in equation 3.101.

The characteristics illustrated in *Figure 3.11* are reflected in the corresponding deflection function, *Figure 3.12* (which has been drawn for a case with $E < E_{crit}$). As the value of J decreases, $\Theta(J)$ is pulled below the Coulomb value, equation 3.102, until it goes into the negative singularity at $J = J_0$ corresponding

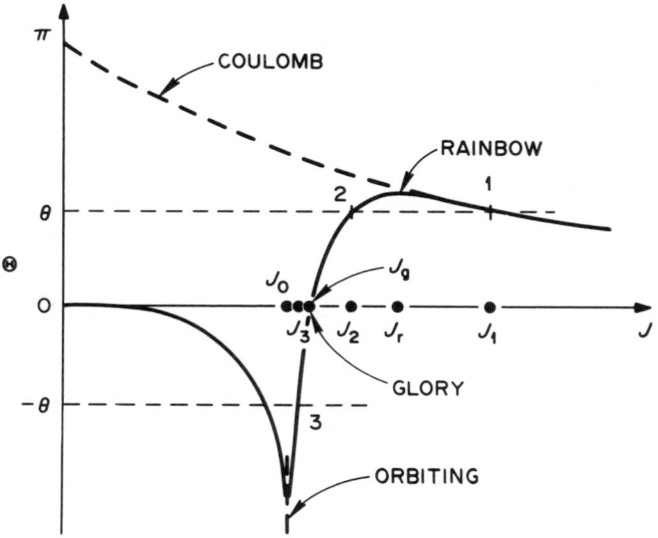

Figure 3.12 Deflection function for the potential shown in Figure 3.11. The angular momentum J_r corresponds to the rainbow orbit, J_g to glory scattering and J_0 to orbiting. J_1, J_2 and J_3 correspond to the three orbits labelled in *Figure 3.11*. When $E > E_{crit}$, the orbiting singularity at $J = J_0$ is replaced by a simple minimum

to orbiting. (If $E > E_{crit}$, this orbiting singularity is replaced by a smooth minimum.) The plunging orbits with $J < J_g$ which emerge on the opposite side correspond to negative values of Θ, while a completely head-on collision, $J = 0$, leads to no deflection, $\Theta = 0$. The fact mentioned above that three or more orbits may lead to the same scattering angle is seen to result from the non-monotonic behaviour of Θ with J; there are an infinite number of orbits contributing for the $E < E_{crit}$ case shown in *Figure 3.12*, although the number is finite for $E > E_{crit}$. Two of these orbits (labelled nos. 1 and 2 in *Figures 3.11, 3.12*) have $\Theta = \theta$ and the remainder have $\Theta = \theta - 2\pi, \theta - 4\pi \ldots$ and $\Theta = -\theta$, $-\theta - 2\pi, -\theta - 4\pi \ldots$, etc., which result from the two negative branches of $\Theta(J)$ near the orbiting value $J = J_0$. (Of these, only that with $\Theta = -\theta$ and labelled 3 is shown in *Figures 3.11, 3.12*.) Successive terms correspond to successively increasing numbers of revolutions performed in a right-handed or left-handed sense respectively before the particle escapes again.

3.10.1.2 Rainbows and glories

There is a non-zero value of $J = J_r$ for which $d\Theta/dJ = 0$ and hence the cross-section equation 3.100 is singular. This is called a ***rainbow*** since the corresponding phenomenon in physical optics for the scattering of light from water droplets is responsible for rainbows. In the vicinity of the rainbow angle $\Theta = \Theta_r$ we may use a parabolic approximation

$$\Theta = \Theta_r - q^2 (J - J_r)^2 \tag{3.103}$$

where the constant q is determined by the curvature of $\Theta(J)$ near $J = J_r$. From equations 3.100 and 3.101 this gives the differential cross-section near $\theta = \Theta_r$ as

$$\begin{aligned} \frac{d\sigma}{d\Omega} &= \frac{J_r}{2qp^2(\theta_r - \theta)^{1/2} \sin\theta_r}, & \theta < \theta_r \\ &= 0, & \theta > \theta_r \end{aligned} \tag{3.104}$$

This shows explicitly the divergence at the rainbow angle (*see Figure 3.13*). In general the cross-section will not be zero for $\theta > \theta_r$ because of the contributions from the negative branches of $\Theta(J)$ which lead to the orbiting singularity. However, because $|d\Theta/dJ|$ is large near $J = J_0$, these contributions are small and decrease exponentially with increasing θ (Ford and Wheeler, 1959; Newton, 1966).

The rainbow shown in *Figure 3.13* is the so-called ***Coulomb rainbow***. As discussed in the preceding section, the deflection function may have another extremum when $E > E_{crit}$, namely the minimum which replaces the orbiting singularity. This gives rise to the so-called ***nuclear rainbow*** which in general will be manifest at a different scattering angle. (Evidence for the existence of nuclear rainbows in measurements of the scattering of α-particles has been cited by Goldberg *et al.*, 1974.)

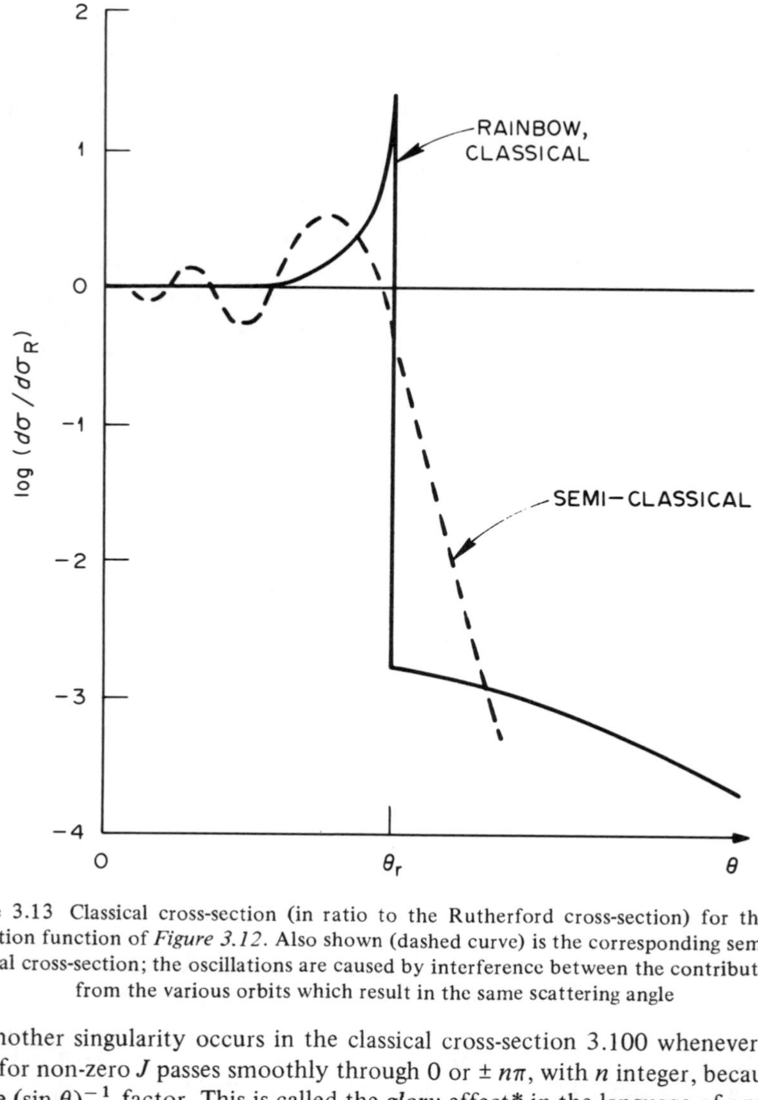

Figure 3.13 Classical cross-section (in ratio to the Rutherford cross-section) for the deflection function of *Figure 3.12*. Also shown (dashed curve) is the corresponding semi-classical cross-section; the oscillations are caused by interference between the contributions from the various orbits which result in the same scattering angle

Another singularity occurs in the classical cross-section 3.100 whenever $\Theta(J)$ for non-zero J passes smoothly through 0 or $\pm n\pi$, with n integer, because of the $(\sin\theta)^{-1}$ factor. This is called the *glory* effect* in the language of optics or meteorology. For charged particles the glory at $\theta = 0$ is overwhelmed by the singularity in the Rutherford cross-section, but a backward glory at $\theta = \pi$ may be observed.

*The backscattering ($\theta = \pi$) of solar radiation from the water droplets in a mist may give rise to a glory; sometimes this can be observed from an aeroplane flying over cloud, just around its shadow. Photographs of glories are reproduced by Bryant and Jarmie, 1974.

3.10.2 *Semi-classical treatments*

According to the correspondence principle, the results of quantum theory must approach those of classical mechanics when very many quanta are involved. When this limit is approached but account is still taken approximately of quantum effects such as interference, we have a semi-classical theory. The cross section in the semi-classical approximation may be similar to, or it may be very different from, the corresponding classical cross-section depending upon the properties of the interaction potential, the bombarding energy and the angle of observation. The quantal and the classical cross-sections for the scattering of (non-identical) point charges (Rutherford or Coulomb scattering) are the same, but for many interactions the quantal value converges non-uniformly to the classical one. There may always be some scattering angle where the two values differ significantly.

Such a semi-classical description can be valid only when the wavelength $\bar{\lambda}$ of the particles is sufficiently short that they may be localised (by building a wave packet) in a region small compared to some characteristic length of the interacting system such as the nuclear radius* ($\bar{\lambda} \ll R$ say). We still retain the interference properties which are characteristic of a wave theory; instead of equation 3.101 we use the coherent form

$$\frac{d\sigma}{d\Omega} = \left| \sum_i \left(\frac{d\sigma_i}{d\Omega}\right)^{1/2} \exp(i\beta_i(\theta)) \right|^2 \tag{3.105}$$

where $(d\sigma_i/d\Omega)$ is the classical cross-section for the ith contributing trajectory and β_i is the corresponding phase.†

When we consider non-elastic collisions leading to transitions between discrete quantum states (i.e. particular energy levels) of the two colliding systems, the non-elastic event has to be treated quantally. However the two particles may be regarded as moving along classical trajectories before and after this event. This will be valid if the non-elastic event represents only a small perturbation of the relative motion. This may occur for an inelastic scattering where the excitation energy is small compared to the bombarding energy (Alder and Winther, 1975) or for the transfer of one or a very few nucleons between large nuclei (Broglia and Winther, 1981). If the non-elastic cross-section for the reaction $\alpha \to \beta$ is small compared to the elastic cross-section, it may then be expressed as

*When the bombarding energy is below the height of the Coulomb barrier, the appropriate characteristic length is one-half the distance of closest approach for a head-on collision, d_0 of equation 2.13, so that then we require $\lambda \ll \frac{1}{2} d_0$. We also require that $\bar{\lambda}$ be short compared to a distance over which the potential V varies appreciably; this may be expressed as $(\bar{\lambda}/V)\,(dV/dr) \ll 1$.

†The full classical theory, equation 3.101, is only restored if we allow equation 3.105 to be averaged over small intervals in the bombarding energy E and the scattering angle θ; it may be shown (Newton, 1966) that the phases β_i vary very rapidly with E and θ in the true classical limit for macroscopic particles so that the interference terms are thus averaged to zero.

$$\frac{d\sigma_\beta}{d\Omega} = P_{\beta\alpha} \frac{d\sigma_{el}}{d\Omega} \qquad (3.106)$$

where $(d\sigma_{el}/d\Omega)$ is the corresponding *elastic* scattering cross-section and $P_{\beta\alpha}$ is the probability of making the transition from the initial internal states α to the final states β. The elastic cross-section may be calculated semi-classically, while $P_{\beta\alpha}$ is derived quantally. It can be expressed as the integral along the classical trajectory, from time $t = -\infty$ to time $t = +\infty$, of the matrix element of the interaction energy $\mathcal{H}(t)$ with the appropriate time-dependent phase

$$P_{\beta\alpha} = \left| \frac{1}{\hbar} \int_{-\infty}^{\infty} \langle \beta | \mathcal{H}(t) | \alpha \rangle \exp(i\omega t) dt \right|^2 \qquad (3.107)$$

where $\hbar\omega = E_\beta - E_\alpha$ is the excitation energy (that is, the change in *internal* energies of the two nuclei or, equivalently, minus the change in the kinetic energy of their relative motion). (Hence the phase factor $\exp(i\omega t)$ arises because the initial and final wave functions are oscillating in time with different periods.)

3.10.2.1 The WKB approximation

Technically the semi-classical theory is realised through the use of the so-called WKB approximation,* the replacement of the discrete partial wave sum, equation 3.50, by an integration over the angular momentum ℓ, and the use of asymptotic expressions (valid for large ℓ) of the Legendre polynomials $P_\ell(\cos\theta)$. Clearly the last two steps require that (i) a large number of partial waves contribute to the collision, and that (ii) the scattering amplitude varies slowly with ℓ so that large changes do not occur when ℓ changes by one quantum unit.

It may be shown that in the WKB approximation the ℓth partial wave function beyond the turning point $r = r_0$ is given by

$$w_\ell^{WKB}(r) = \frac{A}{K_\ell^{1/2}(r)} \sin\left\{ \frac{\pi}{4} + \int_{r_0}^{r} K_\ell(r')dr' \right\}$$

where (3.108)

$$K_\ell(r) = \left\{ \frac{2\mu}{\hbar^2} [E - V(r)] - \frac{\ell(\ell+1)}{r^2} \right\}^{1/2}$$

We see that $\hbar K_\ell(r)$ is just the $P(r)$ introduced in the previous section if we identify $J^2 = \ell(\ell+1)\hbar^2$. (In the spirit of the approximation it is usual to replace $\ell(\ell+1)$ by $(\ell+\frac{1}{2})^2$ so that we have $J \approx (\ell+\frac{1}{2})\hbar$.) The asymptotic behaviour of $w_\ell(r)$ from equation 3.108 then gives the WKB expression for the scattering phase shift defined by equations 3.42, 3.43

*Named after Wentzel, Kramers and Brillouin. It is sometimes called the JWKB method since it was first applied to quantum theory by Jeffreys: *see* any standard text on quantum theory such as Messiah, 1962.

$$\delta_{\ell}^{\mathrm{WKB}} = (\ell + \tfrac{1}{2})\,\frac{\pi}{2} - kr_0 + \int_{r_0}^{\infty} [K_{\ell}(r') - k]\,\mathrm{d}r' \qquad (3.109)$$

where $\hbar^2 k^2 = 2\mu E$. If we now treat the angular momentum $J = (\ell + \frac{1}{2})\hbar$ as a continuous variable, we see immediately from equation 3.98 that the classical deflection function, equation 3.99, is given by

$$\Theta(J) = 2\left(\frac{\mathrm{d}\delta_{\ell}^{\mathrm{WKB}}}{\mathrm{d}\ell}\right)_{\ell=(J-\frac{1}{2})/\hbar} \qquad (3.110)$$

This is the basic expression connecting the classical and quantum descriptions of the scattering.

For large ℓ values and for $\theta > \ell^{-1}$ we may use the asymptotic expression

$$P_{\ell}(\cos\theta) \approx \left[\frac{2}{\ell\pi\sin\theta}\right]^{1/2} \sin\left[(\ell + \tfrac{1}{2})\theta + \frac{\pi}{4}\right] \qquad (3.111)$$

and replace the sum Σ_{ℓ} in expression 3.50 for the scattering amplitude by an integral $\int \mathrm{d}\ell$. If the integral is evaluated by the method of stationary phase, it may be shown (Newton, 1966; Mott and Massey, 1965) that we then obtain the classical expression 3.100 for the scattering cross-section, except that the interferences indicated in equation 3.105 remain if more than one branch of $\Theta(J)$ contributes. [The phases $\beta_i(\theta)$ of the interfering amplitudes also result explicitly from the calculation (Newton, 1966; Ford and Wheeler, 1959).]

For just two contributing values of J (or ℓ), since the relative phase $\beta_1(\theta) - \beta_2(\theta)$ depends upon θ, the cross-section 3.105 will oscillate with θ (as indicated in *Figure 3.13*) between the limits

$$\frac{\mathrm{d}\sigma_{\max}}{\mathrm{d}\Omega} = \left|\left(\frac{\mathrm{d}\sigma_1}{\mathrm{d}\Omega}\right)^{1/2} + \left(\frac{\mathrm{d}\sigma_2}{\mathrm{d}\Omega}\right)^{1/2}\right|^2$$

and (3.112)

$$\frac{\mathrm{d}\sigma_{\min}}{\mathrm{d}\Omega} = \left|\left(\frac{\mathrm{d}\sigma_1}{\mathrm{d}\Omega}\right)^{1/2} - \left(\frac{\mathrm{d}\sigma_2}{\mathrm{d}\Omega}\right)^{1/2}\right|^2$$

It may be shown that the distance between maxima and minima is

$$\Delta\theta \approx \pi/|\ell_1 - \ell_2| \qquad (3.113)$$

It should be remembered that the method just outlined is not valid whenever $\Theta(J)$ is large, such as near the orbiting situation, for then $\delta_{\ell}^{\mathrm{WKB}}$ is varying rapidly with ℓ and the stationary phase assumption cannot be made. Such regions must be treated separately (Ford and Wheeler, 1959).

3.10.2.2 The eikonal approximation

It is by means of the eikonal method that geometrical optics is obtained as a limiting approximation to physical optics (Goldstein, 1950; Newton, 1966). This method has been applied to nuclear scattering particularly by Glauber (1959) and is sometimes referred to as the Glauber method.

The eikonal approximation is most often deduced from the integral equation 3.22 for the scattering wave function (for example, Newton, 1966) but may also be obtained by further approximating the WKB results of the last section. The underlying physical assumption is that the energy of the incident particle is sufficiently high that its trajectory is little deflected from a straight line. Then integrals like those in equations 3.108 and 3.109 may be performed along straight-line paths. In that case, the distance of closest approach r_0 becomes the same as b, the impact parameter. It is convenient to choose the coordinates shown in *Figure 3.14* with the z-axis along the incident direction; we may write the position vector $\mathbf{r} = \mathbf{b} + z\hat{\mathbf{k}}$, with the vector $\mathbf{b}$ in the (x, y) plane and $\hat{\mathbf{k}}$ the unit vector along the z direction.

High energy also implies $V/E \ll 1$ so that we may expand in this small quantity. When we do this in equation 3.109, after some manipulation we obtain the eikonal approximation for the phase shifts

$$\delta_{\ell}^{E} = -\frac{1}{\hbar v}\int_{0}^{\infty} V(\mathbf{b} + z'\hat{\mathbf{k}})\mathrm{d}z' \tag{3.114}$$

Here v is the velocity of the particle and the angular momentum is related to the impact parameter in the usual way, $(\ell + \frac{1}{2})\hbar = pb$. The eikonal approximation to the total (not partial) elastic wave function is most easily obtained from the integral equation 3.22; it is (Newton, 1966; Jackson, 1970)

$$\chi^{E}(\mathbf{k}, \mathbf{r}) = \exp\left\{ik\left[z - \int_{-\infty}^{z} \frac{V(\mathbf{b} + z'\hat{\mathbf{k}})}{2E}\,\mathrm{d}z'\right]\right\} \tag{3.115}$$

The first term in the exponent is simply the incident plane wave while the second term represents the phase shift accumulated while travelling along the straight-line path with impact parameter b (*Figure 3.14*) through the potential V up to the point z. (The *local* wave number is given by the local kinetic energy which at position $\mathbf{r}$ is $(E - V(r))$

$$K(\mathbf{r}) = \left[\frac{2\mu(E - V)}{\hbar^2}\right]^{1/2} \approx k\left(1 - \frac{V}{2E}\right) \tag{3.116}$$

if $V/E \ll 1$, so that equation 3.115 gives the integrated effect of this refraction.) The corresponding scattering amplitude can be obtained by inserting the approximate wave function 3.115 into the integral form 3.24

$$f^{E}(\theta, \phi) = -\left(\frac{\mu}{2\pi\hbar^2}\right)\int \exp(i\mathbf{q}\cdot\mathbf{r})\; V(\mathbf{r})\; \exp\{i\phi(b, z)\}\,\mathrm{d}\mathbf{r}$$

where $\mathbf{q} = \mathbf{k} - \mathbf{k}'$ and

$$\phi(b, z) = -k\int_{-\infty}^{z} \frac{V(\mathbf{b} + z'\hat{\mathbf{k}})}{2E}\,\mathrm{d}z' \tag{3.117}$$

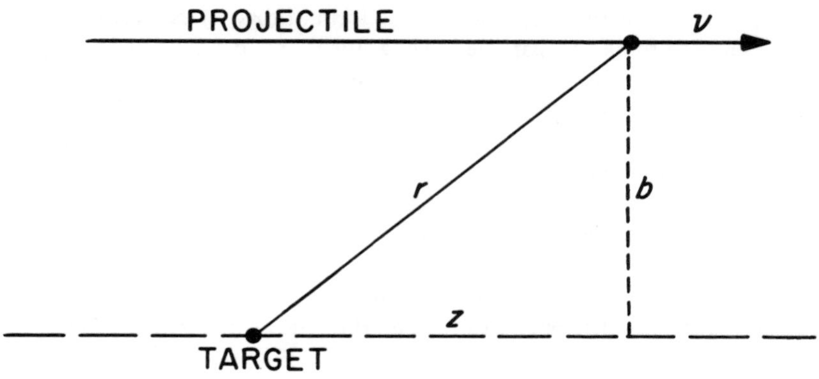

Figure 3.14 Coordinates for use with the eikonal approximation

is the phase shift between χ^E, equation 3.115, and the incident plane wave* at the point z along the straight-line path with impact parameter b.

These formulae may be developed further and some refinements included; for these the reader is referred to standard texts (e.g. Glauber, 1959; Newton, 1966; Jackson, 1970).

The basic approximation of using a straight-line path to evaluate the path integrals in this theory results in it being most suitable for application to the forward scattering of light particles (such as nucleons) with high energies. In most cases of heavy-ion scattering the long-range Coulomb field is strong and makes a non-linear Coulomb trajectory a more plausible approximation.

3.10.3 Diffraction and the effects of strong absorption

The classical and semi-classical descriptions just outlined are for scattering by real potentials; hence they describe elastic scattering only. When non-elastic channels are open they remove flux from the elastic channel and this effect may be represented by making the potential absorptive by adding a (negative) imaginary term to it (*see* sections 3.6.1 and 4.5). If this imaginary part is small, the absorption is weak and the corresponding trajectories are little affected; we may use the trajectory obtained under the influence of the real part of the potential alone and simply allow the absorption to slowly reduce the probability amplitude of finding the particle as it progresses along this path. This view has been applied successfully to atomic and molecular collisions; it also underlies the semi-classical approximation 3.106 for calculating non-elastic transitions.

However, nuclear collisions are usually characterised by the presence of *strong absorption* and this has more dramatic consequences. No longer can we

*Without this phase shift, $\phi(b, z) = 0$, the eikonal approximation 3.117 reduces to the Born approximation 3.25. At large distances, $z \to \infty$, $\phi(b, z)$ becomes the asymptotic phase shift 3.114.

ignore the influence of the absorption on the form of the trajectory; indeed it makes no sense to think of a particle moving along some orbital path in a region where there is strong absorption. Rather we should expect to see the wave nature manifest itself by diffraction of the waves at the edge of the absorptive region since, in nuclear collisions, the onset of absorption as the nuclei approach one another is quite abrupt.

The oscillations in the elastic cross-section indicated in *Figure 3.13* were there caused by interference between the contributions from the more than one classical trajectory which result in the same scattering angle (*Figure 3.11*). Oscillations like these also occur when there is strong absorption but their interpretation is quite different. The purely Coulomb orbits (no. 1 in *Figure 3.11*) may

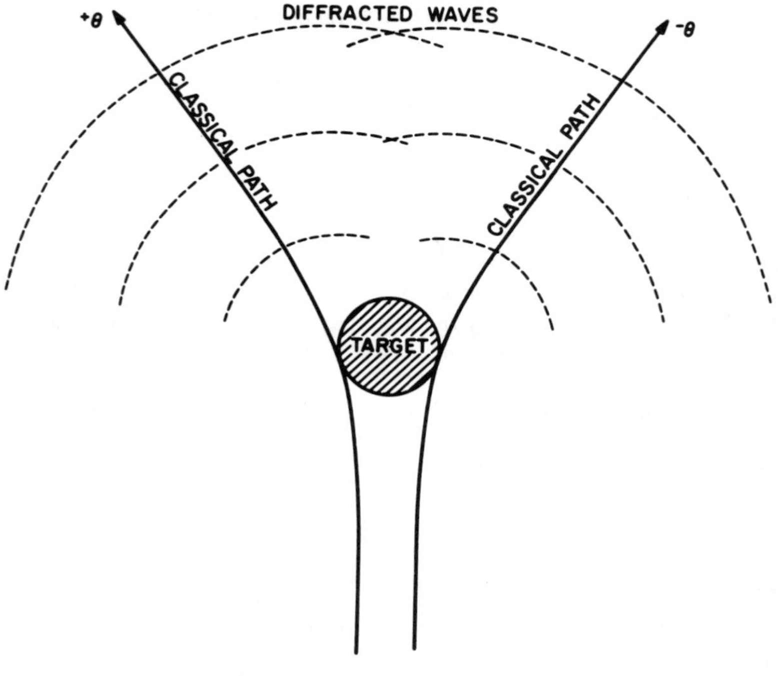

Figure 3.15 Diffraction along the classical trajectories which graze the scattering object and the deflection due to the Coulomb repulsion

still be visualised classically but it is no longer valid to do this for the grazing orbits (no. 2 in *Figure 3.11*) when there is strong absorption. Instead of simple refraction of a grazing ray we have diffraction; diffracted waves fan out from the grazing point as indicated schematically in *Figure 3.15*. In the absence of the Coulomb field the classical trajectories would be undeflected; interference of the diffracted waves from opposite sides of the target nucleus then gives rise to a

Fraunhöfer diffraction pattern like that from two slits.* As we introduce the Coulomb field, the classical grazing trajectories are deflected away from the forward direction and the group of diffracted waves moves to larger angles. Eventually the groups diffracted from opposite sides of the nucleus no longer overlap significantly; at a given angle one sees only the scattering from the near side and the angular distribution pattern is like that of Fresnel diffraction by the edge of a screen.† This latter situation characterises much of heavy-ion elastic scattering. These diffraction phenomena have been studied in detail by Frahn (1984, 1985).

The rapid decrease in cross-section beyond the rainbow angle seen in *Figure 3.13* also appears when there is strong absorption, but again the explanation is quite different. For weak or no absorption, the cross-section is small in this region because the particles which would have been scattered to these angles by the Coulomb field approach close enough to experience the attractive nuclear potential which then pulls their classical trajectories back to smaller angles (*see Figure 3.11*). However, when the nuclear interaction is also associated with strong absorption, particles following these close-approach orbits are simply absorbed and removed from the elastic channel rather than being refracted forward.

The semi-classical theories outlined in the previous section have been modified in various ways to accommodate strong absorption; the reader is referred to the original works for details (Knoll and Schaeffer, 1976; Harney *et al.*, 1974; Koeling and Malfliet, 1975; Brink, 1985; Frahn and Gross, 1976). Although these theories retain, to varying degrees, the language of the classical scattering theory such as deflection function and trajectory, it must be remembered that these quantities may differ greatly from the classical analogues we have discussed. In general they become complex; for example, the turning point (the radius r_0 at which the particles kinetic energy vanishes) moves into the complex plane. Further the quantal deflection function (defined in analogy to equation 3.110 but using the fully quantal phase shift) does not represent the relation between deflection angle and angular momentum as given by the classical equations of motion. These quantal deflection functions do not, when there is strong absorption, behave like the classical one shown in *Figure 3.12*. Rainbows, glories and orbiting become modified. Only when the amount of absorption is reduced to a weak level do we recover from these theories the semi-classical forms discussed previously.

*Strictly speaking, waves are diffracted from all around the rim of the nucleus and the pattern is that for diffraction by a disc rather than by two slits; *see* section 4.3.

†In optics, these two types of diffraction are usually distinguished by saying Fraunhöfer diffraction occurs when both light source and viewing screen are infinitely distant from the diffracting object, whereas if either is at a finite distance we have Fresnel diffraction. We note that in our case, the effect of the Coulomb repulsion is to deflect the incoming rays so that when they strike the rim of the nucleus they appear to have originated from a finite distance even though asymptotically they were parallel. The Coulomb field behaves like a diverging lens (Frahn, 1984).

3.11 THE IMPULSE APPROXIMATION

Consider the scattering of a nucleon by a system of bound nucleons (i.e. a target nucleus). The interaction between the projectile and the target nucleus is then a sum of two-body interactions, $V = \sum_i v_i$, one for each target nucleon i. The formal expression for the scattering amplitude may, by using the techniques outlined in section 3.3.4, be rearranged as a multiple scattering series (Newton, 1966; Jackson, 1970; Kerman *et al.*, 1959); that is, a series of terms like those indicated in *Figure 3.16* involving scattering from one target nucleon, two, three . . ., etc. Each individual scattering event is described by a scattering amplitude, g_i say, which differs from that for scattering from a free nucleon because the scattering occurs imbedded in a medium of other nucleons. These other nucleons exert an influence through the Pauli Exclusion Principle and because they generate a potential in which the scattering pair are moving.

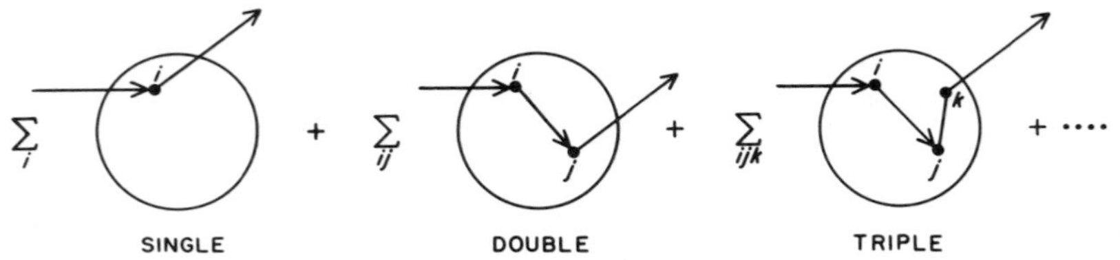

Figure 3.16 Schematic picture of the multiple scattering series. The projectile successively scatters from one, two, three . . . target nucleons

If we retain only the first term of this multiple-scattering series, we have a single-scattering approximation. Further, if their relative kinetic energy is very high (at least 100 MeV), two nucleons inside a nucleus scatter as if they were in free space; the presence of the other nucleons has little effect. Then we may replace the amplitude g_i by f_i, the amplitude for scattering from a *free* nucleon. This is called the *impulse approximation* and it entails assuming that the projectile bombarding energy is much greater than the binding energies of the struck nucleons so that binding effects can be neglected. Thus it is a high-energy approximation.

It is important to note that this is not a Born approximation (section 3.3.3); we do not treat the interaction v_i with the ith nucleon just to first order. Rather we assume that scattering from two or more nucleons is unimportant, but we use the exact scattering amplitude f_i for the single scattering that we retain. The amplitude f_i can be measured by studying the scattering of the projectile from a free nucleon. Then the amplitude of the wave scattered from the whole nucleus is simply the sum of the wavelets scattered from each target nucleon, each with the appropriate phase. The phase takes into account the differences in the path lengths for scattering from the various nucleons; *Figure 3.17* shows two typical paths for a projectile with momentum $\mathbf{k}$ being scattered with momentum $\mathbf{k}'$. If the target nucleons were rigidly fixed at positions $\mathbf{r}_i$, $\mathbf{r}_j$, . . . etc., the total scattering amplitude would be simply

$$f(\theta) = \sum_i f_i(\mathbf{k}, \mathbf{k}') \exp(i(\mathbf{k} - \mathbf{k}')\cdot\mathbf{r}_i) \tag{3.118}$$

However, the target nucleons are not rigidly located but distributed over the nuclear volume with a probability distribution given by their wave functions. They are also moving with definite momentum distributions which are also determined by their wave functions. Consider a transition in which a single target nucleon is excited from a state ψ_A to a state ψ_B. The corresponding momentum distributions are given by the Fourier transforms (Messiah, 1962)

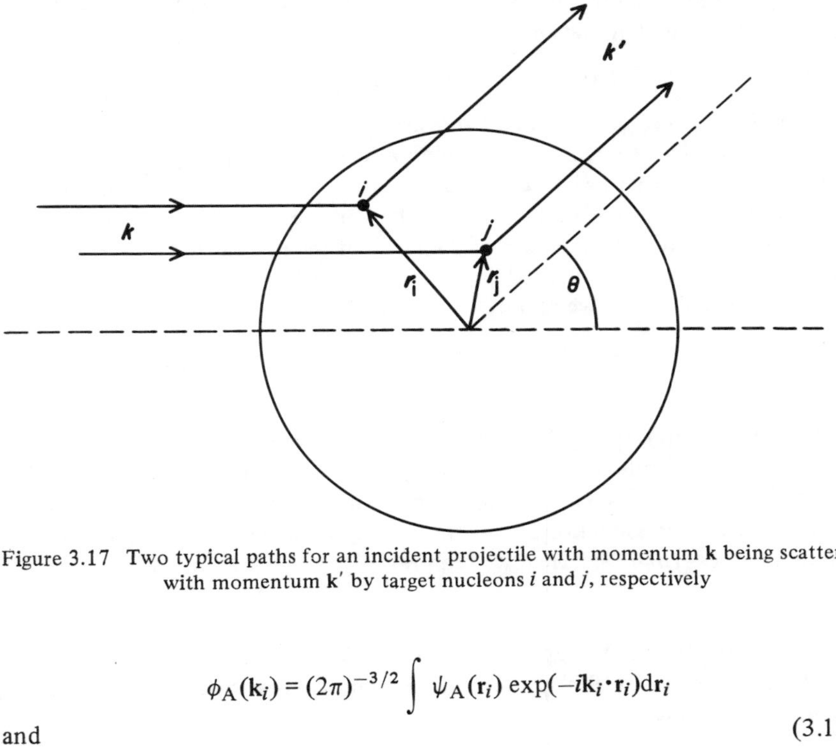

Figure 3.17 Two typical paths for an incident projectile with momentum **k** being scattered with momentum **k**′ by target nucleons *i* and *j*, respectively

$$\phi_A(\mathbf{k}_i) = (2\pi)^{-3/2} \int \psi_A(\mathbf{r}_i) \exp(-i\mathbf{k}_i\cdot\mathbf{r}_i)d\mathbf{r}_i$$

and (3.119)

$$\phi_B(\mathbf{k}_i') = (2\pi)^{-3/2} \int \psi_B(\mathbf{r}_i) \exp(-i\mathbf{k}_i'\cdot\mathbf{r}_i)d\mathbf{r}_i$$

The transition amplitude for the projectile with momentum **k** scattering to momentum **k**′ is then proportional to

$$f_\beta(\theta) = \int \phi_B^*(\mathbf{k}_i')\phi_A(\mathbf{k}_i) \langle \mathbf{k}', \mathbf{k}_i' | f_i | \mathbf{k}, \mathbf{k}_i \rangle \, d\mathbf{k}_i \, d\mathbf{k}_i' \tag{3.120}$$

Momentum conservation demands $\mathbf{k}' + \mathbf{k}_i' = \mathbf{k} + \mathbf{k}_i$. Then we assume that k_i may be neglected compared to k, since this is a high-energy approximation. Further, it is usual to assume that, at a given energy E, f_i can be treated as a function of momentum transfer q alone, where $\mathbf{q} = \mathbf{k} - \mathbf{k}'$; then $f_i \simeq f_i(q, E)$. (A dependence upon scattering angle θ remains because $q^2 = k^2 + k'^2 - 2kk' \cos\theta$.) Then, with equations 3.119, equation 3.120 becomes

$$f_\beta(\theta) = f_i(q, E) \int \psi_B^*(\mathbf{r})\psi_A(\mathbf{r}) \exp(i\mathbf{q}\cdot\mathbf{r})d\mathbf{r} \qquad (3.121)$$

For elastic scattering $\psi_B = \psi_A$ and summation of $|\psi_A|^2$ over all the target nucleons is just the target density distribution, $\rho(r)$; then

$$f_\alpha(\theta) = f_i(q, E)\, F(q), \quad F(q) = \int \rho(r) \exp(i\mathbf{q}\cdot\mathbf{r})d\mathbf{r} \qquad (3.122)$$

where $F(q)$ is a *form factor* similar to that introduced in section 2.11 on electron scattering; here it is the Fourier transform of the ground-state *mass* distribution rather than the charge distribution. Equation 3.121 may be written in a similar form by defining an inelastic form factor

$$F_\beta(q) = \int \psi_B^*(\mathbf{r})\, \psi_A(\mathbf{r}) \exp(i\mathbf{q}\cdot\mathbf{r})d\mathbf{r}$$

The quantity $\psi_B^*(r)\psi_A(r)$ is called a *transition density.*

We may consider the conditions under which the impulse approximation could be valid. In order to be *sudden*, the collision time τ must be small compared to the characteristic period ω^{-1} of the bound target nucleon, $\omega\tau \ll 1$. Now $\tau \approx a/v$, where a is a measure of the range of the nuclear force ($a \sim 2$ fm) and v is the projectile velocity, while $\hbar\omega = B$, where B is the binding energy of the target nucleon. Then we require

$$\frac{aB}{\hbar v} \ll 1$$

For the single scattering approximation to be valid, we require the mean free path Λ of the projectile within the nucleus to be larger than the nuclear radius R, $\Lambda > R$. The mean free path (at high energy) may be related to the average nucleon–nucleon cross-section σ and the density of nuclear matter ρ_0 (*see* section 2.7), $\Lambda \simeq (\rho_0\sigma)^{-1}$. Then we require

$$\sigma\rho_0 R \lesssim 1$$

A review of the impulse approximation applied to nuclear scattering was given by Kerman *et al.* (1959), where the complications arising from spin and isospin were considered.

REFERENCES

Abramowitz, M. and Stegun, I. A. (1970). *Handbook of Mathematical Functions.* New York; Dover Publications

Alder, K. and Winther, A. (1975). *Electromagnetic Excitation: Theory of Coulomb Excitation with Heavy Ions*. Amsterdam: North-Holland

Austern, N. (1970). *Direct Nuclear Reactions.* New York; Wiley

Blatt, J. M. and Weisskopf, V. F. (1952). *Theoretical Nuclear Physics.* New York; Wiley

Brink, D. M. and Satchler, G. R. (1968). *Angular Momentum*, 2nd edn. Oxford; Oxford University Press

Brink, D M. (1985). *Semi-classical Methods in Nucleus–Nucleus Scattering.* Cambridge; Cambridge University Press

Broglia, R. A. and Winther, A. (1981). *Heavy Ion Reactions*. Reading, Mass.; Benjamin
Bryant, H. C. and Jarmie, N. (1974). *Scientific American*. Vol. 231, No. 1, July, 60
Cooper, M. D. and Johnson, M. (1976). *Nucl. Phys.* Vol. A260, 352
Ford, K. W., Hill, D. L., Wakano, M. and Wheeler, J. A. (1959). *Annals of Physics.* Vol. 7, 239
Ford, K. W. and Wheeler, J. A. (1959). *Annals of Physics*. Vol. 7, 259
Frahn, W. E. (1984). In *Treatise on Heavy Ion Science*, Vol. 1. Ed. D. A. Bromley. New York; Plenum Press
Frahn, W. E. (1985). *Diffractive Processes in Nuclear Physics*. Oxford; Oxford University Press
Frahn, W. E. and Gross, D. H. E. (1976). *Annals of Physics*. Vol. 101, 520
Glauber, R. J. (1959). In *Lectures in Theoretical Physics*. Vol. 1. Eds. W. E. Brittin and L. G. Danham. New York; Interscience
Goldberg, D. A., Smith, S. M. and Burdzik, G. F. (1974). *Phys. Rev.* Vol. C10, 1362
Goldstein, H. (1950). *Classical Mechanics*. Reading, Mass.; Addison-Wesley
Harney, H. L., Braun-Munzinger, P. and Gelbke, C. K. (1974). *Z. Phys.* Vol. 269, 339
Hirschfelder, J. O., Curtiss, C. F. and Bird, R. B. (1954). *Molecular Theory of Gases and Liquids*. New York; Wiley
Jackson, D. F. (1970). *Nuclear Reactions*. London; Methuen
Kerman, A. K., McManus, H. and Thaler, R. M. (1959). *Annals of Phys cs*. Vol. 8, 55
Knoll, J. and Schaeffer, R. (1976). *Annals of Physics*. Vol. 97, 307
Koeling, T. and Malfliet, R. A. (1975). *Phys. Reports.* Vol. 22C, 181
Lane, A. M. and Thomas, R. G. (1958). *Rev. Mod. Phys.* Vol. 30, 257
Merzbacher, E. (1961). *Quantum Mechanics*. New York; Wiley
Messiah, A. M. (1962). *Quantum Mechanics I and II*. Amsterdam; North-Holland
Mott, N. F. and Massey, H. S. W. (1965). *The Theory of Atomic Collisions*, 3rd edn. Oxford; Oxford University Press
Newton, R. G. (1966). *Scattering Theory of Waves and Particles.* New York; McGraw-Hill
Preston, M. A. (1962). *Physics of the Nucleus*. Reading, Mass.; Addison-Wesley
Preston, M. A. and Bhaduri, R. K. (1975). *Structure of the Nucleus*. Reading, Mass.; Addison-Wesley
Rasmussen, J. O. (1965). In *Alpha-, Beta- and Gamma-Ray Spectroscopy*. Ed. K. Siegbahn. Amsterdam; North-Holland
Schwarzschild, A. Z., Auerbach, E. H., Fuller, R. C. and Kahana, S. (1976). *Proc. Symp. on Macroscopic Features of Heavy-Ion Collisions*. Argonne National Laboratory Report, ANL-PHY-76-2
Wong, C. Y. (1973). *Phys. Rev. Lett.* Vol. 31, 766
Yoshida, S. (1974). *Ann. Rev. Nucl. Sci.* Vol. 24, 1

EXERCISES FOR CHAPTER 3

3.1 Write down the many-body Schrödinger equation for a + A nucleons and their mutual (two-body) interactions v_{ij}. Transform the equation in a way parallel to the discussion in section 3.1 of its wave-function solutions. Separate the full Hamiltonian into terms governing the internal states of two nuclei, their relative motion under the influence of their mutual interaction and the motion of the centre of mass of the system. Compare this to the equations in section 3.3. Identify the terms H_a, H_A and V_α. (*See*, for example, Messiah, 1962, page 365.)

3.2 Derive explicitly the set of coupled equations represented by equation 3.17 by inserting the expansion 3.14 into the Schrödinger equation 3.13.

3.3 Consider a structureless particle like a nucleon scattering from a target nucleus which has only one excited state. Write down the pair of coupled equations for this system corresponding to equation 3.17. Using the ideas mentioned in section 3.3.4, eliminate the equation for the excited state so as to leave an equation for elastic scattering by an effective potential which includes the effects of the inelastic scattering. Show that this effective potential is complex if the bombarding energy is above the threshold for inelastic scattering.

3.4 Derive an expression for the *s*-wave phase shift for a particle with energy E incident upon a square-well potential of strength V and radius R. Calculate the wave function for the cases illustrated in *Figure 3.4*, namely $V = \pm \frac{1}{2} E$ and $R = 6 \lambda\!\!\!^{-}$.

What is the *s*-wave contribution to the total cross-section? Under what circumstances will the cross-section vanish for zero bombarding energy (Ramsauer–Townsend effect)?

What are the consequences of making the potential complex ($V \rightarrow V + iW$)? Derive equation 3.75 of section 3.6.1. Consider the limits (i) $V \rightarrow -\infty$, and (ii) $W \rightarrow -\infty$.

3.5 Apply the Born approximation to obtain the differential cross-section for the scattering of a particle by a square-well potential

$$V(r) = -V_0, \quad r \leqslant R$$
$$= 0, \quad r > R$$

Deduce an expression for the ℓth-partial wave phase shift. [Hint: use the expansion

$$\frac{\sin(qr)}{qr} = \sum_{\ell=0}^{\infty} (2\ell + 1) P_\ell(\cos\theta) \, [j_\ell(kr)]^2$$

where $q = 2k \sin(\frac{1}{2}\theta)$.] Compare the result for *s*-waves to the exact expression obtained in the previous exercise.

Discuss the conditions under which the Born approximation may be valid for this potential. [*Hint:* see Messiah (1962), pages 813–4.]

3.6 Calculate the differential cross-section for a particle scattering from a potential

$$V(r) = -V_0 \frac{\exp(-\alpha r)}{\alpha r}$$

using the Born approximation. (This is the spatial dependence of the Yukawa interaction between two nucleons. In atomic physics, this form reappears as a screened Coulomb potential.) What is the total cross-section?

Deduce from your results the Born approximation to the cross-section for the scattering of a charged particle by the Coulomb field resulting from a point charge. Compare this with the exact expression derived by Rutherford.

3.7 Neutrons with a bombarding energy of 50 MeV were scattered from protons. The differential cross-section for scattering through 90° in the CMS was found to be one-half of that for 0° scattering. Suppose that the n–p interaction potential has the Yukawa form; use the results of the preceding exercise to estimate the range α^{-1} of the potential.

Is the Born approximation expression for the differential cross-section compatible with the measured cross-sections shown in *Figure 2.22*? [*Hint:* the approximate symmetry about $\theta_{cm} = 90°$ seen in this figure is evidence for the existence of an 'exchange' force between the neutron and proton – see Blatt and Weisskopf (1952), for example.]

3.8 Using the results of Exercise 3.6, show that, in Born approximation, the cross-section for the scattering of a point charge by an extended charge distribution $\rho(r)$ can be written in a form analogous to equation 2.29, namely as the product of the Rutherford cross-section times the square of the Fourier transform $F(q)$ of the distribution $\rho(r)$.

3.9 Verify from equation 3.40 that the amplitude of the ℓth component of the elastically scattered wave (in the sense of equations 3.5 and 3.39) is

$$C_\ell = \tfrac{1}{2} i\,(1 - \eta_\ell)$$

(Note that $C_\ell = 0$ when $\eta_\ell = 1$, i.e. when there is no scattering.) These coefficients C_ℓ are sometimes used instead of the η_ℓ. Rewrite the expressions for the elastic scattering amplitude and the integrated elastic, absorption and total cross-sections in terms of the amplitudes C_ℓ.

3.10 A particle with spin-$\frac{1}{2}$ and momentum $\mathbf{k}$ is scattered from a target nucleus with zero spin to a state of final momentum $\mathbf{k}'$. The requirements of invariance under time-reversal ($\mathbf{k} \to -\mathbf{k}', \mathbf{k}' \to -\mathbf{k}, \sigma \to -\sigma$) and parity conservation (invariance under inversion of the spatial coordinates, $\mathbf{k} \to -\mathbf{k}, \mathbf{k}' \to -\mathbf{k}', \sigma \to \sigma$) place constraints on the forms that the expression for the scattering amplitude

$f(\theta)$ may take. Deduce the most general form for $f(\theta)$. (Here, the vector σ represents the Pauli spin matrices—*see* Appendix A1.)

What can you conclude about the polarisation vector for the particles after scattering?

Generalise your expression to the case where both target and projectile have spin-$\frac{1}{2}$.

3.11 Verify explicitly the second form for the absorption cross-section σ_{abs} in equation 3.62 by using the quantal expression for probability current and the partial wave expansion of the *elastic* wave

$$\exp(i\mathbf{k}_\alpha \cdot r_\alpha) + f_\alpha(\theta)\ [\{\exp(ik_\alpha r_\alpha)\}/r_\alpha]$$

[Hint: the absorption is given by the net flux entering a sphere surrounding the target.]

3.12 Using equations 3.100 and 3.102, derive the Rutherford expression for the scattering by a Coulomb field.

Examine the properties of equation 3.100 for Θ near 0 and π in this case. Show in this way that the cross-section does not diverge at $\Theta = \pi$.

What changes would occur in section 3.10 if the two particles had charges of opposite sign?

3.13 Derive the eikonal approximation for the phase shift, equation 3.114, from the WKB expression of equation 3.109.

3.14 Using *Figure 3.17*, verify the expression 3.118 for the scattering from more than one fixed target nucleon.

Show carefully the steps leading to equation 3.121, the impulse approximation to the scattering amplitude.

4

Models of Nuclear Reactions

The preceding chapter outlined some of the elements of quantum scattering theory. Almost nothing was said about the ways in which the physics of the nucleus influence the parameters which enter the formal theory, although some general features of nuclear reactions were discussed briefly in Chapter 2. We now intend to describe various physical models in more detail. Some models are of more general applicability than others but, as is the way with a model, the more general its nature the less physical content it tends to have. No single physical model describes all the phenomena of nuclear reactions, but the models in use do tend to be restricted to one of two broad categories: direct reactions and compound nucleus reactions (*see* section 2.18).

First we shall turn our attention to simple ways of describing elastic and inelastic scattering when there is strong absorption.

In general terms (sections 2.18.11, 2.18.12) strong absorption means a high probability of removal from the elastic channel because of non-elastic reactions. A more precise operational definition in terms of partial wave amplitudes is given in section 4.1 below. In this situation the scattering is like diffraction by a black sphere. In general the scattering of any complex nucleus by another is characterised by strong absorption. It is also a characteristic of the scattering by nuclei of high-eneıgy (> 100 MeV) nucleons, mesons and hyperons.

Since these systems do not interpenetrate appreciably without being disrupted, we do not expect to learn about the interior region of nuclei from strong absorption scattering. However, just because it is a peripheral process, we can expect to learn about surface properties such as the sizes of nuclei, how diffuse their surfaces are, etc. Further, strong absorption does lead to certain simplicities in the description of the scattering; these are exploited in the next few sections.

4.1 PARTIAL WAVES AND STRONG ABSORPTION

4.1.1 Sharp cut-off model

This simple but suggestive strong absorption model is based upon the idea that nuclei have relatively sharp edges and that any contact between two colliding nuclei inevitably leads to their removal from the elastic channel through the occurrence of inelastic scattering or other reactions. If contact occurs when the centres are separated by R (R is then 'the nuclear radius' for this particular collision) and if the wave number for their relative motion is $k = \lambdabar^{-1}$, then there is complete absorption for relative angular momenta $\ell \leqslant kR$. The amplitude η_ℓ of the outgoing elastic wave describing particles which leave the collision with angular momentum ℓ is*

$$\begin{aligned} \eta_\ell &= 0, \quad \ell \leqslant \ell_c = kR \\ &= 1, \quad \ell > \ell_c = kR \end{aligned} \tag{4.1}$$

This constitutes the sharp cut-off model.

We have seen (equations 3.62, 3.63) that the integrated cross-section for elastic scattering and the absorption cross-section are

$$\sigma_{el} = \pi\lambdabar^2 \sum_\ell (2\ell + 1)|1 - \eta_\ell|^2$$

and

$$\sigma_{abs} = \pi\lambdabar^2 \sum_\ell (2\ell + 1)[1 - |\eta_\ell|^2] \tag{4.2}$$

respectively. Also, from equation 3.66 we have seen that $\eta_\ell = 0$ gives the maximum possible partial absorption cross-section which is then also equal to the partial elastic cross-section. Using 4.1 the sum over ℓ can be performed easily, giving

$$\sigma_{el} = \sigma_{abs} = \pi\lambdabar^2 (\ell_c + 1)^2 = \pi(R + \lambdabar)^2 \approx \pi R^2 \tag{4.3}$$

This is just the geometric cross-section for a disc of radius R. The total cross-section is twice this

$$\sigma_{tot} = 2\pi(R + \lambdabar)^2 \approx 2\pi R^2 \tag{4.4}$$

The apparently anomalous result, 4.4, that the total cross-section is *twice* the geometrical area is related to the result that one cannot have absorption without

*This η_ℓ is the elastic amplitude $\eta_{\ell,\alpha}$ used previously (section 3.4.3). We drop the label α here for simplicity.

The reader may be concerned because, strictly, the angular momentum is $L\hbar = [\ell(\ell + 1)]^{1/2}\hbar \approx (\ell + \frac{1}{2})\hbar$ rather than simply $\ell\hbar$. However, because ℓ may only have integer values, the sharp cut-off approximation can be expected to be valid only if $\ell_c \gg 1$; consequently the uncertainty of $\frac{1}{2}$ is negligible. An equivalent way of saying this is that the approximation will only be valid if the size R is much greater than the wavelength $\lambdabar$, or $kR \gg 1$.

elastic scattering (compare sections 3.1.4, 3.4.10, and *Figure 3.5*). An absorption cross-section, 4.3, equal to the geometrical area is understandable; however, this absorption also implies that the target casts a shadow behind it. The modification of the incident beam which is needed to create this shadow constitutes elastic scattering in the sense in which it is usually defined. An elastically scattered wave [$\psi_{\text{scatt},\alpha}$ in equation 3.2] is produced behind the target which just cancels the incident plane wave in the shadow region.

4.1.2 Comparison with experiment

The step-function behaviour, 4.1, is a rough approximation to the η_ℓ often found for the actual scattering of particles from nuclei. *Figure 4.1* shows examples, typical of many measurements, for protons of 10 and 40 MeV, α-particles of 40 MeV, and ^{16}O ions of 160 MeV, scattering from ^{58}Ni. The behaviour of the $|\eta_\ell|$ for the α-particles and the ^{16}O ions is typical of the scattering of other strongly absorbed particles like deuterons, ^3He and heavier ions at energies high enough that a reasonable number of partial waves contribute; i.e. when $\ell_c \approx kR \gg 1$. In these cases, the main departure from equation 4.1 is a rounding of the step function, with the appearance of a brief transition region in which $|\eta_\ell|$ increases from zero to unity. This is due to the quantum uncertainty in the position of the particles (of order $\lambda\!\!\!^-$), to the finite range of the nuclear forces and to the diffuseness of the nuclear surface, all of which means that there is a short distance near $r = R$ over which the interaction of the colliding pair decreases from full strength to zero.

Protons of 40 MeV involve fewer partial waves than α-particles of the same energy (their wave number is only one-half as large) and are less strongly absorbed. Their $|\eta_\ell|$ show some vestige of similarity to the sharp cut-off model but those for small ℓ are not close to zero. This last feature tends to reduce σ_{abs} below the estimate 4.3, although there is a compensating increase owing to the surface diffuseness. These features appear even more strongly for 10-MeV protons. These results are characteristic of nucleons of low and medium energies. At higher energies, their $|\eta_\ell|$ approach more closely the simple strong absorption form shown by the α-particles in *Figure 4.1.*

4.1.3 Smooth cut-off models

The behaviour of actual nuclei just described has led to the use of functional forms for the η_ℓ which are more general than the step function 4.1 and which include the effects of rounding of the nuclear surface. Some include the possibility of $|\eta_\ell| \neq 0$ for small ℓ. The earliest form used was

$$\eta_\ell \approx \eta(\ell) = \frac{1}{\exp\left(\dfrac{\ell_c - \ell}{\Delta}\right) + 1} \tag{4.5}$$

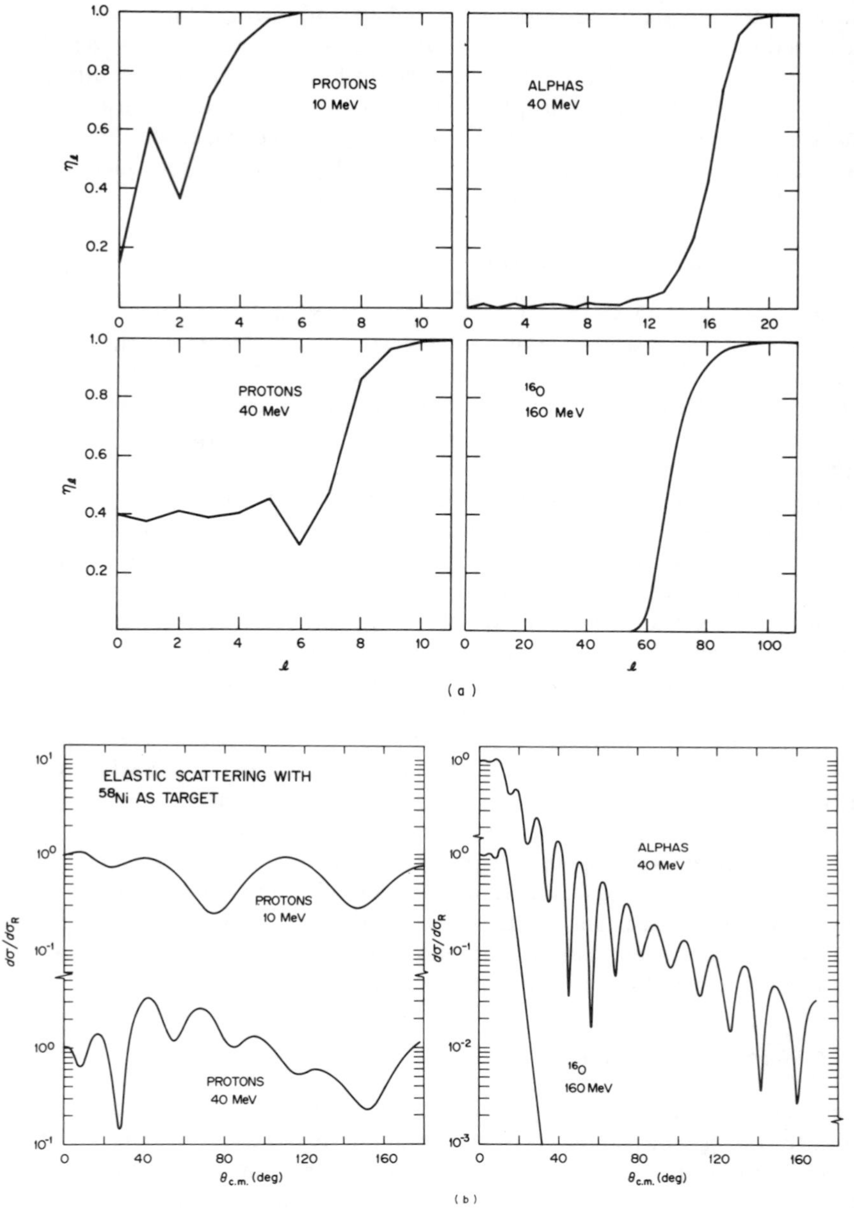

Figure 4.1 Typical examples of (a) the reflection coefficients $|\eta_\ell|$ and (b) the differential cross-sections (in ratio to the Rutherford cross-section) for the elastic scattering of protons, α-particles and ^{16}O ions with energies of 10 MeV per nucleon, and also protons with 40 MeV, all scattering from ^{58}Ni. Note the changes in scale for the angular momentum ℓ

The magnitude of $\eta(\ell)$ is zero for small ℓ, then rises to unity over a short range (about 5Δ) of ℓ values centred on $\ell = \ell_c$. As before, $\ell_c \approx kR$. The sharp cut-off model is then a special case with $\Delta = 0$. The form 4.5 assumes that η_ℓ is real; in practice it has often been found necessary to include an imaginary part also. In some cases the models have been generalised to include the effects of spin. Various smooth cut-off models and their application to experimental data have been reviewed by Frahn (1984, 1985) and Frahn and Rehm (1978) and a quite general discussion has been given by Frahn and Gross (1976). We should note that this approach uses the exact, fully quantal, expression 3.49 for the scattering amplitude. The approximation (or 'model') enters when we assume simple smooth functional forms for η_ℓ such as equation 4.5.

The values of ℓ_c and Δ in equation 4.5 or the quantities which enter similar parameterisations are adjusted to optimise the agreement between the corresponding theoretical differential and absorption cross-sections and the measured ones. Good fits can be obtained for the scattering of most types of particles. The poorest fits are obtained for nucleons of low and medium energies; the example of the $|\eta_\ell|$ for protons shown in *Figure 4.1* indicates that simple analytic forms like equation 4.5 are too smooth for these cases.

4.1.4 The nuclear radius and surface thickness

It is clear that the value obtained for ℓ_c when we analyse elastic scattering data gives a measure of R, the sum of the radii of the two particles, if we put $R = \ell_c/k$ (except for a correction due to the effects of the Coulomb field, *see* section 4.2 below). *Figure 4.2* shows nuclear radii obtained from applying the sharp cut-off model to some measurements of the scattering of α-particles (Kerlee *et al.*, 1957). Analyses of scattering measurements for ^{3}He, ^{4}He and heavier ions using a smooth cut-off model (Frahn 1984, 1985) have yielded R values which are given approximately by

$$R = r_0\,(A_1^{1/3} + A_2^{1/3}) \tag{4.6}$$

where A_1 is the mass number of the target and A_2 that of the projectile, and with r_0 between 1.4 and 1.5 fm. (The results shown in *Figure 4.2* are in agreement with this relation with $A_2 = 4$.) The large value of r_0 obtained reminds us not to identify R as simply the sum of the radii of the density distributions of the two nuclei, for this would suggest a much smaller value, $r_0 \approx 1.1$ (*see* sections 1.7 and 2.11). The present R is an *interaction radius* which measures the separation of the nuclear centres when 'contact' has definitely occurred; i.e. when there is appreciable interaction between the two nuclei which leads to absorption or scattering out of the elastic channel. This distance R depends not only on the matter radii of the two nuclei but is also increased by the finite range of the nuclear forces and the diffuse surfaces of the nuclei. We may also express the rounding parameter Δ in terms of a 'surface thickness' a by writing $\Delta = ka$; values of $a \approx 0.3 - 0.4$ fm are found. Here again we must be careful not to

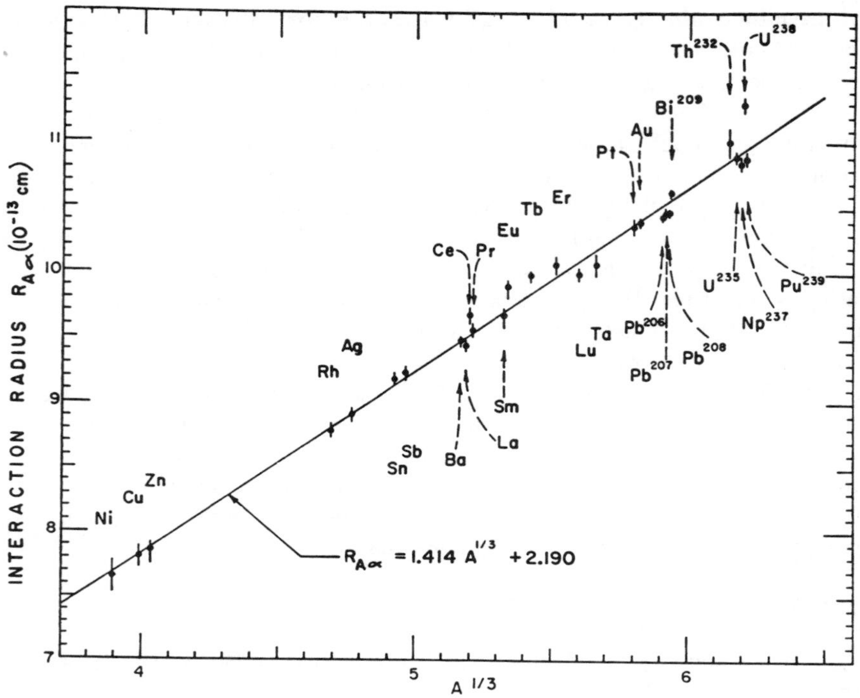

Figure 4.2 Nuclear radii (or 'interaction' radii) obtained from measurements of α-particle scattering when analysed using the sharp cut-off model. The straight line is the result of a least-squares fit. (From Kerlee *et al.*, 1957)

interpret this distance a too literally in terms of the thickness of the diffuse edge of the matter distribution (sections 1.7 and 2.11). However, it is clear that the scattering of strongly absorbed projectiles can give us information about the properties of the nuclear surface.

4.2 EFFECTS OF THE COULOMB FIELD

So far we have ignored the fact that strongly absorbed particles usually carry an electric charge. When there is a Coulomb interaction (*see* section 3.4.7) we must use the appropriate expressions 3.54–3.56 for the elastic scattering amplitude which include the Rutherford amplitude itself. More importantly, however, the partial wave amplitudes η_ℓ are themselves modified by the presence of the Coulomb field. In the strong absorption picture, the effect can be seen very simply; *Figure 4.3* shows two typical trajectories, with and without the deflection due to the Coulomb repulsion. The impact parameter b' associated with a

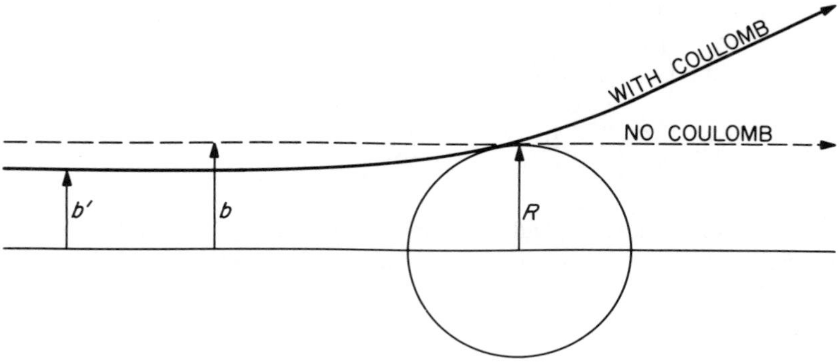

Figure 4.3 The effect of the Coulomb repulsion on a trajectory and how it leads to an effective impact parameter b' for grazing collisons which is smaller than that $b = R$ in the absence of a Coulomb field

glancing collision in the presence of the Coulomb field is smaller than that $b = R$ for an uncharged particle. Now the angular momentum $\ell = kb$, so the critical (or cut-off) value ℓ_c associated with a glancing collision is reduced by the Coulomb field.

Classically (*see* section 2.10), the angular momentum ℓ of a particle with momentum $p = \hbar k$ is related to the distance d of closest approach of its trajectory in a repulsive Coulomb field by equation 2.27

$$\ell = kd\left[1 - \frac{2n}{kd}\right]^{1/2}, \quad \frac{n}{k} = \frac{Z_a Z_A e^2}{2E} \tag{4.7}$$

where n is the Coulomb parameter of equation 2.21. (A classical description of Rutherford scattering is justified provided $n \gg 1$.) If we take for d the strong absorption radius R, we see that the associated angular momentum ℓ_c is now smaller than the value kR because of the Coulomb deflection. Hence we can apply the sharp cut-off or smooth cut-off models as before provided we include the Rutherford amplitude and relate ℓ_c to R through the expression 4.7 with $d = R$. (This was done in arriving at the empirical results expressed by equation 4.6 as well as those shown in *Figure 4.2*.)

With light nuclei or at high energies, or both, so that n is small, the effects of the Coulomb field are only large at small scattering angles. With heavier targets or at lower energies, the Coulomb scattering may dominate over an appreciable range of angles (*see*, for example, the case of $^{16}O + {}^{58}Ni$ in *Figure 4.1* where the ratios of the cross-sections to the Rutherford cross-sections are plotted). This can be understood from the classical relation between the distance d of closest approach of a trajectory and its consequent angle θ of deflection (or scattering);

rearranging equation 2.28 we have

$$d = \left(\frac{Z_a Z_A e^2}{2E}\right)\left(1 + \operatorname{cosec}\frac{\theta}{2}\right) \equiv \frac{n}{k}\left(1 + \operatorname{cosec}\frac{\theta}{2}\right)$$
$$\approx \frac{n}{k}\left(1 + \frac{2}{\theta}\ldots\right) \quad \text{if } \theta \text{ small} \tag{4.8}$$

This determines a critical angle θ_c corresponding to $d = R$, the strong absorption radius. Particles approaching closer than this are removed, consequently the

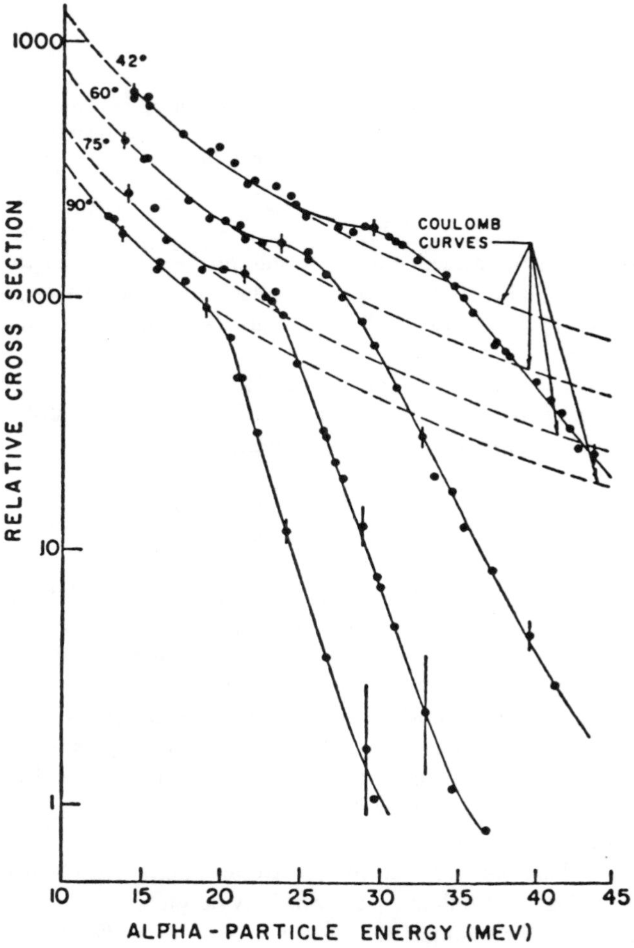

Figure 4.4 Variation of the elastic differential cross-section with bombarding energy at several angles for the scattering of α-particles from ^{208}Pb nuclei. The dashed curves represent the Coulomb or Rutherford scattering. (From Kerlee *et al.*, 1957)

scattered intensity would be zero for scattering angles θ greater than θ_c in this classical approximation. In practice, wave diffraction does allow some intensity to appear in this region but, as *Figure 4.1* shows, the cross-section falls off almost exponentially for $\theta > \theta_c$; see also *Figure 2.12*.

Because the relation 4.8 between d and θ also depends upon the bombarding energy E, a similar phenomenon can be observed at a fixed angle θ as the energy is increased. Below the energy $E = E_c$ for which $\theta = \theta_c$ the cross-section is Rutherford, but above this energy the cross-section will decrease rapidly. *Figure 4.4* shows this for α-particles scattering from lead at several angles. (The radii plotted in *Figure 4.2* were obtained from this type of measurement.)

The role of the Coulomb field in determining the kind of diffraction pattern seen in strong-absorption scattering was discussed briefly in section 3.10.3 and is returned to below.

4.3 DIFFRACTION MODELS AND STRONG-ABSORPTION SCATTERING

A simple scattering process familiar from classical physics is the diffraction of waves at the edges of opaque objects. Two limiting cases are Fraunhöfer and Fresnel diffraction. They may be distinguished by saying that Fraunhöfer diffraction is seen when both source and detector are infinitely distant from the diffracting object, whereas Fresnel diffraction is seen if either is at a finite distance. Both limits have analogues in nuclear scattering.

Diffraction theories of nuclear scattering also start from the idea that strong absorption occurs when two nuclei come into contact. They do not use a partial wave expansion but are based upon the classical diffraction theory of the scattering of waves by a black or totally absorbing sphere.

4.3.1 Fraunhöfer diffraction

When the effects of the Coulomb field may be neglected, the Fraunhöfer diffraction approximation may be derived from the eikonal approximation (section 3.10.2.2) by using a strongly absorbing (i.e. complex) potential V or by following Kirchoff's formulation of Huyghen's principle (Blair, 1966; Jackson, 1970). This gives a scattering amplitude for scattering from an incident momentum $\mathbf{k}$ to an outgoing $\mathbf{k}'$ with scattering angle θ

$$f(\theta) = \frac{ik}{4\pi}\,(1 + \cos\theta)\int_S \exp(i\mathbf{q}\cdot\mathbf{r})\,\mathrm{d}S \tag{4.9}$$

where S is a surface bounding the absorber (i.e. the region, which is assumed to have a sharp edge, where the strongly absorbing interaction is non-zero) and $\mathbf{r}$ is a point on this surface. Also $\mathbf{q} = \mathbf{k} - \mathbf{k}'$ or $q = 2k \sin(\frac{1}{2}\theta) \approx k\theta$ if θ is small. Further, we may put $(1 + \cos\theta) \approx 2$ since we only expect the approximation to be valid for relatively small θ.

When the absorber is spherical or a disc of radius R perpendicular to the beam, equation 4.9 can be easily evaluated to give* for small θ

$$f(\theta) \approx ikR^2 \frac{J_1(x)}{x}, \quad x = kR\theta \tag{4.10}$$

Here $J_1(x)$ is the Bessel function of first order (Abramowitz and Stegun, 1970). The differential cross-section for elastic scattering is then simply

$$\frac{d\sigma}{d\Omega} = k^2 R^4 \left[\frac{J_1(x)}{x}\right]^2, \quad x = kR\theta \tag{4.11}$$

(This expression is a generalisation to three-dimensions of the result for the more familiar problem in two-dimensions of diffraction by a slit.) Since $J_1(x) \approx \frac{1}{2}x - \frac{1}{16}x^3 + \ldots$ for small x, the cross-section has a maximum at $x = 0$ or $\theta = 0$; it falls to zero for $x \approx 3.8$ and then oscillates with successive peaks diminishing rapidly in magnitude. The second maximum occurs for $x \approx 5.1$ and there the cross-section is about 57 times smaller than at $\theta = 0$; the third maximum is reduced by another factor of about 4.2 and occurs for $x \approx 8.4$. The zero between these peaks appears at $x \approx 7.0$. When $x \gg 1$, $[J_1(x)/x]^2 \approx (2/\pi x^3) \sin^2 (x - \frac{1}{4}\pi)$, hence successive peaks or zeros are separated by $\delta\theta \approx \pi/kR$.

We also note that the angular distribution of equation 4.11 depends only upon $x = kR\theta$, so that in this model $(d\sigma/d\Omega)/k^2R^4$ is a universal function of x. These characteristics are illustrated in *Figure 4.5.*

All the radiation which falls on the disc is absorbed and the cross-section for absorption is just the geometrical area of the disc

$$\sigma_{abs} = \pi R^2$$

The total elastic cross-section σ_{el} is just the integral of the differential cross-section 4.11 over all scattering angles. If the ratio $R/\lambdabar = kR \gg 1$ then the diffraction scattering is concentrated at small angles and the integral over θ can be approximated by replacing $\sin\theta d\theta$ by $\theta d\theta$ and the upper limit of integration by ∞ instead of π so that

$$\sigma_{el} \approx 2\pi R^2 \int_0^\infty \left[\frac{J_1(x)}{x}\right]^2 x dx = \pi R^2$$

The total cross-section $\sigma_{tot} = \sigma_{el} + \sigma_{abs} = 2\pi R^2$. These results are the same as were obtained with the sharp cut-off model, section 4.1.1.

The radius R is the only parameter in this model. Provided the measurements show an oscillatory angular distribution like equation 4.11, R can be determined

* The reader will sometimes find used $x = 2kR \sin\frac{1}{2}\theta$ or $x = kR \sin\theta$. These correspond to slightly different choices of the shadow plane when evaluating equation 4.9. At small θ, where the theory may be valid, both prescriptions reduce to $x = kR\theta$.

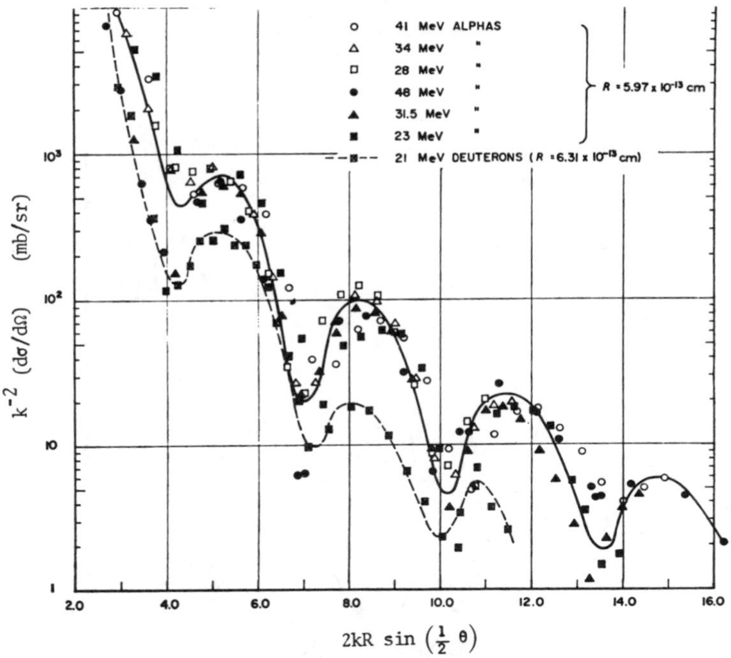

Figure 4.5 An example of an angular distribution for the elastic scattering of deuterons and α-particles showing the typical Fraunhöfer-like diffraction pattern. The cross-sections are plotted against $2kR\sin(\frac{1}{2}\theta) \approx kR\theta$ with R chosen so that their oscillations match those of equation 4.11; the values of R are indicated. (From Blair *et al.*, 1960)

accurately from the spacing of the oscillations. *Figure 4.5* includes some measured cross-sections for α-scattering plotted as a function of $kR\theta$ with R chosen so that their oscillations match those of equation 4.11. The radii R obtained from such analyses are close to those given by equation 4.6 with $r_0 \approx 1.4$–1.5 fm; this supports the idea that it is indeed the same characteristic radius which enters both the Fraunhöfer diffraction and the partial-wave cut-off models. An instructive example of the sensitivity of this type of measurement and analysis is given in *Figure 4.6* which shows precise measurements (Fernandez and Blair, 1970) of the elastic scattering of 42-MeV α-particles by some isotopes of Ca. In this case, the minima near 35° are identified with the zero in equation 4.11 for $x = 10.17$. Since k is known we can then precisely determine how the radius changes as we move along the chain of isotopes; the movement of the minimum by just over 1° between ^{40}Ca and ^{48}Ca corresponds to an increase in R of just over 0.2 fm.

We have previously emphasised that actual nuclei do not interact as though they had sharply defined edges. The effect of a grey edge to the black sphere is like superposing diffraction amplitudes from a number of spheres with slightly different radii. These will all be in phase in the forward direction, $\theta \approx 0$, and will

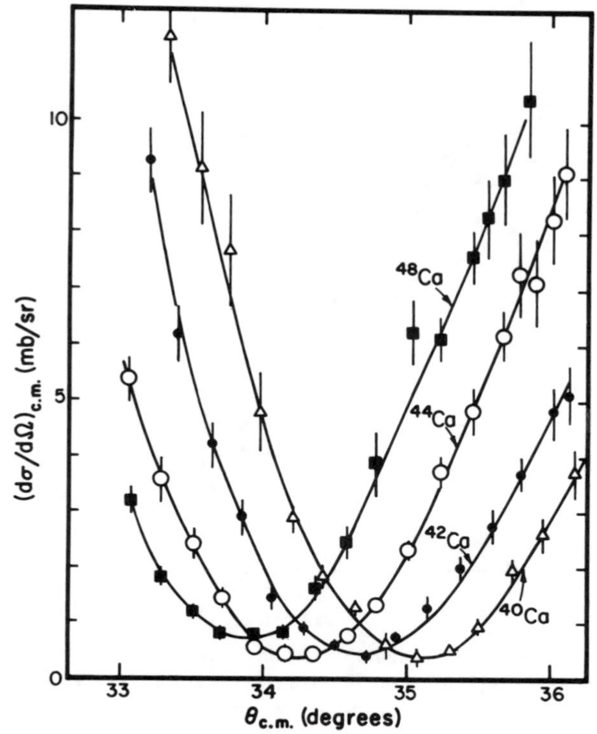

Figure 4.6 A minimum in the differential cross-section for the elastic scattering of α-particles from different isotopes of Ca. This shows how the position of the minimum varies with the mass number A thus allowing the determination of the variation in the radius of the isotopes. (From Fernandez and Blair, 1970)

interfere constructively. As θ increases, however, the oscillations in the various amplitudes will begin to get out of phase and the interference will be less constructive. As a result, a 'fuzzy' edge will lead to scattering intensities whose peak values decrease more rapidly with increasing angle than equation 4.11 implies. Further, the minima may be filled in and no longer go to zero as equation 4.11 predicts. The measured cross-sections shown in *Figure 4.5* (*see* also *Figure 2.39*) deviate from the predictions of the simple diffraction model in just these two ways: clearly diffraction by a sphere with a rounded or fuzzy edge will give better agreement with the measurements. The simplest model of such a rounding results in merely multiplying the expression 4.11 by a damping factor, $F(\theta)$ say, which slowly decreases with increasing θ; the rate of decrease is a measure of the surface diffuseness (*see* Blair, 1966).

The diffraction model as discussed so far ignores the effects of the Coulomb interaction. When these are small ($n \ll kR$, where n is the Coulomb parameter of equation 2.21 or 4.7), they may be corrected for approximately without changing the basic model. If R' is the radius extracted by fitting the Fraunhöfer form 4.11 to the measurements, the 'true' or corrected radius R is given by

$$R = \frac{n}{k} + \left[(R')^2 + \left(\frac{n}{k}\right)^2\right]^{1/2}, \quad \frac{n}{k} = \frac{Z_a Z_A e^2}{2E} \tag{4.12}$$

This is obtained from equation 2.27 by assuming that the characteristic radius R' of the diffraction pattern 4.11 is equal to the impact parameter b' which leads to a glancing collision (*see Figure 4.3*) and corresponds to the use of equation 4.7 in the sharp cut-off model.

4.3.2 Fresnel diffraction

As was discussed in section 3.10.3, the Coulomb interaction between two colliding charged particles acts somewhat like a diverging lens placed in the beam (*see Figures 3.15, 4.3*). This effect and its consequences are discussed in detail by Frahn (1984, 1985). If this effect of the Coulomb field is strong, as occurs for heavy ions or for energies close to or below the Coulomb barrier (situations characterised by a Coulomb parameter $n \gg 1$), the waves diffracted from opposite sides of the nucleus have been deflected sufficiently away from the forward direction that they no longer interfere significantly (*see Figure 3.15*). The diffraction pattern then seen is like that of Fresnel diffraction by the edge of an opaque screen (or, rather, a disc of radius $R \gg \lambda\!\!\!^{-}$).

Classically, complete absorption when two charged particles approach more closely than a separation R would give a differential cross-section of the form

$$\begin{aligned} \frac{d\sigma}{d\Omega} &= \frac{d\sigma_R}{d\Omega}, \quad \theta < \theta_c \\ &= 0, \qquad \theta > \theta_c \end{aligned} \tag{4.13}$$

Here $d\sigma_R/d\Omega$ is the Rutherford cross-section of equation 2.16 and θ_c is the scattering angle for the Rutherford orbit whose distance of closest approach is $d = R$ (*see* section 2.10). In the corresponding wave-mechanical theory we should expect to see oscillations in $d\sigma/d\Omega$ owing to diffraction in the vicinity of the cut-off angle θ_c. Indeed it has been shown (Frahn, 1984, 1985) that this is the case and that the diffraction pattern is of the Fresnel form. The case of $^{16}O + {}^{58}Ni$ in *Figure 4.1* is an example. For $\theta < \theta_c$, the ratio $d\sigma/d\sigma_R$ oscillates about unity with an amplitude which decreases as θ decreases. The largest rise above unity occurs just before θ_c after which $d\sigma/d\sigma_R$ falls rapidly. It continues to fall monotonically for $\theta > \theta_c$. It may be shown that $d\sigma/d\sigma_R = 1/4$ at $\theta = \theta_c$ itself; this result allows one to identify θ_c easily and hence extract a value of R from measured cross-sections by using the relation 4.8 with $d = R$

$$R = \frac{n}{k}\,[1 + \operatorname{cosec} \tfrac{1}{2}\theta_c] \tag{4.14}$$

This is known as 'the quarter-point recipe'. (Allowing the nuclei to have fuzzy edges changes the simple sharp-edge Fresnel pattern somewhat but this quarter-point property remains unchanged: Frahn, 1984, 1985.)

Observed angular distributions of this Fresnel-type, such as exhibited by $^{16}O + {}^{58}Ni$ in *Figure 4.1*, have sometimes been explained in semi-classical terms as being due to the interference between contributions from particles following two classical paths (orbits 1 and 2 in *Figures 3.11, 3.12*) which result in the same deflection close to the rainbow angle. However, although refraction does play an important rôle in determining the angular distribution, this simple refractive picture is only valid if the absorption is weak (see section 3.10.3).

4.3.3 Relation between diffraction and partial wave descriptions

The physical content of the diffraction models just described and the sharp cut-off or smooth cut-off models of section 4.1 would seem to be the same. It is not surprising then that one is able to show mathematically that they are equivalent (*see*, for example, Blair, 1966; Jackson, 1970). For example, if the sharp cut-off values of η_ℓ from equation 4.1 are used in the partial wave expression for the scattering amplitude (equation 3.49 with $\beta = \alpha$), and the cut-off is sufficiently large, $\ell_c \gg 1$, one obtains the Fraunhöfer expression 4.10 when the angle θ is small.

The mathematical trick is to use the relation between Legendre polynomials and the Bessel function of order zero

$$P_\ell(\cos\theta) \approx J_0[(\ell + \tfrac{1}{2})\theta] \tag{4.15}$$

which holds for large values of ℓ and small values of θ. We then replace the sum over ℓ by an integral and use the property

$$\int J_0(x)\, x \, dx = x\, J_1(x)$$

to obtain

$$f(\theta) \approx \frac{iL}{k} \frac{J_1(L\theta)}{\theta}, \quad L = \ell_c + \tfrac{1}{2} \tag{4.16}$$

Since $\ell_c \gg 1$, we have $L \approx kR$ and equation 4.16 becomes the same as the Fraunhöfer diffraction expression 4.10 if the two radii are the same.

Consequently the two models predict the same cross-sections and are, in fact, simply different representations of the same physical situation. This correspondence still holds between smooth cut-off models such as equation 4.5 and diffraction by nuclei with fuzzy edges (Blair, 1966). Further, when there is in addition a strong Coulomb interaction (so that $n \gg 1$), the cut-off partial-wave models are equivalent to Fresnel diffraction (Frahn, 1984, 1985).

4.4 STRONG-ABSORPTION MODELS FOR INELASTIC SCATTERING

Many measurements of the inelastic scattering of strongly absorbed particles which excite low-lying states (either in the target or in the projectile) show large cross-sections (comparable to the elastic cross-sections except at small angles) and angular distributions which are strongly peaked in the forward hemisphere and have the same kind of oscillatory or diffraction structure as the elastic angular distributions (*see Figure 4.7*, for example). These properties are characteristic of *direct reactions* (*see* section 2.18.7) and we can understand them in simple terms like those we have used in the preceding sections for elastic scattering.

4.4.1 Adiabatic approximation

The two basic features are the use of the collective model (*see* section 1.7.4 and, for example, Preston and Bhaduri, 1975) and the adiabatic approximation. We saw that for the application of simple semi-classical or diffraction ideas to be valid requires relatively short wave-lengths or equivalently high energies, so that $kR \gg 1$. Under these circumstances the excitation energies of the low excited states of nuclei are much smaller than the bombarding energy; the adiabatic approximation consists of neglecting them altogether, so that the ground and excited states are taken to be approximately degenerate and the elastic and inelastic scattering can be treated on the same footing. The physical significance of this for nuclear structure can be easily seen if we use the collective rotational model. Then the excitations are simple rotations of a non-spherical nucleus and negligible excitation energy means that the rate of rotation is much slower than the passage of the colliding pair. Hence we can 'freeze' this rotational motion of the non-spherical nucleus, calculate the scattering for each particular orientation and average over the orientation afterwards.

Thus, for example, in an application of the diffraction model we need to calculate the amplitude for diffraction scattering from a non-spherical black object, such as an ellipsoid. This can be done in a variety of approximations (Blair, 1966). Let the coordinates describing the orientation of the nucleus be denoted by ξ. Then we calculate the amplitude $f(\theta, \xi)$ for scattering at an angle θ for a particular value of the ξ. Each quantum state of the nucleus A consists of a particular (in general non-random) distribution of orientations whose probabilities are given by the square modulus of its wave function, $|\psi_A(\xi)|^2$. In order to obtain the transition amplitude between two particular quantum states A_1 and A_2 say, we need to take the *matrix element* of f between these two states

$$f_{1,2}(\theta) = \int \psi^*_{A_2}(\xi)\, f(\theta, \xi)\, \psi_{A_1}(\xi)\mathrm{d}\xi \tag{4.17}$$

The particular case of elastic scattering is given by the diagonal matrix element

$$f_{\mathrm{el}}(\theta) \equiv f_{0,0}(\theta) = \int |\psi_{A_0}(\xi)|^2 f(\theta, \xi)\mathrm{d}\xi \tag{4.18}$$

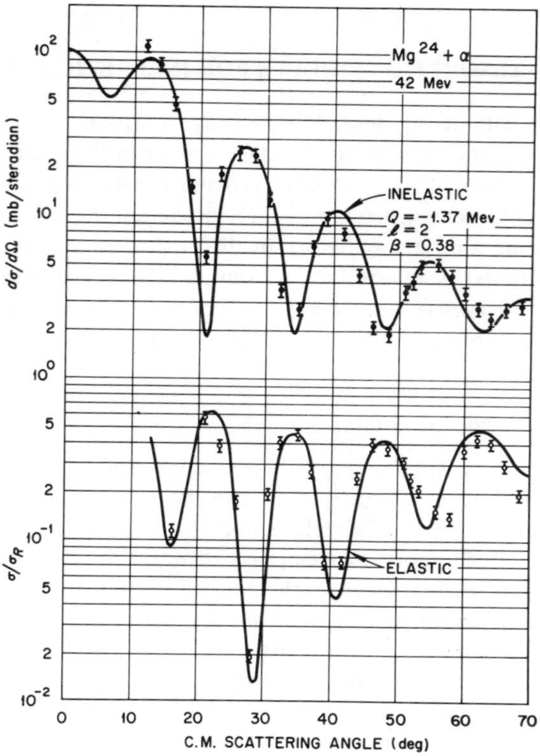

(a)

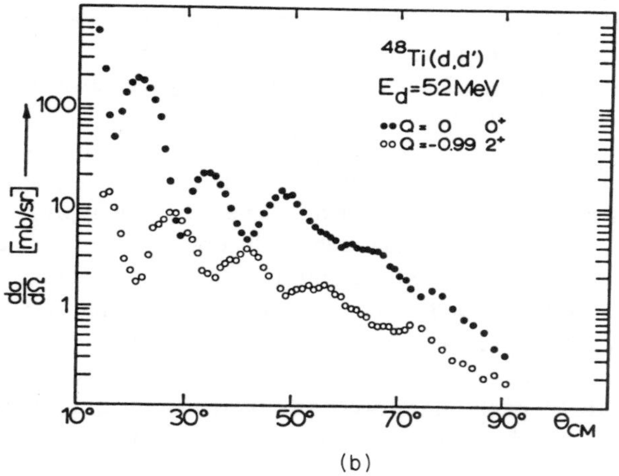

(b)

Figure 4.7 Examples of angular distributions for the inelastic scattering of strongly absorbed systems. Except for ^{90}Zr + t, the elastic cross-sections are shown also for comparison. The curves drawn for the elastic scattering were obtained using the optical model while those for the inelastic scattering represent direct reaction calculations using the DWBA (*see* sections 3.3.3 and 4.7.2); the full and dashed curves are the result of using two different optical model potentials. (a) ^{24}Mg + α at 42 MeV (from Rost, 1962); (b) ^{48}Ti + d at 52 MeV (after Hinterberger *et al.*, 1968); (c) ^{90}Zr + t at 20 MeV (from Park and Satchler, 1971); (d) ^{28}Si + ^{16}O at 142 MeV (from Satchler *et al.*, 1978)

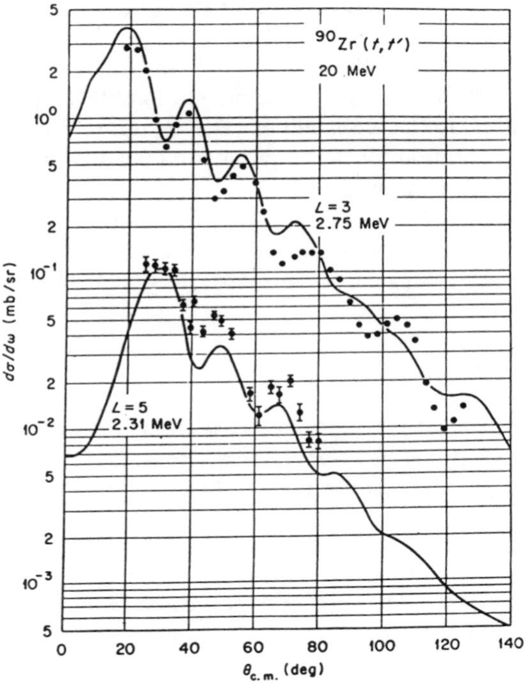

^{90}Zr (t, t′)
20 MeV
L = 3
2.75 MeV
L = 5
2.31 MeV
dσ/dω (mb/sr)
θc.m. (deg)

(c)

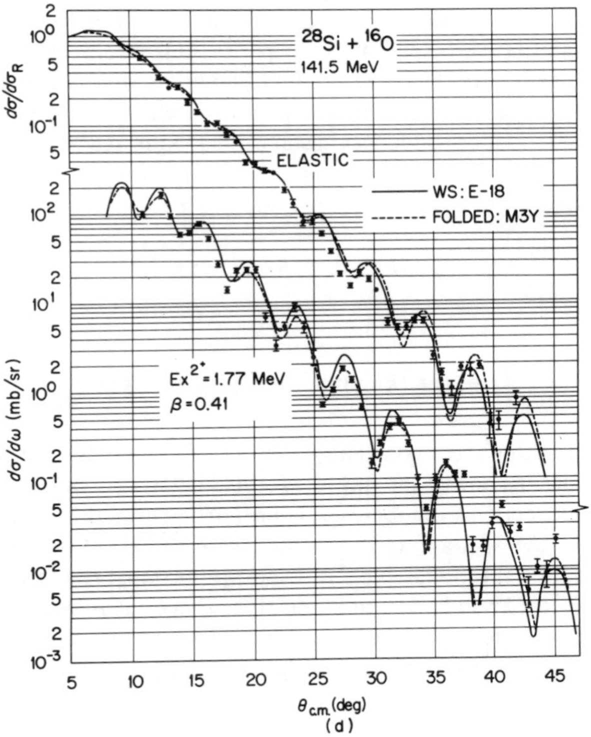

^{28}Si + ^{16}O
141.5 MeV
ELASTIC
WS: E-18
FOLDED: M3Y
Ex^{2+} = 1.77 MeV
β = 0.41
dσ/dσR
dσ/dω (mb/sr)
θc.m. (deg)

(d)

if ψ_{A_0} is the wave function for the ground state. The interpretation of equation 4.18 as a simple average of $f(\theta, \xi)$ over the orientations of the nucleus in its ground state, weighted by the probabilities $|\psi_{A_0}|^2$, is immediate. Equation 4.17 is the quantum generalisation for the amplitude to take one from one distribution $|\psi_{A_1}|^2$ to another distribution $|\psi_{A_2}|^2$.

A completely parallel argument applies to the excitation of collective *vibrational* states. These consist of oscillations in shape about a spherical mean. We calculate the scattering from a particular shape, then take the matrix element corresponding to 4.17.

4.4.2 Fraunhöfer diffraction

We may define the shape of the scatterer by assuming it is strongly absorbing within the non-spherical surface $r = R(\theta, \phi)$, where we can always expand R as (for example, *see Figure 1.2* and Preston and Bhaduri, 1975)

$$R(\theta, \phi) = R_0 + \sum_{L,M} \delta_{LM} Y_L^M(\theta, \phi) \tag{4.19}$$

where $Y_L^M(\theta, \phi)$ are spherical harmonics (Appendix A) and the δ_{LM} may be called *deformation lengths*. The corresponding adiabatic scattering amplitude $f(\theta, \xi)$ can be expanded in powers of the deformation lengths δ_{LM}; the first-order term which is proportional to δ_{LM} itself corresponds to transfer to the nucleus of angular momentum L with z-component M, and parity $(-)^L$. If the target ground state has zero spin and even parity, this means excitation of a state with spin L and parity $(-)^L$. When the Fraunhöfer approximation 4.9 is used, taking the integral over the non-spherical surface defined by equation 4.19, the resulting differential cross-sections again can be expressed in terms of Bessel functions (Blair, 1966). Two typical examples of cross-sections obtained from amplitudes which are first order in the δ_{LM} are

$$\begin{aligned} \frac{d\sigma}{d\Omega}(L = 2) &= (kR_0)^2\, C_2\, [\tfrac{1}{4} J_0^2(x) + \tfrac{3}{4} J_2^2(x)] \\ \frac{d\sigma}{d\Omega}(L = 3) &= (kR_0)^2\, C_3\, [\tfrac{3}{8} J_1^2(x) + \tfrac{5}{8} J_3^2(x)] \end{aligned} \tag{4.20}$$

where $x = kR_0\theta$. Each C_L is a constant which is a characteristic of a given transition in a given nucleus; one of the motives for making measurements of inelastic scattering is to extract this number. C_L is proportional to the square of the matrix element of the δ_{LM} averaged over the z-components M of the spin of the excited nucleus; for a rotational nucleus, C_L is proportional to the square of the magnitude β_L of the 2^L-pole deformation of its shape (as in *Figure 1.2* for example), so that the inelastic cross-section can give direct information on the shape of the nucleus. Just as with the elastic scattering (section 4.3.1), the effect of a fuzzy edge on the scatterer is to modulate the cross-sections 4.20 with a factor $F(\theta)$ which decreases slowly with increasing θ; the same factor F is needed for both the elastic and the inelastic scattering.

For large x, $J_n^2(x) \approx (2/\pi x)\sin^2(x - \frac{1}{2}n\pi + \frac{1}{4}\pi)$, so that the expressions 4.20 oscillate with θ with the same period as the elastic cross-section 4.11. Further, because of the term $\frac{1}{2}n\pi$, the inelastic cross-sections for odd-L (odd parity) oscillate in phase with the elastic ones, while those for even-L (even parity) are out of phase with the elastic. (This is known as the Blair phase rule.) Hence comparison of the measured inelastic angular distribution with that for the elastic

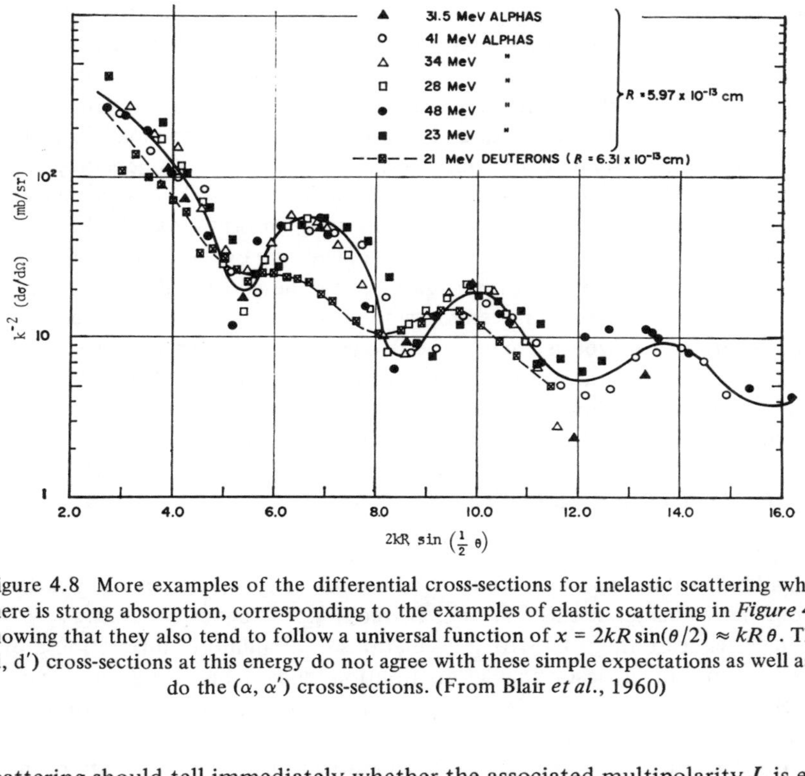

Figure 4.8 More examples of the differential cross-sections for inelastic scattering when there is strong absorption, corresponding to the examples of elastic scattering in *Figure 4.5*, showing that they also tend to follow a universal function of $x = 2kR\sin(\theta/2) \approx kR\theta$. The (d, d′) cross-sections at this energy do not agree with these simple expectations as well as do the (α, α') cross-sections. (From Blair *et al.*, 1960)

scattering should tell immediately whether the associated multipolarity L is even or odd. Only at relatively small scattering angles θ do the theoretical angular distributions for the various L values show differences other than this which are sufficiently marked to allow the particular L to be identified. Further, just as for elastic scattering, the angular distributions for a given L at different energies (i.e. different k) or to different final states (either in different nuclei or the same nucleus), should all fall on a universal curve when plotted against $x = kR_0\theta$. These properties are illustrated in *Figures 4.7, 4.8.*

4.4.3 *Applications of the Fraunhöfer model*

This diffraction model has been compared with many measurements of inelastic α-scattering, generally showing agreement of the quality seen in *Figures 4.5, 4.8.* Applications to the inelastic scattering of deuterons, mass-3 and heavier ions have also been successful, provided the energy was sufficiently high that the conditions $kR \gg 1$ and $n \ll kR$ were satisfied. However, whenever measurements have not satisfied these conditions, the agreement between them and the simple diffraction theory has not been very good.

When judging the quality of fits between theory and experiment obtained (as for example in *Figures 4.5, 4.8*) one must remember the simplicity of the theory. The sharp cut-off results such as equations 4.20 depend only upon two parameters, C_L and R_0. Allowing a rounded edge introduces one other surface thickness parameter, d say. The same values of R_0 and d should apply to all transitions, including the elastic scattering, for a given nucleus. The angular distribution depends only upon R_0 and d, while the cross-section magnitude is proportional to C_L. The predicted angular distribution can be immediately compared with experiment; if it fits with the same values of R_0 and d as we needed for the elastic scattering, our assumption of strong absorption is confirmed. Whether the value of C_L required to give the magnitude of the observed cross-section is reasonable is not known *a priori*. It can be compared with values obtained from scattering at other energies or by other projectiles, in order to check the internal self-consistency of the strong absorption picture. With suitable care (Blair, 1966), it can be compared to an analogous quantity (for example, the deformation of the charge distribution) obtained from electromagnetic excitation or decay of the same transition.

4.4.4 *Coulomb effects and Coulomb excitation*

We saw earlier that the effect of a strong Coulomb interaction on elastic scattering angular distributions was to convert a Fraunhöfer-type of diffraction pattern into one more characteristic of Fresnel diffraction. Analogous effects are seen with inelastic scattering; the strong oscillations of Fraunhöfer diffraction are washed out and the inelastic angular distribution tends to become a simple peak (or 'bell-shape') centred on the scattering angle $\theta = \theta_c$ for the Rutherford orbit which corresponds to a grazing collision (i.e. for which $d = R$).

Another feature which the Coulomb field introduces is the possibility of inelastic excitation by the Coulomb interaction itself; this is called *Coulomb excitation* (*see* section 2.12, also McGowan and Stelson, 1974; Biedenharn and Brussard, 1965; Alder and Winther, 1975). When the Coulomb parameter n is large this process may be comparable to the excitation by the specifically nuclear forces, particularly for the low multipoles (small L values). Of course these two contributions are coherent and interfere. Because the Coulomb interaction is long-ranged ($\sim 1/r^{L+1}$ for the Lth multipole), the Coulomb excitation tends to occur for larger impact parameters than the grazing one and hence it

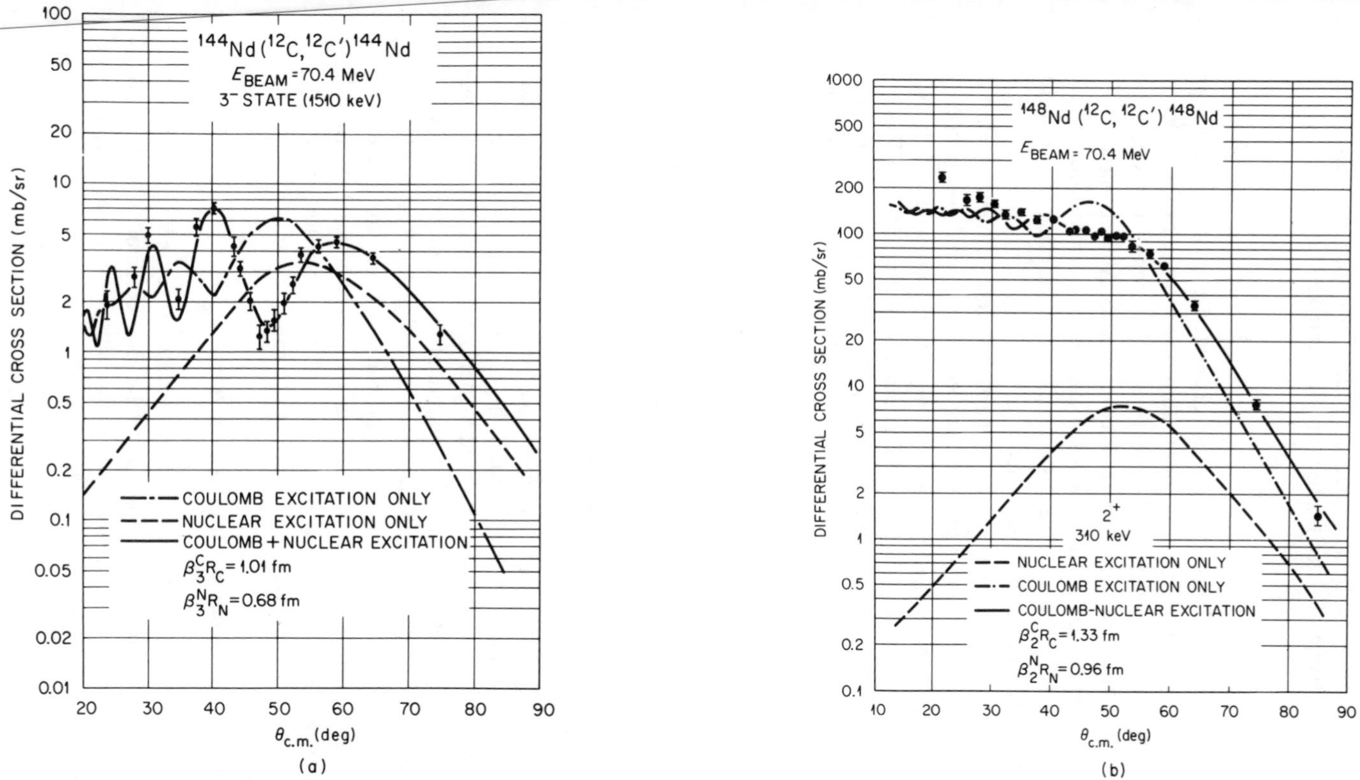

Figure 4.9 Examples of the differential cross-sections for the inelastic scattering of heavy ions, showing the effects of Coulomb excitation and its interference with excitation by the nuclear forces. (a) An octupole (L = 3) excitation where the Coulomb and specifically nuclear contributions are similar in magnitude. Note that the Coulomb–nuclear interference reverses the phase of the oscillations in the angular distribution (solid curve) compared to those for Coulomb excitation alone (dash–dot curve). (b) A quadrupole (L = 2) excitation which is dominated by Coulomb excitation. This occurs primarily because the L = 2 Coulomb interaction ($\sim 1/r^3$) has a longer range than that ($\sim 1/r^4$) for L = 3. Although the Coulomb–nuclear interference is less important here, it still affects the details of the angular distribution. (From Hillis *et al.*, 1977)

peaks at a smaller scattering angle than the nuclear term. The interference between these two terms can be an important source of information about the nuclear interaction. Some of these features are illustrated in *Figure 4.9.*

4.4.5 Extensions of the model

These appear in two ways. First, the general characteristics of the scattering depend upon strong absorption but do not require the validity of the collective model *per se*. This model provides a clear picture of the excitation process, but excitation of other kinds of nuclear states is covered also. All that is required is that the adiabatic approximation be satisfied and that the transition be sufficently strong that it can be described as a simple direct reaction (and not, for example, as a compound nucleus reaction).

The other extension is analogous to the partial wave treatment of elastic scattering and the assumption of phenomenological functional forms $\eta(\ell)$ for the η_ℓ (section 4.1.3). A detailed discussion is beyond the scope of this book (*see* Blair, 1966; Austern and Blair, 1965; Frahn and Rehm, 1978). This approach is closely related to the use of a non-spherical optical potential (*see* below) and provides a link between potential models and strong absorption models. An important consequence is that cross-sections of the form 4.20 result if the excitation process is confined to the nuclear surface so that only the partial waves corresponding to glancing collisions contribute significantly. In the simplest form of this treatment, the contribution to the inelastic scattering from the ℓth partial wave is proportional to $\partial\eta(\ell)/\partial\ell$. The surface nature is then apparent if we use a form like 4.5 for $\eta(\ell)$; the inelastic scattering will be due to a group of ℓ values centred on $\ell = \ell_c$ and with a width of about 3.5Δ.

It should also be remarked that we have only described briefly the inelastic scattering resulting from the terms of first order in the deformation length δ_{LM}. Both the Fraunhöfer diffraction treatment and the partial wave theory just mentioned can be extended to treat terms of higher order. Some nuclear transitions can only be excited by the second and higher order terms; for example, the excitation of a state corresponding to two quanta of the quadrupole vibration requires the quadrupole deformation treated to second order. The excitation of other transitions may involve comparable contributions from first and second orders; for example, the 4^+ state of a rotational band may be reached from the 0^+ ground state either directly (first order) by an $L = 4$ excitation or by two successive $L = 2$ excitations via the 2^+ state of the band.

4.4.6 Strong absorption and other direct reactions

Only peripheral or grazing collisions can lead to direct reactions such as nucleon transfer when there is strong absorption; more head-on collisions lead to more complicated reactions such as the formation of a compound nucleus, while those with larger impact parameters do not interact at all. Consequently, the products of a direct reaction originate from a narrow ring around the strong absorption surface just as in the case of inelastic scattering already discussed. Hence we expect and indeed often do see angular distributions for other direct reactions with characteristics similar to those seen with inelastic scattering when there is strong absorption. Theories for these have been developed either using a diffraction approach (*see*, for example, Henley and Yu, 1964) or postulating the corresponding properties of the partial wave amplitudes (*see*, for example, Frahn, 1985; Chasman *et al.*, 1973). However, while the latter approach has found some recent applications to heavy-ion reactions, currently it is more common to analyse measurements on these direct reactions using the distorted-wave and coupled-channel theories to be described later.

4.5 THE OPTICAL MODEL FOR ELASTIC SCATTERING

4.5.1 Introduction

This is one of the simplest and most successful of nuclear models. The models described in the preceding section were based upon the existence of minimal characteristics such as strong absorption and a fairly well-defined nuclear edge. These were found to imply certain simple properties of the scattering amplitudes and the corresponding differential cross-sections. The results could often be expressed in terms of just two or three physical parameters. The optical model, however, characterises the interaction of two nuclei in terms of a potential. A potential description may give more freedom; for example, it allows one to describe the scattering when the absorption is *not* strong. Indeed, as we shall see, the physical significance of the potential is most easily understood when the absorption is weak. In this respect, the strong absorption models and the optical models are somewhat complementary.

As an example of this increased freedom, we refer back to *Figure 4.1* which shows that the scattering amplitudes η_L need not be smooth functions of L. It is possible to find a potential to reproduce this non-smooth behaviour for the protons as well as one which will give the smooth, strong-absorption behaviour for the α-particles. At the same time we are describing the physical system in more detail. In the strong absorption model we only specify the scattering amplitudes; these allow us to construct the wave function at distances beyond the range of the nuclear forces (*see* section 3.1), for example, at large distances where our detectors measure the scattering, but they do not tell us the wave function in the region where the nuclear forces are acting. A description using a potential, on the contrary, allows us to construct the wave function everywhere by solving the Schrödinger equation.*

*However, there is a question, which is still not properly resolved, as to the physical significance of this wave function when we are dealing with strongly absorbed particles at close distances of approach. The wave function ascribes a certain probability to our finding a composite like a deuteron or α-particle propagating within the target nucleus. Other considerations lead us to expect such composites to disintegrate and lose their identity as soon as they cross the nuclear boundary.

Since an optical potential for neutrons and protons is closely related to the shell model potential for the nucleons bound in a nucleus, we can have much more confidence in the significance in the nuclear interior of the wave functions that we obtain.

The interaction of two nuclei (even if one of them is a single nucleon) is a complicated many-body problem. For the purpose of describing elastic scattering, the optical model attempts to replace this problem by the much simpler problem of two structureless bodies interacting via a simple potential, U say. Except for a possible dependence on the spins of the two nuclei, this model potential is usually assumed to depend only upon the distance r between the centres of mass of the two nuclei, $U = U(r)$. In any case, it is presumed to represent some smoothed average of the actual, complicated interaction.* For a nucleon incident upon a nucleus, the physical significance of the optical potential U is essentially the same as that of the shell model potential for a bound nucleon with negative energy. There is one important difference however; whenever non-elastic scattering is possible, there is a loss of flux from the elastic channel. Hence the optical potential must be absorptive. This is accomplished by making the potential *complex* (*see* below, section 4.5.7).

Because of the parallel between the optical model and the scattering of light waves by a refracting sphere (hence the name 'optical'), it was at one time called the 'crystal ball model'. The absorptive properties required were recognised by referring to it as a 'cloudy crystal ball'. The use of a complex potential is analogous to the introduction of a complex index of refraction for the propagation of light through an absorbing medium.

Historically, a great impetus was given to the development of the optical model by the discovery (around 1950) that the scattering of neutrons from nuclei shared many of the characteristics of the scattering of light by almost transparent (but refracting) spheres. At first, this came as a surprise. The interactions between nucleons were expected to be violent because the nucleon--nucleon force was known to be strong and have a short range. Hence the mean free path of a nucleon in the nuclear fluid was expected to be very short; i.e. nuclear matter was expected to be almost opaque or 'black' so that the strong absorption pictures described in earlier sections would be appropriate. This expectation was reinforced by the early discovery that the cross-sections for the scattering of low-energy neutrons showed very many sharp and closely spaced resonances. Thus the optical behaviour just referred to means that after averaging over these resonances some residuum of the independent-particle motion survives and the effects of the strong forces have been suppressed. The nuclear shell model (*see* Preston and Bhaduri, 1975, for example) was being developed at about this same time and was beginning to have some successes in describing

*There are a number of ways (*see*, for example, Jackson, 1970; Austern, 1970; and other texts on scattering theory) in which the many-body problem of elastic scattering can be formally reduced to an equivalent two-body problem. The general result is a complicated optical *operator* which represents the effective two-body interaction. It is this which we model with our simple potential, U. As in other areas, it is too difficult to start from these first principles and go directly to the experimental measurements. Rather, we construct models to represent the data and then attempt to understand the general features of the models in terms of the formal theory.

nuclear structure; this provided additional evidence that the effects of independent motion in an average field were not completely washed out by internucleon collisions.

Almost simultaneously, the concept of an optical model potential was being applied successfully to the scattering of protons and also α-particles. Nowadays it has even found use in describing the scattering of two heavy ions, such as ^{40}Ar scattering from ^{238}U.

In this section we describe some general properties of the optical model *per se* and its characteristics for various particles. We return to its relationship to an average over resonances after the section on the compound nucleus (section 4.10).

4.5.2 'Echoes' in neutron cross-sections

Plots of measured total cross-sections σ_T for neutrons on a given target versus neutron energy show broad maxima several MeV in width. Plots of σ_T at a given energy against A, the mass number of the target, show similar maxima (*see*, for example, *Figure 2.6*). A composite three-dimensional plot for neutron energies up to 30 MeV is shown in the left side of *Figure 4.10*. The first remarkable feature is that the cross-sections for various nuclei combine to form a smooth surface. Next we see the maxima as ridges on this surface (two can be seen in *Figure 4.10*) whose loci move to higher energy as A increases. These features persist to energies of over 100 MeV, as *Figure 4.11* shows for three nuclei. This is re-emphasised by *Figure 4.12*, which locates these maxima and minima in the energy-A plane.

If nuclei behaved toward incident neutrons like strongly absorbing or black spheres, we would expect from section 4.1.1 that the total cross-section would be $2\pi(R + ƛ)^2$, where R is the nuclear radius and $ƛ$ is the reduced wavelength of the neutrons. This expression implies that σ_T monotonically increases with A and monotonically decreases with energy, and indeed it does reproduce the *average* behaviour of the measured σ_T provided a radius of $R \approx 1.4\,A^{1/3}$ fm is used (*Figure 2.6*). The oscillations about this average value are due to the nuclei not being completely opaque to neutrons. This partial transparency can be represented by a complex potential well (*see* below) through which the neutrons move; that is to say, an attractive complex potential $-(V + iW)$ in the nuclear interior which rapidly falls to zero at the nuclear radius, $r = R$, and for which $W/V \ll 1$ so that the absorption is not strong.

The right side of *Figure 4.10* shows the total cross-sections calculated for such an optical potential which covers the same range of E and A as the observed cross-sections on the left side; it shows clearly the first two ridges. These results can be understood very easily. *Figure 4.13* gives a simplified picture of the neutron waves crossing a cubical region of dimension d in which there is an attractive potential of depth V. If the wavelength is sufficiently short (i.e. the energy is sufficiently high) so that $ƛ \ll d$, we can neglect refraction and diffraction at the edges. Inside the cube the neutron has a larger wave number K than

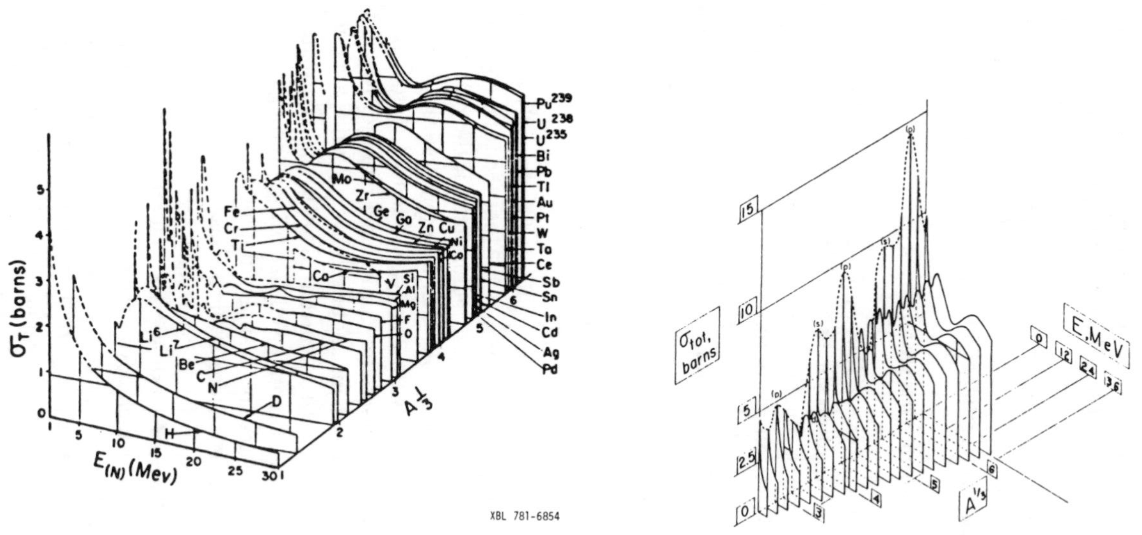

Figure 4.10 Plots showing how the total cross-sections for neutrons incident on nuclei vary with the neutron energy in the range 0–30 MeV and with the size of the target (expressed as $A^{1/3}$). To the left are the measured values (from Peterson, 1962); to the right are the results of calculations using a simple optical potential which represents a partially transparent nucleus (from McVoy, 1967)

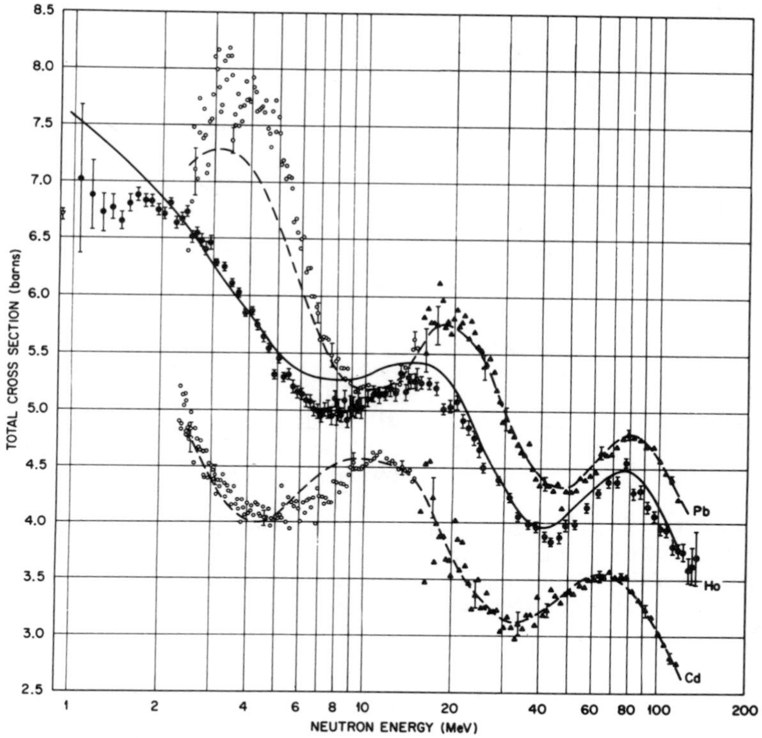

Figure 4.11 The variation with energy of the total cross-section for neutrons with energies from 1 to about 100 MeV for targets of Cd, Ho and Pb nuclei. The Ho nuclei are strongly deformed while the Cd and Pb nuclei are essentially spherical. (From Marshak *et al.*, 1968)

the wave number k outside because the attractive potential increases its kinetic energy. The wave passing through is advanced in phase by

$$\phi = d(K - k) \tag{4.21}$$

compared to that part of the wave which passes by outside. (Such a wave function would be given by the eikonal approximation 3.115 with $\phi = kVd/2E$; this agrees with the phase 4.21 if $V \ll E$.) Remembering that the scattered wave is the *difference* between this and the undisturbed plane wave (sections 2.15, 3.1.4), we see that there is maximum scattering when ϕ is an *odd* multiple of π so that the portion of wave which passed through is exactly minus what would have been there in the absence of the potential. When ϕ is an *even* multiple of π, the wave which passed through the potential appears to have been undisturbed on the far side and there is no scattering (except that due to the edge effects which we have neglected).

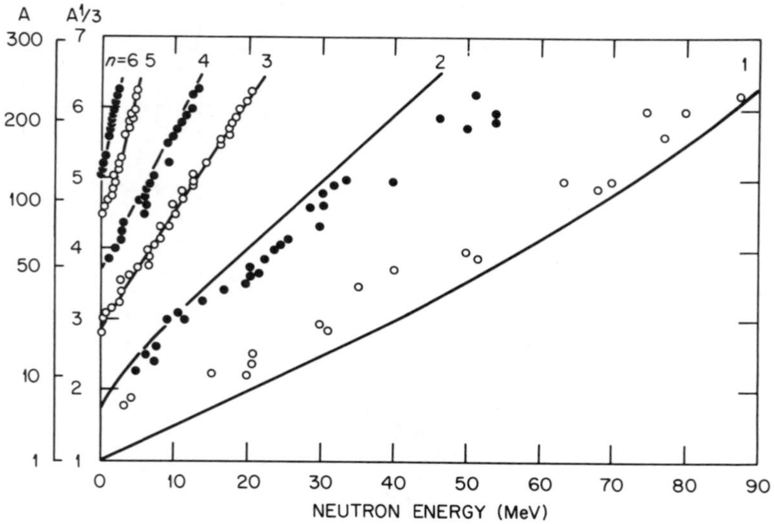

Figure 4.12 The loci of the maxima and minima of the neutron total cross-sections (as in *Figures 4.10, 4.11*) plotted in the energy-A plane. Odd values of n correspond to maxima, even values to minima. There are large uncertainties in the measured positions for $n = 1$ and 2. The curves are the loci for $\phi = n\pi$ calculated using equation 4.21. (After Peterson, 1962)

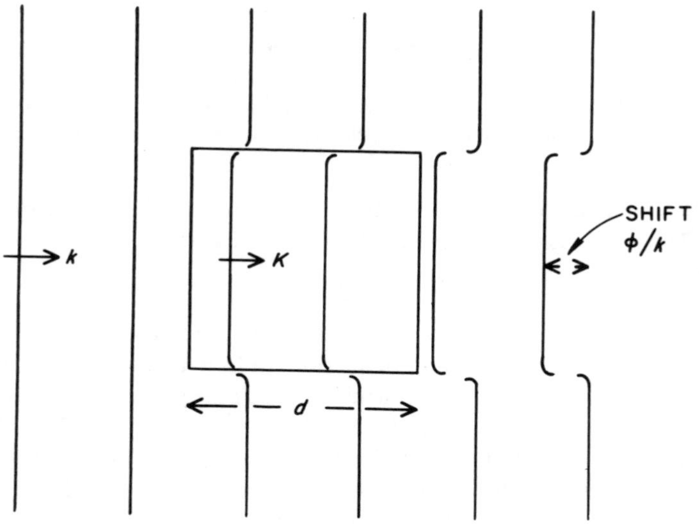

Figure 4.13 Illustrating the appearance of 'echoes' in the total cross-sections for neutron scattering owing to the interference between the retarded wave passing through a slab of nuclear matter and the undisturbed wave passing around it

Now nuclei are not cubes but spheres; however the *average* path length across a spherical nucleus of radius R is $\frac{4}{3}R$ so we may use the same argument with d replaced by $\frac{4}{3}R$. Hence we would expect the cross-section to be a maximum when $\phi = \frac{4}{3}R(K - k) \approx$ (odd integer) $\times\ \pi$ and a minimum when $\phi \approx$ (even integer) $\times\ \pi$. The loci of these values of $\phi = n\pi$ for $n = 1$–6 are plotted as full lines in *Figure 4.12* where as usual R is assumed to be proportional to $A^{1/3}$. Because $(K - k) \sim E^{-1/2}$ when $V \ll E$, the maxima move to larger A as the energy E is increased, or vice-versa. This behaviour is to be contrasted with that resulting from an alternate resonance condition which might have occurred to the reader, namely $KR = n\pi$, corresponding to fitting an integral number of half-wavelengths within the nuclear potential. This would result in the resonances moving to *smaller A* as E increases. Because of this, the broad maxima found in the cross-sections owing to condition 4.21 have been referred to as *echoes* or *anti-resonances* (*see* McVoy, 1967; McVoy *et al.*, 1967; where detailed discussions are given).

The discussion so far has treated the potential as real, so that any absorption of the neutrons as they traverse the nucleus has been neglected. The effect of absorption is to reduce the amplitude of the maxima and minima until with strong absorption no part of the wave traversing the nucleus survives and the cross-sections take on the uniform value of $2\pi R^2$. The measurements clearly indicate that the absorption for neutrons is weak.

The simple classical picture just used to derive the relation 4.21 is only valid for high energies where refraction effects are not too important. However early wave-mechanical studies of total cross-sections for neutrons with energies below a few MeV were made and were able to explain the qualitative features of the data by using a square potential well of depth $V + iW \approx 42\ (1 + 0.03i)$ MeV and with a radius of $1.45\ A^{1/3}$ fm. We see that $W/V \approx 0.03$ is indeed small. The large radius required is due to the use of a well with a sharp edge; more physically reasonable potentials with rounded edges are described below.

4.5.3 *Average interaction potential for nucleons*

If an incident nucleon p interacts with each target nucleon t through a potential $v(r_{tp})$, where $r_{tp} = |\mathbf{r}_t - \mathbf{r}_p|$ is the distance between them, then the overall potential it experiences owing to the target nucleus is

$$U(\mathbf{r}_p) = \int \rho(\mathbf{r}_t)\, v(\mathbf{r}_{tp})\, d\mathbf{r}_t \tag{4.22}$$

where $\rho(\mathbf{r}_t)$ is the density of the nucleus at the position $\mathbf{r}_t$ (*see Figure 4.14*). This expression is familiar when v is the electrostatic potential between two charges and ρ is then the charge distribution, ρ_c. We know that if ρ_c is well localised in some region, outside of that region the total electrostatic potential $U_c(r_p)$ is simply the Coulomb potential Q/r_p where Q is the total charge. The situation is quite different for nuclear forces because then the potential v has a short range; for separations r_{tp} greater than about 1 fm, v decreases exponen-

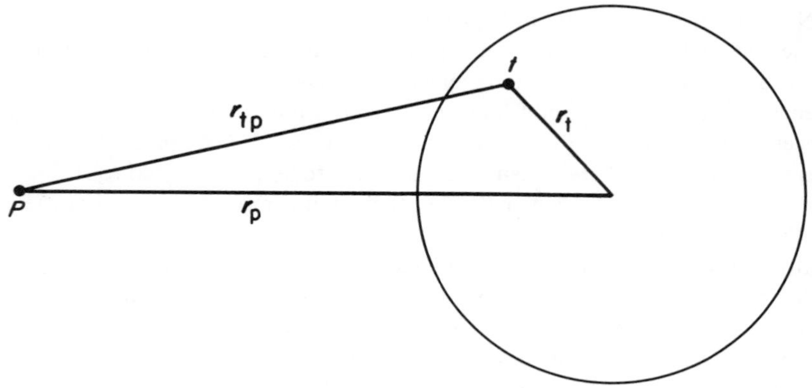

Figure 4.14 The coordinates for the folding integral, equation 4.22

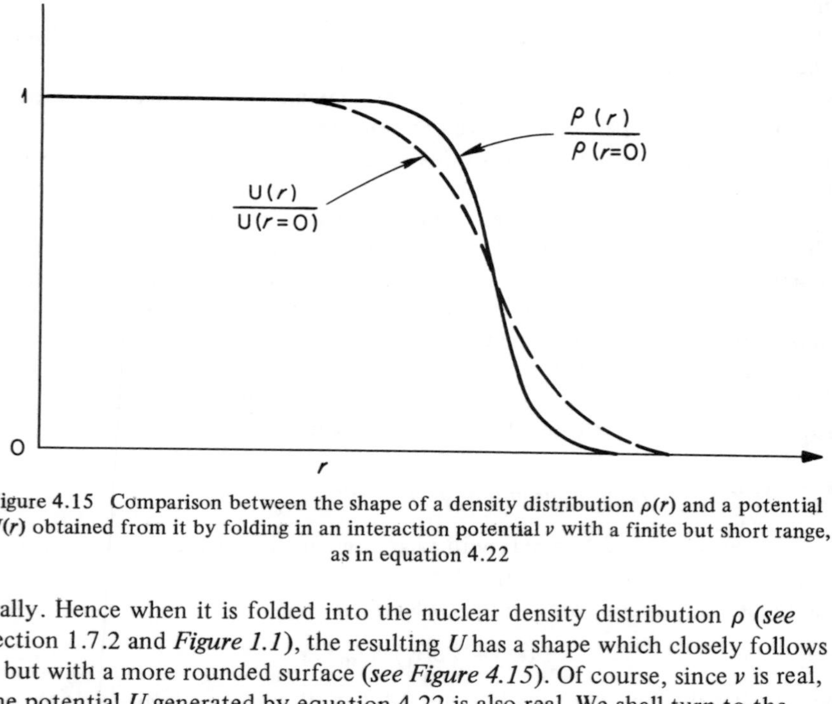

Figure 4.15 Comparison between the shape of a density distribution $\rho(r)$ and a potential $U(r)$ obtained from it by folding in an interaction potential v with a finite but short range, as in equation 4.22

tially. Hence when it is folded into the nuclear density distribution ρ (*see* section 1.7.2 and *Figure 1.1*), the resulting U has a shape which closely follows ρ but with a more rounded surface (*see Figure 4.15*). Of course, since v is real, the potential U generated by equation 4.22 is also real. We shall turn to the imaginary part of U later.

In the centre of a large nucleus, the density is approximately constant at $\rho = \rho_0 \approx \frac{1}{6}$ nucleon fm^{-3}. There the potential U will also be constant with a value

$$U(r_p \approx 0) = \rho_0 \int v(r_{tp})\, d\mathbf{r}_{tp} = \rho_0 J_0, \text{ say} \tag{4.23}$$

Simple potentials v which explain low-energy nucleon–nucleon scattering data have volume integrals $J_0 \approx -450$ MeV fm^{-3}, so we would expect $U(r_p \approx 0) \approx -75$ MeV for the potential felt by a neutron or proton inside a large nucleus. We shall see later that this is too large; the empirical values are some 30% less. However, we have ignored several features which will lead to corrections to this number. For example, the indistinguishability of an incident and a target nucleon means that they can be exchanged and this will count as part of the elastic scattering.

One important modification of the simple theory as we have presented it here occurs because high-energy nucleon–nucleon scattering measurements indicate that the interaction v between two nucleons is very strongly repulsive at short distances ($r_{tp} \lesssim 0.5$ fm). This makes the simple folding formula 4.22 inadequate when v itself is used; however these difficulties are overcome in the theories of Brueckner and Bethe (Preston and Bhaduri, 1975; Bethe, 1971) by transforming v into an effective interaction, usually called a G-matrix, which is better behaved. Besides the short-range repulsion, this G takes care of the effects of the presence of the other nucleons (such as the influence of the Pauli Exclusion Principle) when the interacting pair are imbedded in a nucleus. One result is to make the effective interaction G *density-dependent*; that is to say, the strength of interaction of a nucleon pair depends upon the density of the nuclear matter in which they are imbedded. This makes G weaker inside the nucleus than outside.

When these effects are taken into account, the prescription 4.22 gives good agreement with the real part of the empirical potentials which are deduced from the observed scattering (Jeukenne *et al*., 1977). This implies that the processes which give rise to the absorption (*see* below) do not prevent equation 4.22 from being a good approximation for the optical potential.

4.5.4 Energy dependence of the potential

We can also get a measure of the potential depth for nucleons from the binding energies of nuclei if we treat each of them as a Fermi gas (*see* Preston and Bhaduri, 1975, for example). The energy ϵ of each nucleon in such a gas has a kinetic part T and a potential part U, so $\epsilon = T + U$. For our purpose we ignore the effects of the surface, of any neutron excess and of the Coulomb repulsion between the protons. (This is equivalent to considering the limit of a large nucleus with $N = Z$ and with uncharged protons, a limit usually called 'nuclear matter'.) Then $\overline{\epsilon}$, the average of ϵ, is just the volume term in the Weizsacker mass formula for nuclear binding energies and is about -16 MeV. The average kinetic energy of a nucleon in a Fermi gas is $\overline{T} = \frac{3}{5} T_F$ where T_F is the maximum kinetic energy, or *Fermi energy*. The average potential energy per nucleon is $\frac{1}{2}\overline{U}$, the factor of $\frac{1}{2}$ arising because the total potential energy of the system is a sum of pairwise interactions; if we simply summed U (defined as in equation 4.22) for each nucleon, each interacting pair would be counted twice. Thus the average energy per nucleon is

$$\bar{\epsilon} = \tfrac{3}{5} T_F + \tfrac{1}{2} \bar{U} = -16 \text{ MeV} \tag{4.24}$$

We know from the measured densities of nuclei that $T_F \approx 40$ MeV. Hence equation 4.24 gives the *average* potential to be $\bar{U} \approx -80$ MeV.

Now the energy S needed to separate the least tightly bound nucleon from a large nucleus (thus leaving the residual nucleus in its ground state) must be the same as the average binding energy per nucleon. But S is just minus the energy of a nucleon at the top of the Fermi gas distribution (the *Fermi surface*)

$$S = -\epsilon_F = -T_F - U_F \tag{4.25}$$

In our case, with $S = -\bar{\epsilon} = 16$ MeV, this means that the potential experienced by this nucleon is $U_F = -56$ MeV. That is, the requirement that $\bar{\epsilon} = \epsilon_F$ (a reflection of the saturating property of nuclear forces) cannot be satisfied with a constant potential U; as Weisskopf, 1957, pointed out, this means that the potential felt by a nucleon moving in the nucleus must depend upon its kinetic energy T. (The exchange process mentioned above gives just such an effect, and seems able to account for much of the observed energy dependence.) As the simplest approximation, let us assume a linear dependence, say

$$U(T) = U_0 + \alpha T \tag{4.26}$$

Then equations 4.24 and 4.25 give $U_0 = -116$ MeV and $\alpha = 3/2$ if $S = -\bar{\epsilon} =$ 16 MeV. Hence the potential becomes weaker (less attractive) as the kinetic energy increases.

Now it is clear that in this simple Fermi gas model the potential experienced by a nucleon incident on such a nucleus is also U. However, equation 4.26 gives U in terms of T, the kinetic energy inside the nucleus. But this is greater than the bombarding energy E outside because of the attractive potential. Substituting from $E = T + U$ and rearranging, we get

$$U(E) = \frac{U_0 + \alpha E}{1 + \alpha}$$

$$\approx -46.4 + 0.6E$$

for the potential in terms of the bombarding energy. This estimate of the potential strength is close to the empirical values for low bombarding energies, $E \approx 0$, but the energy dependence is too strong. For example, it predicts $U = 0$ for $E = (46.4/0.6) \approx 77$ MeV; this is not observed experimentally. However, these sorts of defect arise because the model we have been discussing, although it serves a useful didactic purpose, is greatly oversimplified. More sophisticated treatments (for example, Jeukenne *et al.*, 1977) do reproduce the observed values of the potential strength and its energy dependence.

4.5.5 Spin-orbit coupling

According to the Dirac wave equation (Messiah, 1962), an electron moving in an electrostatic potential $U(\mathbf{r})$ 'sees' a magnetic field owing to relativistic effects; the magnetic moment associated with the electron's spin angular momentum $\mathbf{s}$ then interacts with this field resulting in a *spin-orbit coupling* potential of the so-called Thomas type

$$-\tfrac{1}{2}i\left(\frac{\hbar}{m_e c}\right)^2 \mathbf{s}\cdot(\nabla U \times \nabla)$$

When the potential U is central or spherically symmetric (depends only upon the magnitude r and not the direction of $\mathbf{r}$), this reduces to

$$\tfrac{1}{2}\left(\frac{\hbar}{m_e c}\right)^2 \frac{1}{r}\frac{dU(r)}{dr}\,\mathbf{s}\cdot\mathbf{l}$$

The same *form* is usually employed in the optical model potentials for nucleons and other projectiles with non-zero spin (as well as in the shell model potential for nucleons bound in the nucleus—*see* Preston and Bhadhuri, 1975, for example). However, in these cases it certainly does not arise as a simple relativistic effect. Indeed, we find empirically that the coupling needed for nucleons has the opposite sign to the Thomas term and is 20–30 times stronger.

Now the nucleon-nucleon force itself is dependent upon the spins of the nucleons and in particular contains non-central terms of the spin-orbit and tensor type. For example, there is a two-body spin-orbit interaction potential of the form

$$v_{so} = f(|\mathbf{r}_t - \mathbf{r}_p|)\,[(\mathbf{r}_t - \mathbf{r}_p) \times (\mathbf{p}_t - \mathbf{p}_p)\cdot(\mathbf{s}_t + \mathbf{s}_p)]$$

where $\mathbf{p}_t$ and $\mathbf{p}_p$ are the momenta of the target and projectile nucleons, respectively. The expression 4.22 is only valid if the v there represents a central interaction; however, an analogous folding can be performed for the spin-orbit interaction v_{so} when proper attention is paid to the spins of the nucleons (Blin-Stoyle, 1955). If the target nucleus is spin-saturated, that is has closed shells with as many spins 'up' as 'down', the expectation values of $\Sigma_t \mathbf{p}_t$ and $\Sigma_t \mathbf{s}_t$ will vanish leaving an effective one-body spin-orbit potential of the form

$$U_{so} = g_{so}(r_p)\,\mathbf{r}_p \times \mathbf{p}_p\cdot\mathbf{s}_p = g_{so}(r_p)\,\boldsymbol{\ell}_p\cdot\mathbf{s}_p \tag{4.27}$$

which is experienced by the projectile nucleon. (In practice, when $s_p = \frac{1}{2}$, the $\mathbf{s}_p$ in equation 4.27 is often replaced by the Pauli spin operator $\sigma = 2\mathbf{s}_p$.)

When the interaction v_{so} has a very short range, it can be shown (Blin-Stoyle, 1955) that the radial dependence in equation 4.27 becomes

$$g_{so}(r) = \frac{C}{r}\frac{d\rho(r)}{dr}$$

where $\rho(r)$ is the density distribution for the target nucleus and C is a constant. This suggests that the spin-orbit potential, 4.27, is peaked at the nuclear surface, even if it is not exactly proportional to the derivative of the density. The physical reason for this is simple. A nucleon moving through uniform nuclear matter, such as is found in the interior of nuclei, would not experience a spin-orbit coupling because there is no preferred point of reference about which the angular momentum **L** should be measured. Only for finite nuclei can **L** be defined and the finiteness is experienced only as the nucleon crosses the nuclear surface, provided the range of the spin-orbit force is short compared to the nuclear radius.

The operator $\mathbf{l}\cdot\mathbf{s}$ in equation 4.27 conserves the magnitude of the spin and orbital angular momentum but allows their relative orientation to be changed (*see* Appendix A and *Figure A2a*). The plane of the orbit may be tilted by the interaction which at the same time flips over the spin so that the *total* angular momentum $\mathbf{j} = \mathbf{l} + \mathbf{s}$ and its z-component remain constants of the motion. (This also means that the coupling scheme, 3.89, diagonalises the interaction 4.27.) Using $\mathbf{j}^2 = \mathbf{l}^2 + \mathbf{s}^2 + 2\mathbf{l}\cdot\mathbf{s}$, the expectation value is given by

$$\langle \mathbf{l}\cdot\mathbf{s}\rangle = \tfrac{1}{2}\,[j(j+1) - \ell(\ell+1) - s(s+1)]$$

When $s = \frac{1}{2}$

$$\begin{aligned} \langle \mathbf{l}\cdot\mathbf{s}\rangle &= \tfrac{1}{2}\ell & \text{if } j = \ell + \tfrac{1}{2} \\ &= -\tfrac{1}{2}(\ell+1) & \text{if } j = \ell - \tfrac{1}{2} \end{aligned}$$

Empirically, we find that g_{so} in equation 4.27 is negative so that particles with spin and orbit parallel, $j = \ell + \frac{1}{2}$, experience a more attractive total potential than those with spin and orbit antiparallel, $j = \ell - \frac{1}{2}$. Consequently, the two spin states are scattered differently and it is this which gives rise to spin polarisation, or a preferential scattering of one spin orientation compared to the other (section 2.13).

The tensor force between two nucleons may also contribute to the spin-orbit potential, 4.27. Further, when the projectile has a spin $s_p > \frac{1}{2}$ (such as for a deuteron which has $s_p = 1$), there may be more complicated tensor types of spin-orbit coupling; the form 4.27 is then referred to as the vector coupling.

4.5.6 Average potentials for composite projectiles

Potentials for composite projectiles, like deuterons, α-particles and heavier ions, can be constructed in the same way as for nucleons. Now, however, we must integrate over the nucleons in the projectile as well as in the target. If the density distribution in the target is $\rho_t(\mathbf{r}_t)$ and that in the projectile is $\rho_p(\mathbf{r}_p)$, equation 4.22 is generalised to

$$U(R) = \iint \rho_p(\mathbf{r}_p)\,\rho_t(\mathbf{r}_t)\,v(r_{tp})\,d\mathbf{r}_p\,d\mathbf{r}_t \qquad (4.28)$$

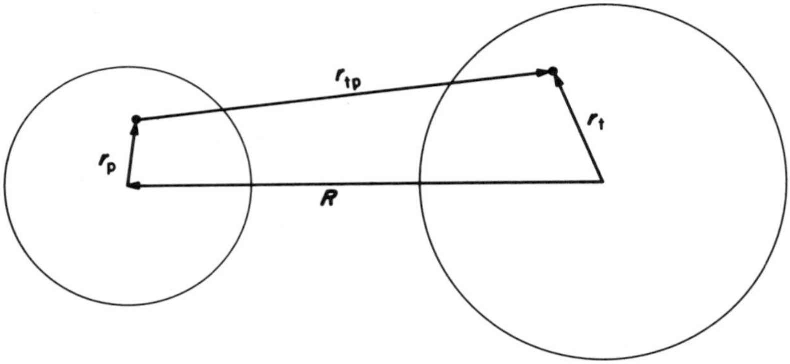

Figure 4.16 The coordinates for the double-folding integral, equation 4.28

where R is the separation between the centres of mass of the target and projectile. (The coordinates are shown in *Figure 4.16*.) Clearly the additional integration over the projectile distribution will lead to a potential much like that from equation 4.22 (*see Figure 4.15*) but with an even more diffuse surface. More important however; if there are n nucleons in the projectile, its potential will be approximately n times stronger than that for a single nucleon. Then, if a nucleon optical potential well is about 50 MeV deep, we would expect depths of about 100, 150 and 200 MeV for deuterons, mass 3 (^{3}He and tritons) and α-particles, respectively.

4.5.7 Imaginary potentials and absorption

Equation 4.22 would be an exact expression for the optical potential if nuclei had no structure and could not be excited, so that elastic scattering were the only possible process. But other reaction channels are possible and their existence affects the elastic scattering. For example, an incident nucleon may first excite the target nucleus and then de-excite it, leaving with its initial energy unchanged. This is illustrated in part (b) of *Figure 4.17*, where an incident nucleon excites a target nucleon to another orbit and then in a subsequent collision de-excites it. Potential 4.22 only describes the scattering shown in part (a) of *Figure 4.17*, where the target is undisturbed. The target may be pictured as responding to the forces exerted on it by the projectile by polarising or deforming, returning to its former state as the projectile leaves. (Of course, if the projectile were itself a composite nucleus, it also could be polarised in this way during the collision.) The optical potential is a simple 'effective' interaction which is designed to describe the whole scattering, including the contributions from these virtual processes, but without referring to them explicitly. To this extent it must then differ from the simple potential, 4.22. (*See* Exercise 3.3 for a simple example.)

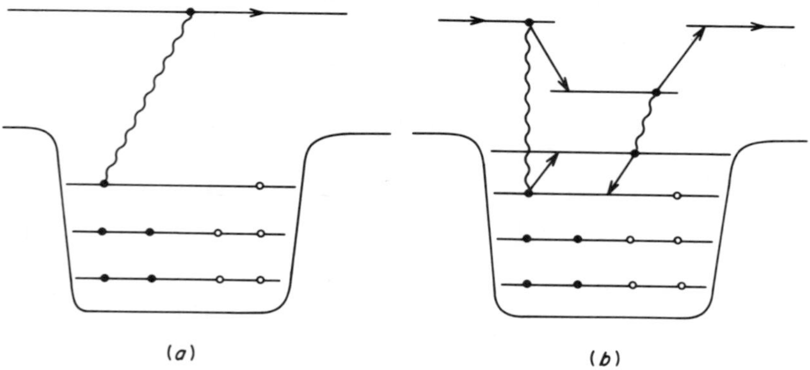

Figure 4.17 Illustrating (a) the scattering of a nucleon by interaction with a target nucleon in which the latter is undisturbed and (b) a double scattering in which a target nucleon is first excited and then de-excited

If an excitation of this type takes place, but the system does *not* return to the entrance channel, we have a non-elastic event. The projectile is lost from the entrance channel and can be thought of as being absorbed. If we are not interested in the details of the non-elastic processes, but only in their effect on the elastic scattering, we can represent this absorption by adding an imaginary term to the optical potential. This is closely related to the use of a complex refractive index for describing the passage of light waves through an absorbing and refracting medium.

Consider a plane wave corresponding to particles of mass m moving along the z-axis in a uniform potential $U = -(V + iW)$. The wave function is

$$\psi = \exp(ikz), \quad k = [2m(E - U)/\hbar^2]^{1/2}$$

Now the wave number k is complex, $k = k_1 + ik_2$ say. Then ψ can be written

$$\psi = \exp(-k_2 z)\exp(ik_1 z)$$

that is, the wave is damped exponentially by the factor $\exp(-k_2 z)$. The particle density also decreases exponentially as one moves along the z-direction

$$|\psi|^2 = \exp(-2k_2 z)$$

The absorption coefficient $2k_2$ is the reciprocal of the mean free path Λ. If W is small, we have

$$\Lambda^{-1} = 2k_2 \approx Wk_1/(E + V), \quad k_1 \approx [2m(E + V)/\hbar^2]^{1/2} \qquad (4.29)$$

so that Λ is inversely proportional to the strength W of the imaginary part of the potential.*

*Note that we must have an *attractive* imaginary potential, $Im U < 0$ or $W > 0$. If $W < 0$, then $k_2 < 0$ and the particle density *increases* exponentially; i.e. we would have particle emission, not absorption.

If we examine the flux $\mathbf{j}(\mathbf{r})$ associated with a solution $\psi(\mathbf{r})$ of the Schrödinger equation with a complex potential $U(\mathbf{r})$, it is straightforward (*see* Messiah, 1962) to show that it is not solenoidal but has a divergence which is proportional to the imaginary part of U and to the particle density

$$\text{div}\,\mathbf{j}(\mathbf{r}) = \frac{2}{\hbar}\,|\psi(\mathbf{r})|^2\, Im U(\mathbf{r}) \tag{4.30}$$

Simple classical considerations lead to an instructive relation between the imaginary potential W for nucleons and the average* nucleon–nucleon scattering cross-section, $\bar{\sigma}$. A nucleon traversing nuclear matter of density ρ_0 nucleons per unit volume with a velocity v will suffer $\rho_0\bar{\sigma}v$ collisions in unit time. Hence its mean free path is $\Lambda = (\rho_0\bar{\sigma})^{-1}$. But we saw that the mean free path is related to the imaginary potential by equation 4.29. Remembering that $\hbar k_1 = mv$, we get

$$W = \tfrac{1}{2}\hbar\, v\, \rho_0 \bar{\sigma} \tag{4.31}$$

For low energies this gives $W \sim 50$ MeV, much larger than the empirical values. This estimate will be in error for a number of reasons. Quantum-mechanical effects are neglected. The average cross-section for collision within nuclear matter will be reduced from that for free space because of the Pauli Exclusion Principle; the two colliding nucleons can only scatter into states not already occupied by the other nucleons. When this Exclusion Effect is included (still in a semi-classical way), W for low-energy nucleons is reduced by an order of magnitude, to about 3 MeV. As the bombarding energy is increased, the Exclusion Principle becomes less effective and W increases. In addition, the density of spectator nucleons is much less in the nuclear surface and they interfere less with the colliding pair; this tends to make the imaginary potential stronger in the nuclear surface.

These various features are in qualitative agreement with what is found empirically. Although the numerical values obtained from the simple formula, 4.31, are not very accurate, it has some didactic value.

4.5.8 Analyses of scattering experiments

In some situations, a set of experimental scattering data only allows us to determine simple properties such as the overall size (radius) of the nucleus. In more favourable circumstances (such as with shorter wave-lengths due to higher energy) the scattering may reveal more detailed shape information. The usual procedure is to adopt a particular functional form for the potential and adjust its parameters until the scattering calculated with it gives a good fit to the

*σ depends upon the relative energy of the two nucleons, so has to be averaged over the energies of the various target nucleons. This is usually done assuming a Fermi gas for the latter. Further, since the σ are different for like and unlike nucleon pairs, these must be averaged also.

measurements. This process is not without ambiguities. However, we are guided by two criteria.

(i) *Reasonableness*. This means that we build into the model the features that we expect from general considerations. For example, we are guided by equation 4.22 and *Figure 4.15* when choosing the shape of the real part of the potential.

(ii) *Continuity*. We expect the potentials for a given type of projectile to be similar for target nuclei with adjacent masses, or for adjacent energies on a given nucleus. Indeed we demand that the potential parameters vary slowly and smoothly (if at all) as we move through the Periodic Table and as we change the bombarding energy.

The latter requirement means that we would not apply the optical model to scattering data when rapid variations with energy or even distinct resonances are present, unless these variations could be averaged over first (*see* section 4.10, for example).

The most widely used functional form for optical model potentials is

$$U(r) = -V f(r, R, a) - iW f(r, R', a') - iW_D g(r, R', a') \tag{4.32}$$

where f is the Woods–Saxon form factor*

$$f(r, R, a) = (e^x + 1)^{-1}, \quad x = \frac{r - R}{a} \tag{4.33}$$

At the origin, $f(r = 0) \approx 1$, while $f = \frac{1}{2}$ at $r = R$. Further, f falls from 9/10 to 1/10 over a distance $4.4a$ centred on $r = R$. The first imaginary term in equation 4.32 has the same shape but may have a slightly different radius R' and surface diffuseness a'. The second imaginary term is peaked at the nuclear surface and is the derivative of the first

$$g(r, R', a') = 4a(d/dr) f(r, R', a') \tag{4.34}$$

These shapes are illustrated in *Figure 4.18*. In addition, if the particle has spin, we may add a spin–orbit coupling term like equation 4.27; this is essential if we are to describe polarisation measurements (section 2.13; *see*, for example, Barschall and Haeberli, 1971). Usually the so-called Thomas form is chosen for the radial dependence with $g_{so}(r)$ proportional to $r^{-1}(d/dr) f(r, R_{so}, a_{so})$.

4.5.8.1 Nucleon scattering

Analyses of nucleon-scattering data have been made for energies from essentially zero to about 1 GeV. Experimentally it is more difficult to measure neutron scattering, and most available data are for energies below about 50 MeV, so much of our knowledge of the optical potential for nucleons comes from studies

*Originally introduced by Eckart, this form factor is usually referred to as 'Woods–Saxon' or 'Saxon' when used for optical potentials. The same shape is often used to describe charge or mass distributions (*see* section 2.11); in that context it is usually called a Fermi distribution!

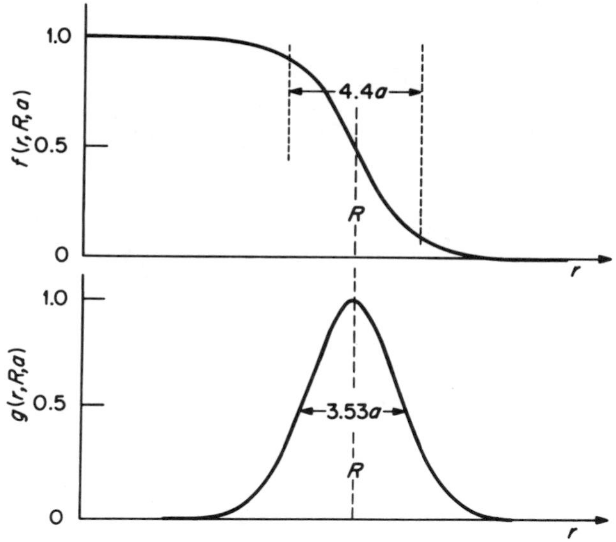

Figure 4.18 The Woods–Saxon, equation 4.33, and Woods–Saxon derivative, equation 4.34, shapes used for optical model potentials. The figure is drawn for $R/a \approx 9$, which corresponds to the potential for a nucleon on a nucleus with $A \sim 100$

with protons. An exception occurs at low energies of a few MeV or less where the Coulomb barrier prevents protons from experiencing the nuclear potential whereas neutrons are not inhibited in this way. (We return to very low-energy ($\lesssim 1$ MeV) neutron scattering in section 4.10.)

An example of some fits to the measured differential cross-sections and polarisations of 100–182 MeV protons are shown in *Figure 4.19*. We note especially how similar are the distributions from target nuclei of similar mass and the regular way in which the pattern changes as the target mass, and therefore size, increases.

The depth of the real potential well, V, is found to be roughly the same for all nuclei and of order 50 MeV. The energy dependence of the depth (compare equation 4.27) is $-\partial V/\partial E \gtrsim 0.3$ for energies below about 20 MeV, but this has decreased to about 0.2 between 40 and 60 MeV, confirming that the energy variation is not truly linear. As equation 4.22 leads us to expect, the radius R is approximately proportional to $A^{1/3}$ just like the density distribution. Its value is roughly $R \approx 1.2\,A^{1/3}$ fm. The surface diffuseness parameter is $a \approx 0.7$ fm, corresponding to a decrease in the potential from 90% to 10% of its full strength over a distance of about 3 fm (*see Figure 4.18*).

At the lower energies, there is no strong evidence for a 'volume' component to the imaginary potential (i.e. we may put $W \approx 0$ in equation 4.32) and the absorptive processes appear to be mostly confined to the nuclear surface. However, a volume term, increasing with energy, can be detected at 30 MeV and

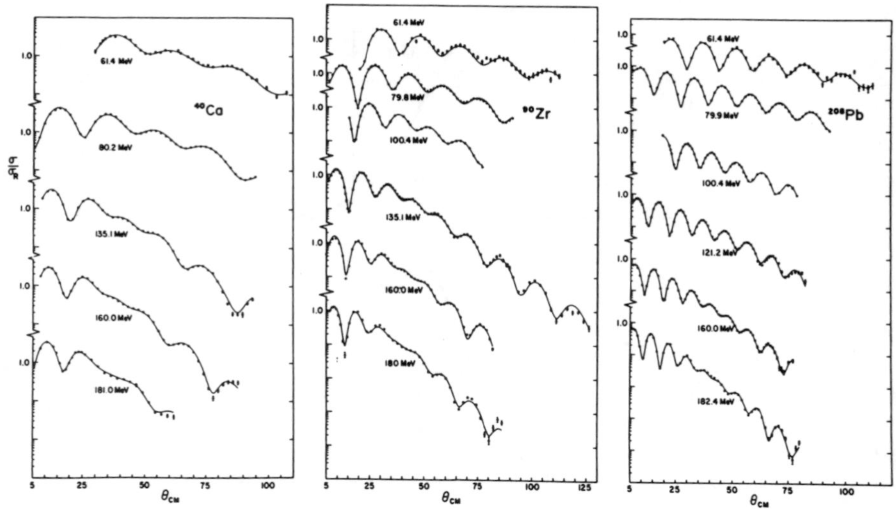

Figure 4.19 (a) The differential cross-sections and (b) the asymmetries for the scattering of polarised protons with several energies from a variety of target nuclei. The dots are the measured values and the curves represent fits obtained using the optical model.
(From Nadasen *et al.*, 1981)

above, whereas the surface absorption steadily decreases in strength as the energy is raised; there seems to be no need for any surface term for energies much above 50 MeV. There is a tendency for the radius R' of the imaginary potential to be somewhat larger than R, that for the real part, while a' is smaller than a; typically, $R' \approx 1.32A^{1/3}$ fm, $a' \approx 0.6$ fm. (*See*, for example, Nadasen *et al.*, 1981 and Satchler, 1983, for more details.)

The numbers just quoted are all subject to some uncertainties. For example, it is often possible to make small changes in V and R without changing the scattering appreciably provided the product VR^2 is kept constant. Indeed, it has been found that two quantities that can be determined with less ambiguity are the *volume integral* J of the real potential and its *mean square radius* $\langle r^2 \rangle_U$

$$J = \int U(r)\mathrm{d}\mathbf{r}, \quad J\langle r^2 \rangle_U = \int U(r)r^2 \,\mathrm{d}\mathbf{r} \tag{4.35}$$

We note that the integral 4.22 has the properties that

$$J = AJ_0, \quad \langle r^2 \rangle_U = \langle r^2 \rangle_\rho + \langle r^2 \rangle_v \tag{4.36}$$

where A is the number of nucleons in the nucleus, and J_0 is the volume integral of the nucleon-nucleon potential defined in equation 4.23. Also $\langle r^2 \rangle_\rho$ and $\langle r^2 \rangle_v$ are the mean square radii of the density ρ and the potential v, respectively.

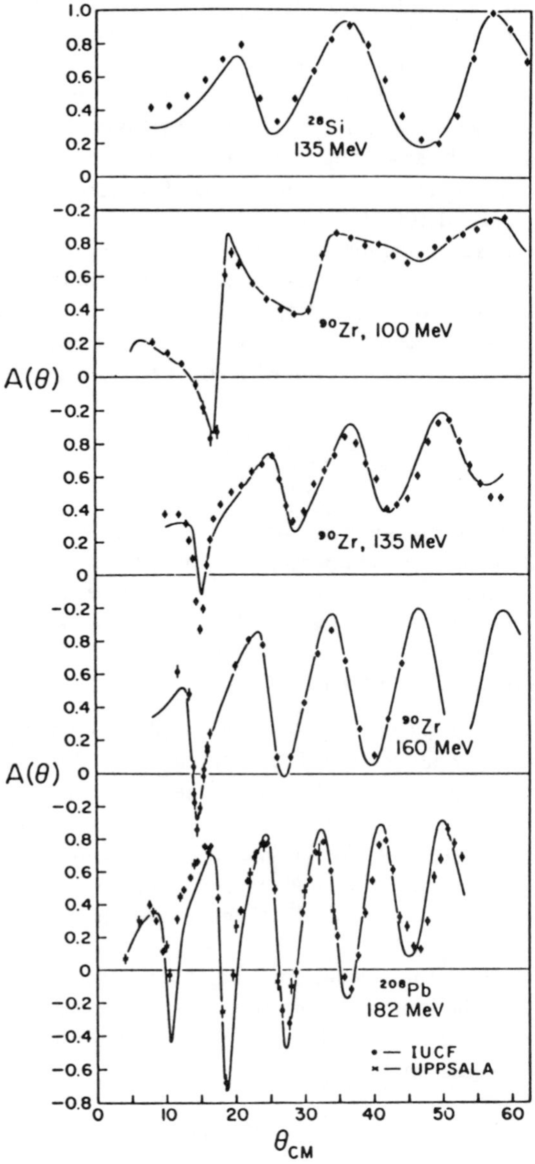

Figure 4.19(b)

(These relations are modified somewhat if v in equation 4.22 is replaced by an effective interaction G which is *density-dependent*.) The empirical potentials have values of J which are closely proportional to A (with $J_0 \approx 450$ MeV fm^3 at zero energy). The values of $\langle r^2 \rangle_U$ fall approximately on a line $\langle r^2 \rangle_U \approx (0.9\,A^{2/3} + 6)$ fm^2 which is roughly that expected from the density distributions measured by electron scattering if $\langle r^2 \rangle_v \approx 6\ \text{fm}^2$.

The depth of the real potential steadily decreases as the bombarding energy is increased until it appears to pass through zero in the vicinity of 300 MeV and then become repulsive.* The scattering tends to be dominated by the imaginary potential. Currently, new data are being obtained with accelerators for protons with energies in the range of 100 MeV to 1 GeV.

Most of the neutron-scattering measurements are for energies $\lesssim 30$ MeV for practical reasons. Analysis of these implies potentials with similar characteristics to those just described for protons (*see* Rapaport, 1982, for example). The potential depths V for neutrons tend to be less than for protons; in particular, this difference increases as we go to nuclei with larger neutron excesses, in such a way that

$$V_{\text{p}} - V_{\text{n}} \approx 50\left(\frac{N-Z}{A}\right)\ \text{MeV} \tag{4.37}$$

This effect is due to the proton–neutron interaction being stronger than the proton–proton or neutron–neutron interactions. When equation 4.22 is suitably generalised to take account of this, the observed difference in neutron and proton potentials can be understood (*see* Satchler, 1969, for example).

4.5.8.2 Scattering of composite particles

This is characterised by strong absorption, as discussed in earlier sections. As a consequence, considerable ambiguity is found in the optical potential parameters needed to fit the scattering data. The data largely determine the values of the potential in the surface region and the scattering is often very insensitive to the potential in the interior. *Figure 4.20a* shows a series of potentials for 40-MeV α-particles, all of which generate the same cross-sections for scattering angles less than about 90° (*see Figure 4.20b*) and all of which have the same value in the surface.

To give 'the surface' a more specific meaning here, *consider Figure 4.21* which shows curves for the sum of the real part of the potential energy $U(r)$ (attractive nuclear plus repulsive Coulomb) and the centrifugal energy for various values of the angular momentum ℓ

$$S(r) = Re\ U(r) + \frac{\hbar^2}{2\mu}\ \frac{\ell(\ell+1)}{r^2} \tag{4.38}$$

*This is a manifestation of the short-range repulsive component of the nucleon–nucleon force which also causes the S-wave phase shift for nucleon–nucleon scattering to go through zero at a similar energy.

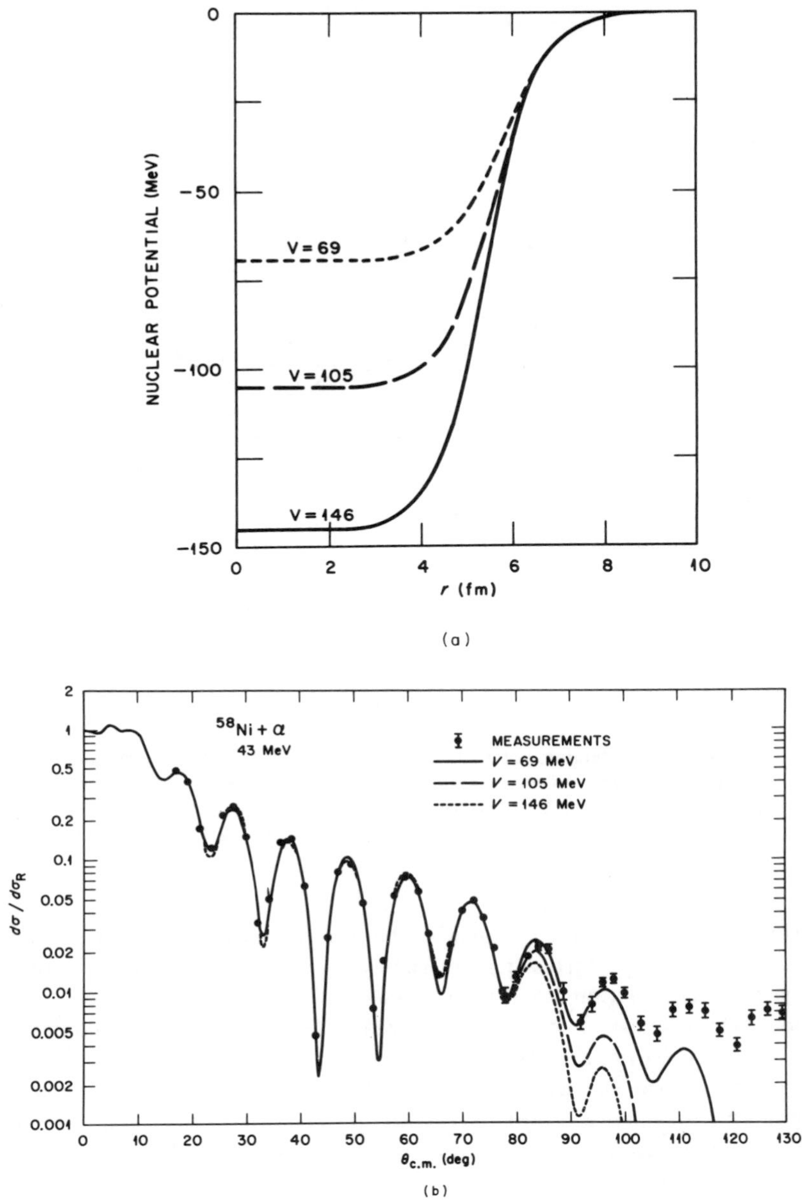

Figure 4.20 (a) Three different optical potentials which fit the cross-sections for the elastic scattering of α-particles with 43 MeV from ^{58}Ni nuclei. (b) The corresponding differential cross-sections in ratio to the Rutherford ones. (After Broek *et al.*, 1964)

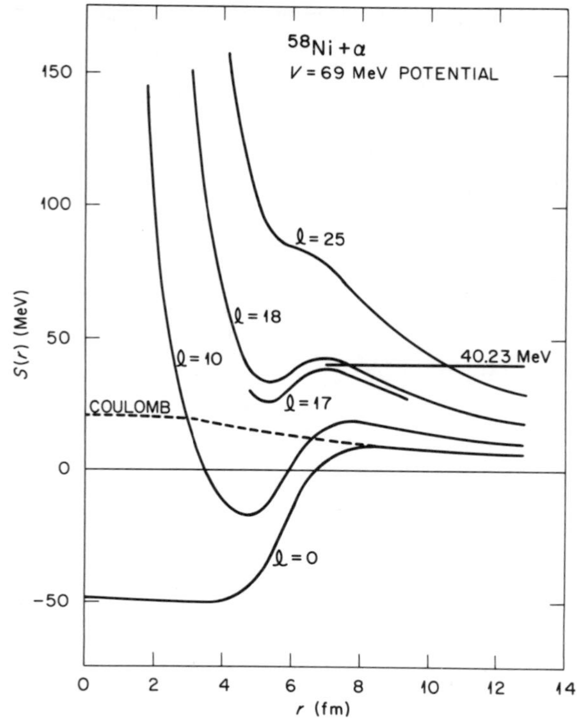

Figure 4.21 The total effective 'potential' $S(r)$ of equation 4.38, nuclear plus Coulomb plus centrifugal, for the potential of *Figure 4.20* with $V = 69$ MeV and for various values of ℓ. α-Particles of 43 MeV bombarding ^{58}Ni have a CMS energy of 40.23 MeV. Classically (*see* section 3.10.1), orbiting would occur for ℓ between 17 and 18

Consider the curve for which a given bombarding energy E just surmounts the hump or barrier in the surface region. The wave-mechanical theory of the tunnelling effect (*see* section 3.6.3) then tells us that the transmission across the barrier in this case is approximately 50%. Since those particles which do penetrate within the barrier are absorbed, this means that $|\eta_\ell|^2$ is approximately 1/2 for this $\ell = \ell_c$. Also, classically, particles with $\ell = \ell_c$ are moving tangentially at the point $r = R_c$ where $E = S(r)$, i.e. the top of the barrier is their point of closest approach on a classical orbit. (We now recognise R_c to be similar to the strong-absorption radius discussed in section 4.3.) Particles with $\ell < \ell_c$ approach more closely and tend to be almost completely absorbed, hence the potential they feel in the interior is not very important. Those with ℓ appreciably greater than ℓ_c do not approach closely enough to feel the optical potential at all. Hence it is clear that, to a good approximation, we need only to get the correct potential in the vicinity of $r = R_c$ in order to reproduce the observed scattering. Conversely it is difficult to extract from the measurements information about the potential in other regions. For while the observed scattering may determine quite well the strengths of the real and imaginary potentials near $r = R_c$ and give some information about their slopes in that region, there are many combinations of the parameters in the potential 4.32 which will reproduce those values, as *Figure 4.20* shows.

The ambiguity just described makes it difficult to test the prediction of section 4.5.5 that the optical potential depth in the nuclear interior should be approximately n times that for a nucleon if the projectile contains n nucleons. However, careful studies, particularly for the higher bombarding energies and for cross-sections at large scattering angles (both features emphasising those particles which have penetrated most deeply into the nucleus), have often shown a preference for potentials which come close to satisfying this requirement (*see*, for example, Goldberg *et al.*, 1974; Satchler, 1983). Indirect evidence sometimes comes from non-elastic reactions. Although a whole range of potentials may give the same elastic scattering, that is generate wave functions with the same asymptotic form, these wave functions are different in the nuclear interior. Hence they may give different results when used in a calculation of some other reaction (*see* below, for example).

A qualitatively new feature, compared to nucleons, which is introduced here is the possibility of dissociation ('break up') of the projectile in the field of the target. This will be influenced by the strength with which the projectile is bound. Loosely bound projectiles like the deuteron ($d \to n + p - 2.23$ MeV), ^3He and triton ($^3He \to d + p - 5.5$ MeV, etc.), ^6Li ($^6Li \to \alpha + d - 1.5$ MeV) or ^9Be ($^9Be \to n + 2\alpha - 1.6$ MeV) are easily broken up in the weak field at some distance outside the target; consequently their absorptive or imaginary potentials extend to large radii (e.g. typically $R' \approx 1.7\,A^{1/3}$ fm for deuterons, $\approx 1.6\,A^{1/3}$ fm for ^3He if we use the volume absorption ($W_D = 0$) form of equation 4.32). On the other hand, the α-particle is tightly bound ($\alpha \to {}^3He + n - 20$ MeV, etc.) and can approach more closely before the field is sufficiently strong to cause its disintegration. For α-particles we find the imaginary potential has about the same radial extent as for a single nucleon plus the size of the α-particle itself.

The optical model has also been applied successfully to describe the scattering of heavy ions. The kind of ambiguity just discussed for α-particles is even more evident here; almost the only information obtained from the scattering is about the potential close to the strong absorption radius. However, the simple folding model of equation 4.28 does seem to reproduce the empirical potential in that region when realistic interactions v are used.

The physical significance of an optical potential for the interaction of two heavy ions of comparable mass is not clear. A prescription like that of equation 4.28 may be meaningful as long as the two density distributions ρ_p and ρ_t do not overlap appreciably. (Fortunately, there is very little overlap at the strong absorption radius which is typically about 3 fm larger than the sum of the radii of the two density distributions.) However, it must become less valid when the centres of the two nuclei approach more closely. The Pauli Principle acting on the nucleons forbids us to simply superpose the two nuclei at rest without highly exciting them. Further, the saturating property of nuclear forces (*see* Preston and Bhaduri, 1975, for example) ensures that this would require a considerable amount of work for the overlapping region would have a density of roughly twice the normal value. If the nuclei approach slowly or *adiabatically*, they will tend to readjust in shape so that their summed density remains about normal. In that case, their interaction potential energy does not simply depend on the coordinate r. If their energy is high so that the collision is *sudden*, abnormally high densities may be achieved momentarily but may be associated with a very positive potential energy. Attempts have been made to include such a repulsive 'core' in the optical potential, but results have so far been inconclusive because of the insensitivity of the scattering to this inner region.

Since analyses of elastic scattering data with optical potentials also yield values for the scattering amplitudes η_ℓ, we can deduce strong-absorption radii R_c. The results are similar to those obtained using the analytic forms for η_ℓ or the diffraction models discussed earlier.

4.6 THE MEANING OF A NUCLEAR RADIUS

The term 'nuclear radius' has been encountered many times in the preceding sections so it is appropriate at this point to review what this expression means. It must by now be evident to the reader that there is no single, unambiguous meaning. We have discussed nuclear radii R whose values range from $\approx 1.0 \times A^{1/3}$ to $\approx 1.5 \times A^{1/3}$ fm. All of these indicate in a rough way the 'size' of a nucleus, but clearly in detail they must be measuring different things.

We measure the size of a nucleus by using some probe, such as a beam of electrons, neutrons or ^{16}O ions, etc. The thing we can first extract from such an experiment is some measure of an *interaction region* within which the probe feels the target.

We may distinguish two kinds of situation. In one we have a long-range but relatively weak force, namely the Coulomb force. For this, the interaction region is of infinite extent (the Rutherford cross-section diverges at $\theta = 0$ and does not have a finite integral). However this result is trivial and well understood. What we are looking for are *deviations* from Rutherford scattering owing to the nucleus not being a point but having a finite size, and this, of course, does occur in a limited region. Because the force is weak, these deviations can be treated using perturbation theory in the way outlined in section 2.11 (*see* also Jackson, 1970; Uberall, 1971) in order to obtain almost directly the *charge distribution* of the nucleus. (This corresponds to unfolding the integral 4.22 when v is the Coulomb interaction potential, $v = \pm e^2/r_{tp}$.) This charge distribution can be characterised by a radius, such as the radius c at which the charge density falls to one-half of its value in the nuclear interior (*see* sections 1.7.2, 2.11, equation

2.33 and *Figure 2.14*). If the electron-scattering measurements only cover small momentum transfers, it may not be possible to deduce the shape of the charge distribution but only to obtain a measure of its ***mean square radius*** $\langle r^2 \rangle$ (*see* equation 2.32). This depends both upon the half-value radius c and the surface diffuseness a; for example, for the Fermi distribution, 2.33, we have

$$\langle r^2 \rangle_{ch} \approx \tfrac{3}{5} c^2 + \tfrac{7}{5} \pi^2 a^2 \tag{4.39}$$

Even then, the charge distribution ρ_{ch} is not the same as the distribution ρ_p of the centres of mass of the protons because each proton itself carries a charge distribution of finite extent with $\langle r^2 \rangle_{proton} \approx 0.7$ fm^2. The two are related by a folding integral like equation 4.22

$$\rho_{ch}(\mathbf{r}) = \int \rho_p(\mathbf{r}') \, \rho_{proton} \, (|\mathbf{r} - \mathbf{r}'|) \, d\mathbf{r}' \tag{4.40}$$

where ρ_{proton} is the charge distribution of an individual proton. The relation corresponding to equation 4.36 is

$$\langle r^2 \rangle_{ch} = \langle r^2 \rangle_p + \langle r^2 \rangle_{proton} \tag{4.41}$$

The main result of the folding 4.40 is that ρ_p and ρ_{ch} have similar half-value radii but ρ_{ch} has a more diffuse surface than ρ_p. The ρ_p distribution is of more fundamental interest because it is that which is given directly by the proton wave functions.

Electron scattering does not tell us about the distribution ρ_n of neutrons within a nucleus. For that we must rely upon the strong, specifically nuclear, interactions which primarily give information about the matter distribution $\rho_m = \rho_p + \rho_n$. If ρ_p is known independently, we can deduce ρ_n. (Some measurements, such as those resulting in equation 4.37, can be used to give information on the *difference* $\rho_n - \rho_p$; we shall not discuss these here but they have been reviewed elsewhere (Satchler, 1969).)

The other kind of situation occurs when the interaction used to probe the nucleus is strong but has a *short range*, which means the specifically nuclear force. Then the interaction region within which the probe feels the nucleus is of finite extent and has a fairly well-defined boundary. It is this boundary and its radius which is involved in strong absorption scattering. This is also the interaction region which is represented in more detail by the optical potential, and which may be approximately related to the matter distribution through expressions like equation 4.22 or 4.28.

The strong absorption radius essentially measures the distance of approach at which contact is made and non-elastic events begin to be important. (The various prescriptions for extracting this radius from either the diffraction models or from the η_ℓ themselves differ in detail, corresponding to slightly different criteria as to what is meant by 'important'.) Because nuclei have diffuse edges and because of the finite range of the force, these radii are much larger than the radii of the density distributions. *Figure 4.22* illustrates this by showing the matter-density distributions along the line of centres for $^{16}O + {}^{208}Pb$ when their

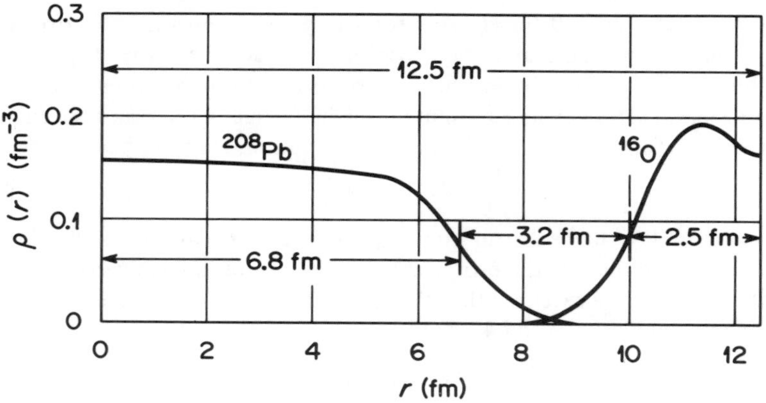

Figure 4.22 The matter-density distributions for the nuclei $^{16}O + ^{208}Pb$ along the line joining their centres when the centres are separated by the strong absorption radius, here about 12.5 fm

centres are separated by the strong absorption radius. Nuclei making this kind of grazing collision have a 50% probability of inducing a non-elastic event, i.e. of being 'absorbed'. We see that the points at which their individual densities have fallen to one-half their central values are separated by about 3 fm; this is typical of heavy-ion collisions whose strong absorption radii are given by $R \approx R_{m1} + R_{m2} + 3$ fm, where R_m is the half-density radius, $R_m \approx 1.1\, A^{1/3}$. This relation also reproduces approximately the strong absorption radii for lighter ions such as α-particles; the differences between this expression and the alternative form, equation 4.6, are not large. In any case, this radius is not a characteristic of a single nucleus but of a particular colliding pair; it is a measure of the sum of two nuclear matter radii with corrections for diffuse edges and forces of finite range.

The radius of an optical potential (R in equation 4.32) is the radius at which the real potential has one-half the depth it has in the interior (at small r values). It does not bear any simple relation to a strong absorption radius; indeed we may use an optical potential when the absorption is weak and there is no such radius. The potential radius may be related to the radius of the nuclear density distribution if simple formulae like 4.22 and 4.28 are valid.

The folding relation, 4.22, is approximately correct for nucleons scattering from nuclei. It predicts a potential radius which is approximately equal to the matter radius (although the diffuseness is greater). Also the mean square radii of the potential and density are related by equation 4.36. Now the mean square radius $\langle r^2 \rangle_U$ of the potential is well determined by the scattering measurements so that we can deduce $\langle r^2 \rangle_\rho$ for the density ρ if the $\langle r^2 \rangle_v$ of the interaction v is known. The radius of an empirical nucleon optical potential, $\approx 1.2 \times A^{1/3}$ fm, is close to but a little larger than the radius of the proton distribution, $\approx 1.1\, A^{1/3}$ fm, deduced from electron scattering. It is believed this is not due primarily to the neutron distribution having a larger radius than the proton one, but rather due to corrections to the model which make equations 4.22 and 4.36 no longer exactly correct. In particular, the effective interaction ν depends upon the density of the target nucleus in such a way that ν is stronger in the low-density nuclear surface than in the higher-density interior region.

The radius of an optical potential for strongly absorbed composite particles has much less physical significance. As we have seen, the measured scattering mainly determines the potential near the strong absorption radius and its insensitivity to the potential in the interior region means that an unambiguous *potential* radius often cannot be determined. In any case, as already discussed, the nuclei must become polarised in each other's field for these close impacts and the simple formula, 4.38, can no longer be used.

An interesting comparison of the radii determined in these various ways from analysis of some α-particle scattering has been made by Fernandez and Blair (1970).

While all this may seem confusing, we can summarise briefly by distinguishing three important nuclear radii. The first is the radius of the matter distribution of a nucleus, which falls to one-half its central value at approximately

$$R_{\mathrm{m}} \approx 1.1\, A^{1/3} \text{ fm} \tag{4.42}$$

The second is the radius at which the depth of the optical potential for a nucleon incident upon a nucleus falls to one-half of its central value

$$R_U \approx 1.2\, A^{1/3} \text{ fm} \tag{4.43}$$

Finally there is the strong absorption radius for the collision of two composite nuclei, the distance of approach at which there is a 50% probability of a non-elastic event occurring

$$R_{\mathrm{SA}} \approx R_{\mathrm{m1}} + R_{\mathrm{m2}} + 3 \text{ fm} \tag{4.44}$$

or

$$\approx (1.4\text{–}1.5) \times (A_1^{1/3} + A_2^{1/3}) \text{ fm} \tag{4.45}$$

This is a characteristic not of a single nucleus but of a colliding pair.

4.7 DIRECT REACTIONS

Direct reactions and compound nucleus reactions were contrasted in section 2.18.1, and the importance of the former was discussed in section 2.18.6. Three types of direct reaction were mentioned: 'collective' inelastic scattering, stripping (or pick-up) and knock-out. We describe here in more detail some models of these three types of reaction. A treatment of inelastic scattering in the strong absorption approximation has already been given (section 4.4). Analogous descriptions can be made for other reactions, such as stripping, when their angular distributions exhibit similar diffraction-like characteristics. However, the most widely used and perhaps the most successful descriptions are based upon potential models such as we shall describe here. (See also Satchler, 1983.)

4.7.1 A semi-classical model

Some of the underlying physical ideas can be understood from simple, semi-classical, considerations (Butler *et al.*, 1958). Because this approach is very crude, we shall only discuss it briefly and in its simplest form.

If there is strong absorption, particles which emerge after only one collision with a target nucleon must have had that collision in the nuclear surface. Deeper penetration would have led to formation of a compound nucleus, so we consider collisions confined to a spherical surface of radius R. In general, the angular momentum or spin I_A of the target nucleus will change because of the non-elastic collision. Let this change be denoted L, so that

$$\mathbf{L} = \mathbf{I}_B - \mathbf{I}_A, \quad \text{with } |I_B - I_A| \leqslant L \leqslant I_B + I_A \tag{4.46}$$

where A, B denote initial and final nuclei, respectively. To conserve angular momentum, this must be balanced by a change in the relative orbital angular momentum of the colliding pair (we ignore the spin, if any, of the projectile and ejectile*). Now if the pair are separated by a distance $\mathbf{r}$ and their relative momentum is $\hbar\mathbf{k}$, their orbital angular momentum is $\mathbf{L} = \mathbf{r} \times \mathbf{k}$ (in units of $\hbar$). Hence the angular momentum $\mathbf{L}$ transferred to the target must be

$$\mathbf{L} = (\mathbf{k}_\alpha - \mathbf{k}_\beta) \times \mathbf{R} = \mathbf{q} \times \mathbf{R} \text{ say} \tag{4.47}$$

where $\mathbf{R}$ is the point on the sphere at which the interaction took place (*see Figure 4.23*) and $\mathbf{k}_\alpha$, $\mathbf{k}_\beta$ are the initial and final relative momenta, respectively.

Now consider the excitation of a definite state in the final nucleus; energy conservation then determines the magnitude of $\mathbf{k}_\beta$. The direction of $\mathbf{k}_\beta$ defines the scattering angle, θ. For a fixed angle of observation θ (and hence fixed magnitude for $\mathbf{q}$), equation 4.47 is only satisfied along two rings of radius L/q on opposite sides of the surface of the sphere (*see Figure 4.23*). Then particles observed in this direction θ which have transferred L units to the target have originated from collisions occurring on these two rings.

In order to remain on the sphere, the ring radius must be less than R, or $L/q \leqslant R$. This may put a limit on the angle of observation. For example, in many cases of interest (e.g. inelastic scattering to low-lying levels) $\mathbf{k}_\alpha$ and $\mathbf{k}_\beta$ have similar magnitudes. Then, if θ is small, we may put $q \approx \bar{k}\,\theta$, where $\bar{k} = \frac{1}{2}(k_\alpha + k_\beta)$. The restriction becomes

$$\theta \geqslant L/\bar{k}R \tag{4.48}$$

*Then the transfer L also determines any change in parity of the nucleus. If the target A has parity π_A (where $\pi = \pm 1$) and the residual nucleus B has parity π_B, then $\pi_A \pi_B = (-)^L$.

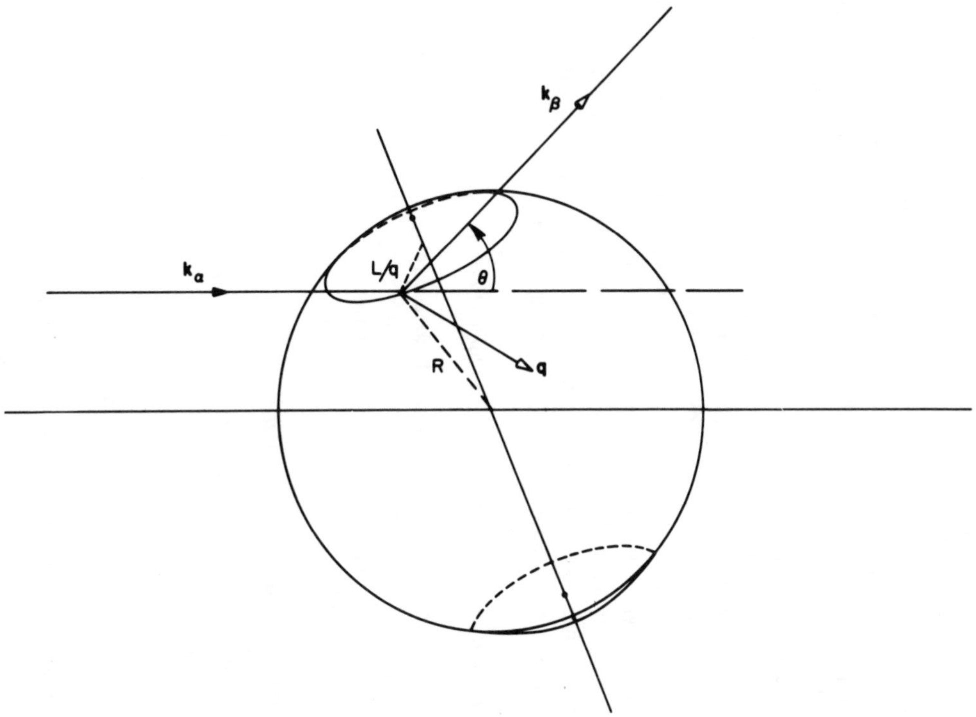

Figure 4.23 Illustrating the active rings, of radius L/q, on opposite sides of the nucleus, from which particles observed in the direction θ may originate, classically, after transfer of L units of angular momentum. (After Butler *et al.*, 1958)

that is, for transfers $L \neq 0$ to occur, scattering through an angle of at least $L/\bar{k}R$ must take place. (These considerations also imply that we must have $L \leqslant 2\bar{k}R$.)

Further, we can expect interference between the reaction-product waves emitted from the two active rings on opposite sides of the nucleus. When the radius of the rings is small, the average phase difference between the two waves is roughly $2\bar{k}R\theta$. This should be $2n$ for constructive interference, with n integer, so we expect a diffraction pattern with maxima at angles θ for which $\bar{k}R\theta = n\pi$ (just as for diffraction at two slits separated by $2R$).

Consequently, we can expect a differential cross-section with an angular distribution showing a diffraction pattern with peaks separated in angle by roughly $\pi/\bar{k}R$. Further, the first peak will occur at an angle of order $L/\bar{k}R$ when L units of angular momentum are transferred. This last feature provides us with a means of measuring the angular momentum change in a direct reaction, and hence deducing from equation 4.46 something about the spin I_{B} of the residual nuclear state if we know the target spin I_{A}.

We note that these general features were deduced simply from the assumption of strong absorption which confined the reactions to the nuclear surface. We did not need to specify the type of reaction, whether it was rearrangement or simply inelastic scattering. Hence the general characteristics of the elastic and inelastic scattering of strongly absorbed particles discussed in sections 4.1–4.4 are seen to be applicable to a much wider range of reactions also. When the absorption is weaker, there can be contributions from the nuclear interior also, although almost always there is still an emphasis on the surface region.

4.7.2 Perturbation theory and the Born approximations

In order to give a more detailed account of surface reactions, as well as to describe reactions involving particles which are not strongly absorbed, it is helpful to use potentials and to invoke the help of perturbation theory. First we deduce some qualitative results.

4.7.2.1 Plane-wave Born approximation

It follows from section 3.3.3 that if we write the differential cross-section for the A(a, b)B reaction as

$$\frac{\mathrm{d}\sigma_\beta}{\mathrm{d}\Omega} = \frac{\mu_\alpha \mu_\beta}{(2\pi\hbar^2)^2} \frac{k_\beta}{k_\alpha} \, |T|^2 \tag{4.49}$$

then the first Born approximation to the transition amplitude T is

$$T^{(\mathrm{BA})} = \int \psi_{\mathrm{B}}^* \psi_{\mathrm{b}}^* \exp(-i\mathbf{k}_\beta \cdot \mathbf{r}_\beta) \, V \, \psi_{\mathrm{A}} \psi_{\mathrm{a}} \exp(i\mathbf{k}_\alpha \cdot \mathbf{r}_\alpha) \, \mathrm{d}\tau \tag{4.50}$$

In this expression, ψ_i is the wave function which describes the internal state of nucleus i, V is the potential of interaction between the colliding pair and τ represents all the variables to be integrated over. Their relative motion before and after the collision has been approximated by plane waves. When the matrix element 4.50 does not vanish, we say we have a direct reaction.

Now nuclear forces have a short range, so let us suppose that it is a reasonable approximation to take $\mathbf{r}_\alpha \approx \mathbf{r}_\beta$, = $\mathbf{r}$ say, in equation 4.50. Rearranging, we can then write

$$T^{(\mathrm{BA})} \approx \int \mathrm{d}\mathbf{r} \exp(i\mathbf{q}\cdot\mathbf{r}) \left\{ \int \psi_\mathrm{B}^* \psi_\mathrm{b}^* \, V \, \psi_\mathrm{A} \psi_\mathrm{a} \, \mathrm{d}\tau' \right\}, \quad \mathbf{q} = \mathbf{k}_\alpha - \mathbf{k}_\beta$$

$$= \int \mathrm{d}\mathbf{r} \exp(i\mathbf{q}\cdot\mathbf{r}) \, F(\mathbf{r}), \text{ say} \tag{4.51}$$

where we have separated the variables τ into $\mathbf{r}$ and the remainder, τ'. The integral over τ' in braces (which may also be written $\langle \mathrm{b}, \mathrm{B} | V | \mathrm{a}, \mathrm{A} \rangle$ as in equation 3.30) remains a function of $\mathbf{r}$, $F(\mathbf{r})$ say, which we may call the form factor for the reaction. The partial wave expansion, equation 3.35 of section 3.4, may be introduced to give

$$T^{(\mathrm{BA})} \approx \sum_{L=0}^{\infty} i^L \, [4\pi(2L+1)]^{1/2} \int j_L(qr) \, Y_L^0 \, (\theta, \phi) \, F(\mathbf{r}) \, \mathrm{d}\mathbf{r} \tag{4.52}$$

In general $F(\mathbf{r})$ will not be spherically symmetric because the various internal functions ψ_i may refer to nuclei with non-zero spins. This means that $L \neq 0$ terms may survive the integration in equation 4.52 and there may be more than one of them. Indeed the L here can be identified as the angular momentum transfer defined in equation 4.46 and all the values allowed by that equation may contribute to T. Usually, however, only one or sometimes two values of L are of importance for a given transition. For example, if the target, projectile and ejectile all have zero spin, as for inelastic α-scattering from even nuclei, then L is limited to one value, $L = I_\mathrm{B}$. In other cases, selection rules imposed by the internal structure of the nuclei involved may limit L. Further, since the parity change is given by $(-)^L$, only L values which are all even or all odd may participate.

The expression for T may be further simplified if we argue that the use of plane waves in equation 4.50 is a fair approximation outside the nucleus, but that these wave functions should be put to zero inside the nucleus (because of absorption). Further, the form factor $F(\mathbf{r})$, being a nuclear property, falls off rapidly beyond the nuclear surface, so that the contributions to the integral should be confined to radii $r \approx R$ at the nuclear surface. This leads us to evaluate equation 4.52 at $r = R$ so that it has the form

$$T^{(\mathrm{BA})} \approx \sum_L C_L j_L(qR) \tag{4.53}$$

where the constants C_L determine the contributions from the various L values. When only one L is important, we see that the differential cross-section is proportional to the square of the Bessel function

$$\frac{d\sigma^{(BA)}}{d\Omega} \propto [j_L(qR)]^2 \tag{4.54}$$

The dependence on scattering angle θ is contained within q, since

$$q^2 = k_\alpha^2 + k_\beta^2 - 2k_\alpha k_\beta \cos\theta \tag{4.55}$$

Remembering the oscillatory nature of the spherical Bessel functions (*see Figure 3.3*), we see we have regained the diffraction-like angular distributions by yet another route. The formula 4.54 contains the characteristics discussed in the previous section and elsewhere; the first (and major) peak occurs at an angle for which qR is a little greater than L and the peaks are separated by those angles which increment qR by π.

In applying equation 4.54, the radius R was usually adjusted to give a good fit to the measured distribution. It was generally found to be much larger than the radius of the density distribution. Indeed, it is clear from our derivation that R here is closely related to the strong absorption radius discussed earlier.

4.7.2.2 Distorted-wave Born approximation

The arguments leading to equations 4.53, 4.54 were inconsistent in that strong absorption was invoked to eliminate the nuclear interior, $r < R$, but the motion at larger radii was described by undisturbed plane waves. Now we know we cannot have absorption without also having elastic scattering; a plane wave with a hole punched in it is not a solution to any Schrödinger equation. However, both phenomena are included in the generalisation of equation 4.50 called the distorted-wave Born approximation (*see* section 3.3.3), usually abbreviated to DWBA. As its name implies, the relative motions before and after the non-elastic event are described not by plane waves but those waves distorted by elastic scattering and its accompanying absorption. Hence this approximation is likely to be valid when the most important single event to take place when two nuclei collide is elastic scattering. Then other reactions can be treated as perturbations, as weak transitions between elastic scattering states. This condition is often satisfied.

Each of the $\exp(i\mathbf{k}_i \cdot \mathbf{r}_i)$ in equation 4.50 is then replaced by a function $\chi(\mathbf{k}_i, \mathbf{r}_i)$ which has the asymptotic form of equations 3.5, 3.23, 3.39, etc.

$$\chi(\mathbf{k}, \mathbf{r}) \rightarrow \exp(i\mathbf{k}\cdot\mathbf{r}) + \psi_{\text{scatt}} \tag{4.56}$$

(Only *elastic* scattering is included in ψ_{scatt}.) These distorted waves χ are usually generated by solving Schrödinger's equation with an optical potential which has been obtained by fitting measurements of the corresponding elastic scattering cross-sections. Then we have the DWBA for the transition amplitude

$$T^{(\text{DWBA})} = \iint \chi^{(-)}(\mathbf{k}_\beta, \mathbf{r}_\beta)^* \langle \text{b, B} | V | \text{a, A} \rangle \chi^{(+)}(\mathbf{k}_\alpha, \mathbf{r}_\alpha)\, d\mathbf{r}_\alpha d\mathbf{r}_\beta \tag{4.57}$$

instead of equation 4.50. In general the nuclear matrix element $\langle \text{b, B} | V | \text{a, A} \rangle$ is a function of both $\mathbf{r}_\alpha$ and $\mathbf{r}_\beta$, but just as before if we can put $\mathbf{r}_\alpha \approx \mathbf{r}_\beta = \mathbf{r}$ say, we can write it $F(\mathbf{r})$ and the generalisation of equation 4.51 is

$$\mathcal{T}^{(\mathrm{DWBA})} = \int \chi^{(-)}(\mathbf{k}_\beta, \mathbf{r})\, F(\mathbf{r})\, \chi^{(+)}(\mathbf{k}_\alpha, \mathbf{r})\, \mathrm{d}\mathbf{r} \tag{4.58}$$

Simple analytic expressions like equations 4.52 or 4.53 are now difficult to obtain; for example the distorted waves cannot be combined in the simple way that led from equation 4.50 to equation 4.51. In practice, usually each distorted wave is expanded into partial waves as in equation 3.34 and an analogous multipole expansion made for the form factor $F(\mathbf{r})$. The integration over the directions of $\mathbf{r}$ may then be done analytically, leaving a sum of partial-wave radial integrals. Each of these is proportional to the corresponding partial-wave scattering matrix element, as in equation 3.49 or 3.91, for example. Usually one has to resort to a high-speed computer in order to calculate these radial integrals and hence the amplitudes 4.57 or 4.58 and the cross-sections. However, what has been lost in

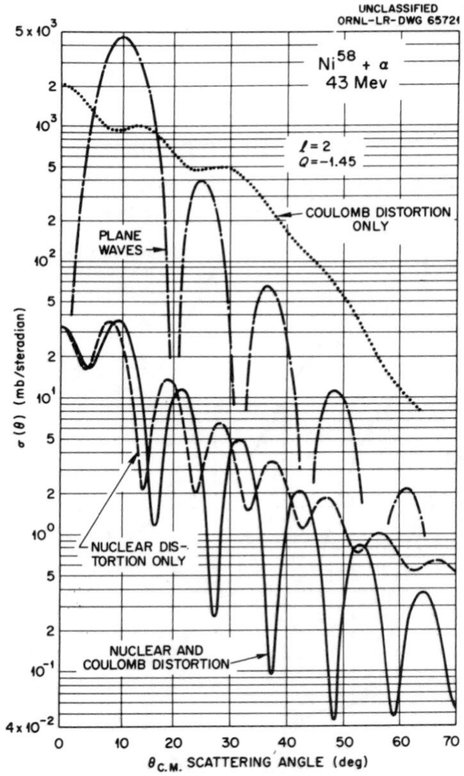

Figure 4.24 Comparisons of the inelastic cross-sections predicted using plane-wave and distorted-wave theories. Curves are given showing the effects of distortion by the Coulomb field alone and the nuclear potential alone, as well as their combined effects. The measured cross-sections are in close agreement with the curve including nuclear and Coulomb distortion effects. (From Bassel *et al.*, 1962)

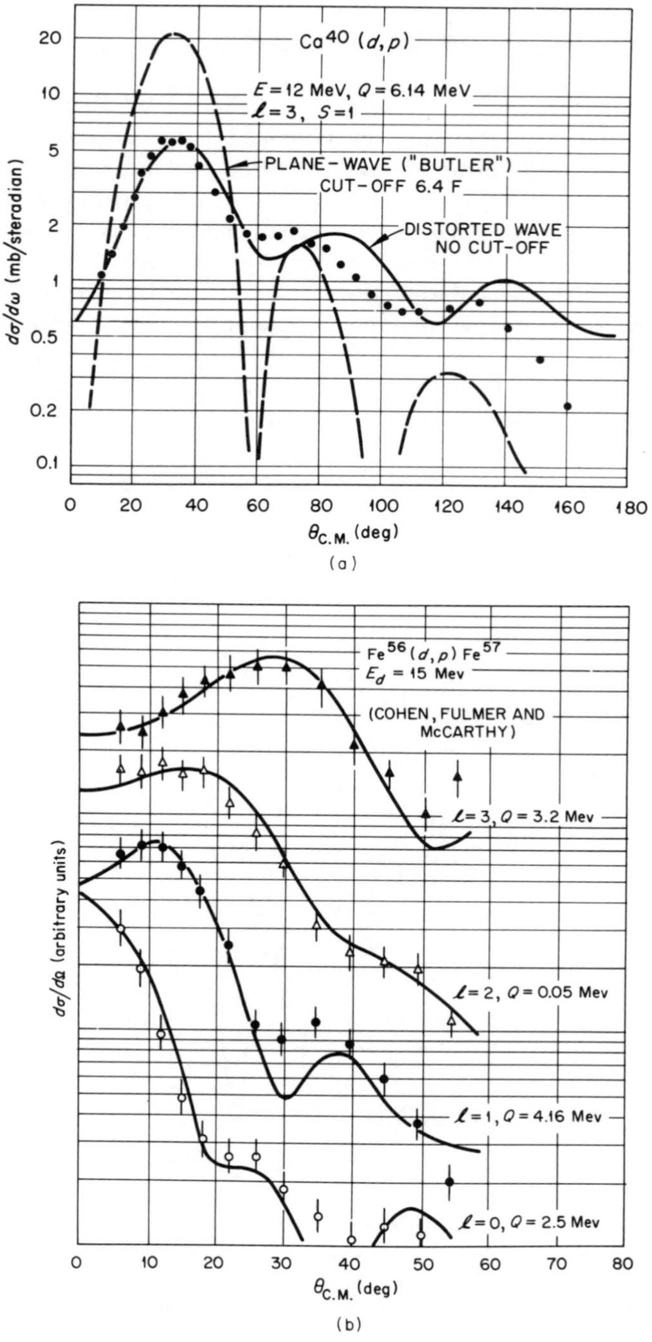

Ca^{40} (d,p)
E = 12 MeV, Q = 6.14 MeV
ℓ = 3, S = 1
PLANE-WAVE ("BUTLER") CUT-OFF 6.4 F
DISTORTED WAVE NO CUT-OFF
dσ/dω (mb/steradian)
θ$_{C.M.}$ (deg)
(a)
Fe^{56} (d,p) Fe^{57}
E$_d$ = 15 Mev
(COHEN, FULMER AND McCARTHY)
ℓ = 3, Q = 3.2 Mev
ℓ = 2, Q = 0.05 Mev
ℓ = 1, Q = 4.16 Mev
ℓ = 0, Q = 2.5 Mev
dσ/dΩ (arbitrary units)
θ$_{C.M.}$ (deg)
(b)

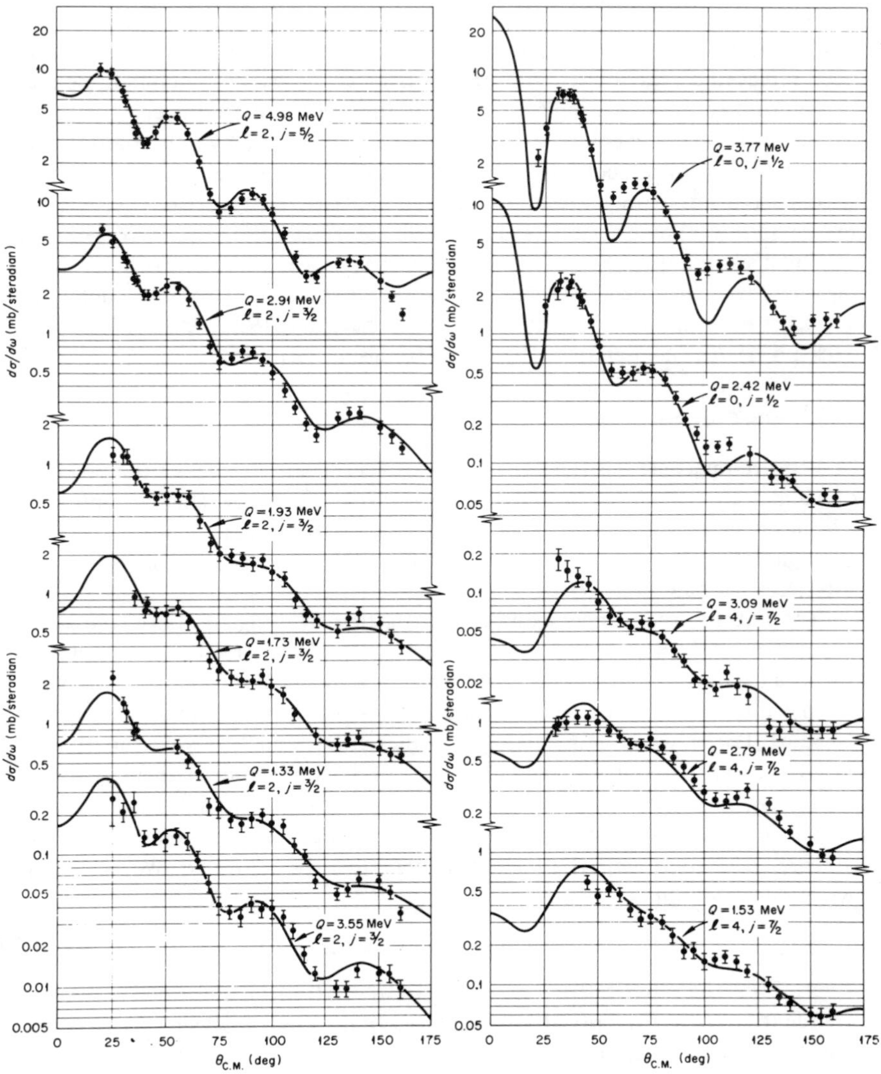

(c)

Figure 4.25 Calculated cross-sections for some (d, p) reactions in comparison with measured values. (a) Comparison of the plane-wave theory of Butler with the distorted-wave theory for one case (from Lee *et al.*, 1964). (b) Illustrating the changes in the angle of the peak cross-section which allow the ℓ of the transferred neutron to be identified (from Cohen *et al.*, 1962). (c) Distorted-wave Born approximation fits to proton angular distributions from ^{90}Zr (d, p) reactions with 12-MeV deuterons to various excited levels of ^{91}Zr with transfers of ℓ = 0–4 (from Silva and Gordon, 1964)

simplicity is regained in the accuracy with which the reaction can be described (*see Figures 4.24, 4.25* for examples). Although the simple plane waves have been replaced by the more complicated distorted waves, in principle these may be determined by measuring the corresponding elastic scattering cross-sections. Then various models of nuclear structure may be tested by constructing the corresponding form factors $F(\mathbf{r})$ (which contain all the structure effects in this theory) and comparing the consequent non-elastic cross-sections with the measured ones. Some specific examples are described in the following sections.

The distorted-wave Born approximation has proved to be one of the most powerful tools for the analysis of nuclear direct reactions. (For more detailed accounts of its application, *see*, for example, Satchler, 1983; Austern, 1970; Jackson, 1970; Glendenning, 1983; also Hodgson, 1971.)

4.7.3 Inelastic scattering

The emitted particle b is the same as the projectile a if we neglect the possibility of exchange between the projectile and the target nucleons. Then we automatically have the coordinates $\mathbf{r}_\alpha = \mathbf{r}_\beta$ in the integral 4.50 or its distorted-wave analogue 4.58. The role of a model is now to define the interaction V and the internal wave functions ψ_i.

The states most strongly excited by inelastic scattering are those recognised as involving collective motion of some kind, vibrational or rotational. Indeed, this has become one of the ways of identifying such states. It suggests that we invoke the collective model (*see* section 1.7.4 and, for example, Preston and Bhaduri, 1975), which immediately provides us with nuclear wave functions ψ_i. For V we may be guided by equation 4.22 or 4.28. If the target density distribution is not spherical, then the corresponding optical potential will not be either. It will follow the density distribution in vibrating in shape about a spherical average or in rotating if the nucleus is permanently deformed. We may then write the potential as

$$U(\mathbf{r}) = U_0(r) + \Delta U(\mathbf{r}) \tag{4.59}$$

where U_0 is the spherically symmetric part and ΔU is the deformed remainder. We identify U_0 as the optical potential we obtain when analysing ***elastic*** scattering data, and treat ΔU as a perturbation which can cause inelastic scattering so that $V = \Delta U$ in equation 4.51 or 4.57. The latter can be expanded in terms of the orthonormal spherical harmonics Y_ℓ^m (Appendix A)

$$\Delta U(\mathbf{r}) = \sum_{L,M} V_{LM}(r)\, Y_L^M(\theta, \phi) \tag{4.60}$$

Then the term with L, M can induce the transfer of angular momentum L (with z-component M) to the nucleus in an inelastic collision. Further, the parity change is $(-)^L$. (This means that transitions involving a so-called 'unnatural' parity change of $(-)^{L+1}$ cannot be excited in this model. They require a spin-dependent interaction.)

It is worth noting that the interaction 4.60 is similar in form to the operator which induces the emission or absorption of electric multipole radiation, except that the latter has a radial dependence proportional to r^L. Indeed we find experimentally a close parallel between inelastic scattering cross-sections and the corresponding EL transition rates.*

A simple way to obtain the expansion 4.60 is to assume that $U(r)$ depends upon a nuclear radius R (as, for example, in equation 4.32) and allow the surface $r = R(\theta, \phi)$ to be non-spherical as in equation 4.19. We then expand U as a Taylor series in the deformation lengths δ_{LM}

$$U(\mathbf{r}, R) = U_0(r, R_0) + \frac{\partial U_0(r, R_0)}{\partial R_0} \sum_{LM} \delta_{LM} Y_L^M(\theta, \phi) + \ldots \quad (4.61)$$

The second term is just ΔU to first order in the δ_{LM}; then we see that in this model the shape of the radial factors $V_{LM}(r)$ is the same for all L, M. If U_0 is assumed to be a function of $(r - R_0)$, as in equation 4.33, then $\partial U_0/\partial R_0 = -\partial U_0/\partial r$ and the $V_{LM}(r)$ are peaked at the nuclear surface (*see* $g(r)$ as defined by equation 4.34 and shown in *Figure 4.18*). Further, once we have obtained U_0 from the values of the parameters of an analysis of elastic scattering measurements, then we know the $V_{LM}(r)$ aside from the deformation length, δ_{LM}. The matrix element of δ_{LM} is a characteristic of a given transition and is the quantity we wish to measure. For a rotational nucleus, it is R_0 times the magnitude β_L of the 2^L-pole deformation of the shape of the potential. Since it appears just as a multiplicative constant in the transition amplitude if we use the Born or distorted-wave approximations, we can obtain its value by matching the magnitudes of the calculated and measured cross-sections.

The fact that, in this model, the interaction is peaked at the nuclear surface might seem to lend support to the surface approximation, 4.53, even in the absence of strong absorption (such as for the scattering of nucleons). However, the radius R_0 appearing here is the radius R_0 of the *potential* U. A radius larger than R_0 is needed in the plane-wave Born approximation expression 4.54 in order to obtain cross-section peaks at the correct angles. This is borne out by

*When the colliding pair are both charged, there is also a Coulomb potential acting between them. This will also be non-spherical if one or both of their charge distributions is non-spherical; consequently there will be a 2^L-pole electric potential term to be added to V_{LM} in equation 4.60. The action of this term is called *Coulomb excitation* (*see* Biedenharn and Brussaard, 1965; Alder and Winther, 1975; McGowan and Stelson, 1974). The Coulomb potential has a long range; the 2^L-pole term experienced by the projectile decreases slowly like $1/r^{L+1}$ compared to the exponential fall-off of the specifically nuclear interaction. Hence if the bombarding energy is below the Coulomb barrier (section 2.18.8 and *Figure 2.35*), the two particles do not approach closely enough for the short-range nuclear forces to act so that Coulomb excitation is the only process possible. Measurements of Coulomb excitation are an important source of information on electric transition rates. As the bombarding energy is increased, the Coulomb and nuclear terms become of comparable importance and then the interference between their contributions can help to elucidate the nuclear interaction (*see Figure 4.9*, for example).

Figure 4.24 which shows a comparison of calculations based on both theories with measurements on the scattering of α-particles. The radius needed in the plane wave theory is an effective interaction radius similar to the strong absorption radius. Even when there is weak absorption, a large value of R is required to represent implicitly the refraction, diffraction and reflection of the incident and emitted waves. The more complete distorted-wave theory includes these effects explicitly and gives very much better agreement with the measurements without the use of an arbitrarily adjusted radius R.

In principle, the deformed potential model just described is valid only for 'collective' transitions, even though it does give results in reasonable agreement with the measurements on many weak or non-collective transitions. It would be more satisfying to start from the nucleon–nucleon forces and use more detailed wave functions of the shell model type to describe the initial and final internal nuclear states. This approach has had many successes (*see*, for example, Satchler, 1983; Glendenning, 1983; Madsen, 1974). However, the simple deformed potential plus collective model remains a very useful way of representing the results of measurement.

4.7.4 Stripping and pick-up reactions

The prototypes of these reactions are the deuteron stripping reactions (d, p) and (d, n) and their pick-up inverses (p, d) and (n, d); we shall discuss these first.

4.7.4.1 Deuteron stripping and pick-up

As illustrated in *Figure 2.28*, one of the nucleons may be stripped from the deuteron as it makes a peripheral collision with the target while the other nucleon proceeds more or less undisturbed. Hence we expect to see the emergent nucleon predominantly in the forward direction. Further, the angular momentum L transferred to the target is carried by the captured nucleon.

Of course, because of detailed balance (sections 2.17, 3.8.4), the inverse pick-up reactions, (p, d) and (n, d), are described by the same amplitudes. The physics involved appears more transparent perhaps if we discuss the pick-up process. Consider the A(p, d)B reaction for definiteness; then d = n + p and A = B + n, and the neutron is transferred. The proton picks up the neutron through their mutual interaction V_{pn}. The transition amplitude $T_{\text{p,d}}$ is given by equation 4.50 or its distorted-wave analogue 4.57, with a → p, b → d and $V = V_{\text{pn}}$. Then $\psi_{\text{a}} \rightarrow \psi_{\text{p}}$ just refers to the spin of the proton, which we shall ignore here, $\psi_{\text{b}} \rightarrow \psi_{\text{d}}(\mathbf{r}_{\text{n}} - \mathbf{r}_{\text{p}})$ is the wave function for the deuteron in its ground state and $\psi_{\text{B}}(\xi)$ is the wave function of the residual nucleus in a given state. (Here ξ represents all the internal variables of nucleus B.) The simplest case occurs when the target A consists of doubly closed shells, plus a neutron in a definite orbit. Then its wave function ψ_{A} can be factored*

$$\psi_{\text{A}}(\xi, \mathbf{r}_{\text{n}}) = \psi_{\text{B}}(\xi)\, \phi_{\text{n}}(\mathbf{r}_{\text{n}}) \tag{4.62}$$

*For simplicity, we are ignoring here the antisymmetrisation between this neutron and the indistinguishable neutrons in the closed shells.

Here ϕ_n is the wave function for the shell model single-particle orbit which the neutron occupies. Removal of this neutron will simply leave the closed-shell nucleus B in its ground state. Since the interaction V_{pn} does not depend upon ξ, the internal coordinates of B, these may be integrated over immediately

$$\int \psi_B^*(\xi)\psi_A(\xi, r_n)d\xi = \phi_n(r_n) \tag{4.63}$$

The position of the centre of mass of the deuteron is $\mathbf{r}_d = \frac{1}{2}(\mathbf{r}_p + \mathbf{r}_n)$. Since the neutron–proton interaction potential $V_{pn}(\mathbf{r}_p - \mathbf{r}_n)$ has a short range, it is reasonable to put $\mathbf{r}_p \approx \mathbf{r}_n$ when evaluating the transition amplitude; i.e. we use

$$V_{pn}(\mathbf{r}_p - \mathbf{r}_n)\psi_d(\mathbf{r}_p - \mathbf{r}_n) \approx D_0\delta(\mathbf{r}_p - \mathbf{r}_n) \tag{4.64}$$

where $\delta(\mathbf{x})$ is the Dirac delta function. This is called the *zero-range approximation*. The plane-wave Born approximation 4.50 then becomes

$$T_{p,d}^{(BA)} \simeq D_0 \int \exp(-i\mathbf{q}\cdot\mathbf{r}_n)\,\phi_n(\mathbf{r}_n)d\mathbf{r}_n \tag{4.65}$$

with $\mathbf{q} = \mathbf{k}_d - \mathbf{k}_p$. Comparison with equation 4.51 shows that the neutron wave function ϕ_n plays the role of the form factor $F(\mathbf{r})$.

There is a simple interpretation of equation 4.65 for $T_{p,d}^{(BA)}$. The proton enters with momentum $\mathbf{k}_p$ (in units of $\hbar$) and the deuteron leaves with $\mathbf{k}_d$. The difference $\mathbf{q} = \mathbf{k}_d - \mathbf{k}_p$ is carried into the deuteron by the picked-up neutron. Thus the proton had to find a neutron in the target with this momentum $\mathbf{q}$. The probability amplitude of finding the neutron with $\mathbf{q}$ when its wave function is ϕ_n is just the Fourier transform of ϕ_n evaluated at $\mathbf{q}$ (Messiah, 1962) which is what we recognise equation 4.65 to be.

If we had not made the approximation $\mathbf{r}_p \approx \mathbf{r}_n$, an additional Fourier transform would have appeared as a factor in $T_{p,d}^{(BA)}$, namely

$$\int \exp(-i\mathbf{K}\cdot\mathbf{s})\, V_{pn}(\mathbf{s})\psi_d^*(\mathbf{s})d\mathbf{s} \tag{4.66}$$

where $\mathbf{s} = \mathbf{r}_p - \mathbf{r}_n$ and $\mathbf{K} = \frac{1}{2}\mathbf{k}_d - \mathbf{k}_p$. This also has a simple interpretation. The proton initially had momentum $\mathbf{k}_p$, the neutron had $\mathbf{q} = \mathbf{k}_d - \mathbf{k}_p$. In the outgoing deuteron, they each have a momentum $\frac{1}{2}\mathbf{k}_d$ owing to the motion of the centre of mass of the deuteron. The remainder, $\pm(\frac{1}{2}\mathbf{k}_d - \mathbf{k}_p)$, is partly to be found as internal motion within the deuteron (described by ψ_d) and partly is transferred back and forth between the neuteron and proton via their mutual interaction, V_{pn}.

Returning now to equation 4.65, suppose that the single-particle orbital ϕ_n has definite angular momentum, L say. It can be factored into radial and angular parts

$$\phi_n(\mathbf{r}_n) = u_{nL}(r_n)\, Y_L^M(\theta_n, \phi_n) \tag{4.67}$$

If we use the partial wave expansion given in equation 4.52, the orthonormality of the spherical harmonics (Appendix A) immediately gives

$$T_{p,d} \approx \text{constant} \times \int j_L(qr_n)\, u_{nL}(r_n)\, r_n^2\, dr_n \tag{4.68}$$

Hence the characteristic L is here the angular momentum of the orbit which the neutron occupied before being picked up. Making the surface approximation and replacing the integral by its value on the surface, $r_n = R$, again returns us to the simple forms 4.53 and 4.54.*

Other states of the nucleus B can be reached by picking up a neutron from the closed shells of A. The more general case in which A does not simply consist of closed shells plus a single neutron, so that the wave function ψ_A cannot be factored as in equation 4.62, can be treated in a similar fashion. The result is the same except for an additional *spectroscopic factor* multiplying the expression for the cross-section. The spectroscopic factor S_L is the probability of finding the target nucleus A in the configuration represented by the right side of equation 4.62, namely the residual nucleus B in the observed state plus a neutron in the particular shell model orbital denoted by ϕ_n. Since the cross-section is proportional to this S_L, its value can be determined experimentally by comparing the calculated and measured cross-sections. This provides a stringent test of the validity of shell model wave functions and is one of the main reasons for the importance of stripping and pick-up reactions as a source of information about nuclear structure. The other reason is that the measured angular distribution allows one to identify the L value. This not only gives information on the spin and parity of the residual nuclear state through the relations 4.46 but tells us which neutron orbital is involved (*see*, for example, Macfarlane and Schiffer, 1974).

Again, the more sophisticated distorted-wave approximation gives much better agreement with measured differential cross-sections than does the use of plane waves. The factorisation 4.62 or its generalisation are used in equation 4.57 and often the zero-range approximation 4.64 is made to simplify the calculation. *Figure 4.25* gives some examples; other examples were shown in *Figure 2.20*. The plane-wave or 'Butler' theory gives peaks at approximately the correct angles if the cut-off radius R is chosen judiciously (again a large value of R is required) but it fails altogether to give the detailed shape of the angular distribution. This is not surprising in view of the simplifying assumptions made; it does, however, give us some understanding of the general features observed, especially for light nuclei.

4.7.4.2 Other stripping and pick-up reactions

A single nucleon can be stripped from other projectiles in reactions such as (^{3}He, d) and (α, t). The theory for these is in essence the same as that we have just described for deuteron stripping. Again a zero-range approximation analogous to equation 4.64 is often made. A corresponding transfer process can be observed with heavier ions also, such as in (^{14}N, ^{13}N) reactions, although with these larger systems the zero-range approximation is never adequate (*see*, for example, von Oertzen, 1974).

*The result is then close to that obtained in Butler's original theory of deuteron stripping (Butler, 1957) which actually carried out the integral for $r \geqslant R$ after replacing $u_{nL}(r)$ by its known asymptotic form. This theory also included the factor 4.66.

In addition one may transfer more than one nucleon. One can hope to learn in this way about any cluster structures in nuclei. For example. (d, ^{6}Li) or (^{12}C, ^{16}O) reactions which may pick up an α-particle-like cluster of four nucleons would be favoured if there were an appreciable probability of preformation of these clusters in the target. Such 'α' transfer reactions are analogous to naturally occurring α-radioactivity. Particularly important are two-nucleon transfer reactions such as (t, p) and (^{16}O, ^{18}O). Single-nucleon transfers like deuteron stripping probe the probabilities of finding a single nucleon in a given shell model orbit. Transfer of two nucleons, however, depends not only upon the individual motions of the two nucleons but also upon the correlation between those motions. For example, the two neutrons picked up in a (p, t) reaction must be found close together, with their spins antiparallel (because of the Pauli Exclusion Principle) and with no relative angular momentum. The transfer L is then the angular momentum of the centre of mass of the pair (*see*, for example, Towner and Hardy, 1969; Glendenning, 1983).

The theoretical description of multi-nucleon transfer again follows similar lines to that for deuteron stripping and the same general characteristics emerge, although in detail the treatment is more complicated.

4.7.5 Knock-out reactions

These consist of reactions A(a, bc)B in which there are three nuclei in the final state. Such reactions may result from successive evaporations following the formation of a compound nucleus, but we are concerned here with reactions of the type (a, a′b) in which particle b is ejected after being directly struck by the projectile a, which then also escapes with reduced energy (*see*, for example, Jacob and Maris, 1966, 1973).

Reactions resulting in more than two fragments do not show a line spectrum for the energy of each fragment even for a transition between definite states of A and B. Energy and momentum can be conserved with a continuous distribution of energies among the fragments; only the summed energies can show a line spectrum. However, the distribution is not arbitrary; it is determined by the reaction mechanism and the structural details of the nuclei involved.

If we measure the energies $E_{a'}$ and E_b of the emitted particles, energy balance then tells us the binding energy that b had in the target nucleus. For example, if b is a nucleon, we find peaks in the summed energy $(E_{a'} + E_b)$ spectrum corresponding to the binding energies of the various shell model single-particle orbitals. This is one of the most direct pieces of evidence for the existence of these orbits, and the measurements are the nuclear analogues of the early Franck–Hertz experiments on the ionisation of atoms. We do not see sharp lines because of configuration mixing; the single-particle character has been partly dissolved and spread into a range of states around the original single-particle energy. Peaks in the spectrum for the ejection of complex particles b (such as α-particles from (p, pα) or (α, 2α) reactions, say) could imply significant multi-nucleon clustering in the target.

The simplest situation is for a target A which can be regarded as the residual nucleus B in its ground state plus the extra particle b in a definite orbital as in equation 4.62. To a first approximation we can regard the target particle b as at rest and neglect its binding to the remainder, B. This is known as 'quasi-free' scattering and would be valid for very high bombarding energies. Then when it is struck by the projectile a, the result will be the same as if the particle b were free. We could use the impulse approximation (section 3.11), replacing the interaction υ between the projectile and the struck-particle by the scattering amplitude for free scattering of this pair of particles. Further, if a and b have the same mass, as in a (p, 2p) or (α, 2α) reaction, the two emerging particles will be moving at right angles in the laboratory system; i.e. $\theta_a + \theta_b = \frac{1}{2}\pi$ in *Figure 4.26* (*see* Appendix B). Further, the two final directions will be coplanar with the incident beam. We could then detect the two particles in coincidence, varying the angle $(\theta_a + \theta_b)$ between the two detectors. If we find a strong preference for emission at the appropriate angle (e.g. 90° for equal masses) we have evidence for a direct knock-out reaction.

The last remark does need some qualifying. The struck particle b will not be completely at rest but will be moving about in the target; this will allow emergence at other relative angles and also allow non-coplanar events; i.e. we shall see a peak centred at the favoured angle but broadened by this initial motion, as in the upper part of *Figure 4.26b*. Further, only if b is initially in an *s*-state, with no angular momentum, can we *ever* find it at rest. If it has a finite amount of angular momentum, it will always be found in motion, although this motion will be slow compared to the projectile a if the bombarding energy is high. Consequently in these cases there will actually be a *minimum* for emission at the $(\theta_a + \theta_b)$ corresponding to free scattering, with a peak on either side. This is illustrated for knock-out from a *p*-state in the lower part of *Figure 4.26b*. Hence by making these angular correlation measurements, we can learn the angular momentum of the state from which b was ejected, as well as its energy. Further, the shape observed for the peak should also tell us about the momentum distribution the nucleon had in the target nucleus before it was struck and ejected.

The situation as we have described it so far can be described in terms of the plane-wave Born approximation or impulse approximation (Jackson, 1970, 1971). However, we have completely ignored refraction and absorption effects. The momentum and direction of motion of the particles will be altered as they cross the nuclear surface, and multiple collisions in the target correspond to absorption. The refraction causes further broadening of the peaks in the energy spectra, and in addition may result in the central minimum being partially filled in. Just as in section 4.7.2.2, these effects can be included by using distorted waves to describe the motions. (If the impulse approximation is also used, this becomes the *distorted-wave impulse approximation*.) The calculations required are considerably more elaborate than those discussed earlier because we have three free particles in the final state to share the energy and momentum (Jackson, 1970, 1971).

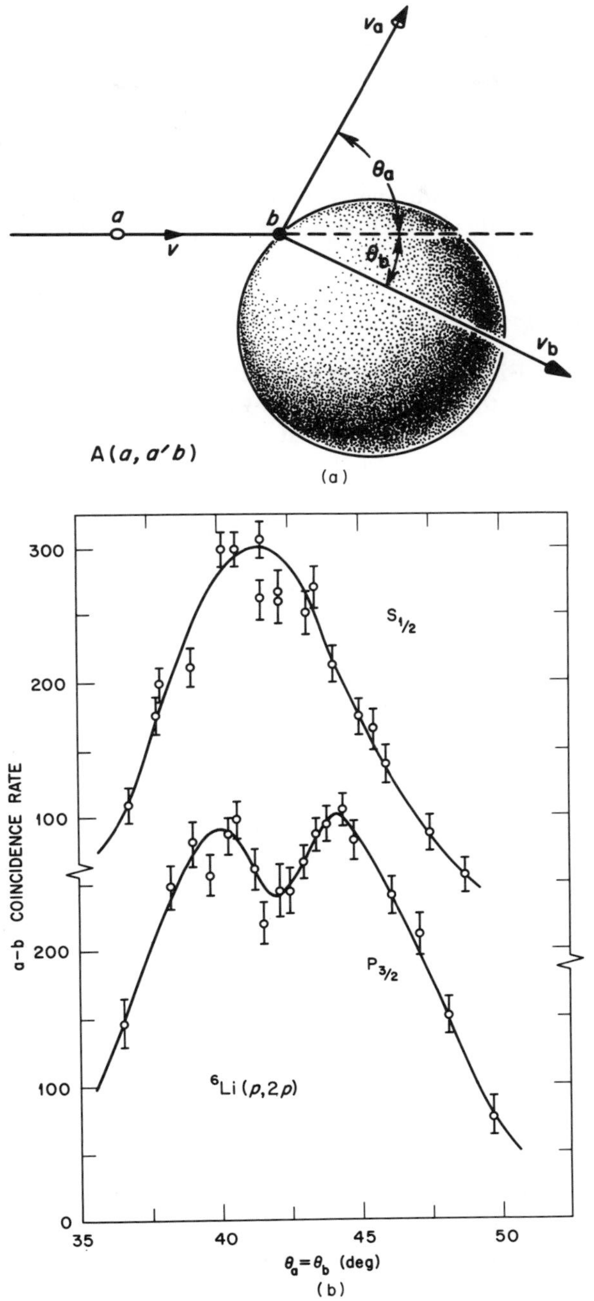

Figure 4.26 (a) Illustrating a knock-out reaction of the type (a, a′b). (b) The angular correlation between protons emitted with equal deflections to left and right (that is, with $\theta_a = \theta_b$) in the ^{6}Li (p, 2p) reaction at 450 MeV bombarding energy for knock-out from the $1s_{1/2}$ and $1p_{3/2}$ orbits. The maximum and minimum, respectively, are shifted slightly from $\theta_a = \theta_b = 45°$ because of refraction of the incident and outgoing protons. (From Jacob and Maris, 1966)

Measurements of knock-out reactions are difficult; high energy is required together with the high intensities needed to make coincidence measurements. In addition we must have good energy resolution if definite final states of the residual nucleus are to be studied. Most of the experiments have been made using the (p, 2p) reaction, with bombarding energies between about 100 MeV and 1 GeV, although there have been studies of (e, e′p), (p, p′α) and (α, 2α) reactions, etc.

4.7.6 Multi-step processes and strong coupling

The descriptions we have given so far of direct reactions have used a first-order perturbation theory, the plane- or distorted-wave Born approximation. That is, the interaction V has been assumed to act only once, implying that terms of higher order are negligible. Indeed, 'first-order (or one-step) process', and 'direct reaction' are often regarded as almost synonymous (*see* also section 2.18.1). However, it is sometimes possible for second-order and other low-order terms to be important, without the process being sufficiently complicated for us to describe it as a compound-nucleus reaction. This can happen if only two or a few channels need be considered explicitly, but the coupling (i.e. the matrix elements of V) between them is sufficiently *strong* that considering them only to first-order is inadequate. In this way, the state of the colliding system may oscillate back and forth between these few strongly coupled configurations without diffusing into the many more complicated configurations needed to make up the localised and relatively long-lived system we would call a compound nucleus.

Another possibility is that the coupling between the entrance channel and some particular exit channel is *weak* for some reason, either because the interaction V itself is weak or because the initial and final nuclear internal states have a poor overlap. Then this exit channel may be fed by a two-step process, via an intermediate state which is more strongly coupled to both the initial and final states than they are to each other. This indirect feeding may be as strong or stronger than the direct one-step process.

We can best understand these possibilities by considering some specific examples of multi-step processes.

Perhaps the simplest (and historically the earliest) example of strong coupling occurs for inelastic scattering. *Figure 4.27a* indicates schematically the A(a, a′)A* reaction between the ground state A and an excited state A*. Each arrow indicates a matrix element of V and its contribution to the scattering amplitude. The number of arrows included equals the number of steps in this multi-step contribution. The first arrow corresponds to the Born approximation described above. However, we may also reach A* by de-exciting back to the ground state A and re-exciting A*; this is third-order in V, or a three-step term. (The de-excitation step to the ground state A represents a second-order contribution to the *elastic* scattering in which the state A* has been *virtually* excited.) If the diagonal

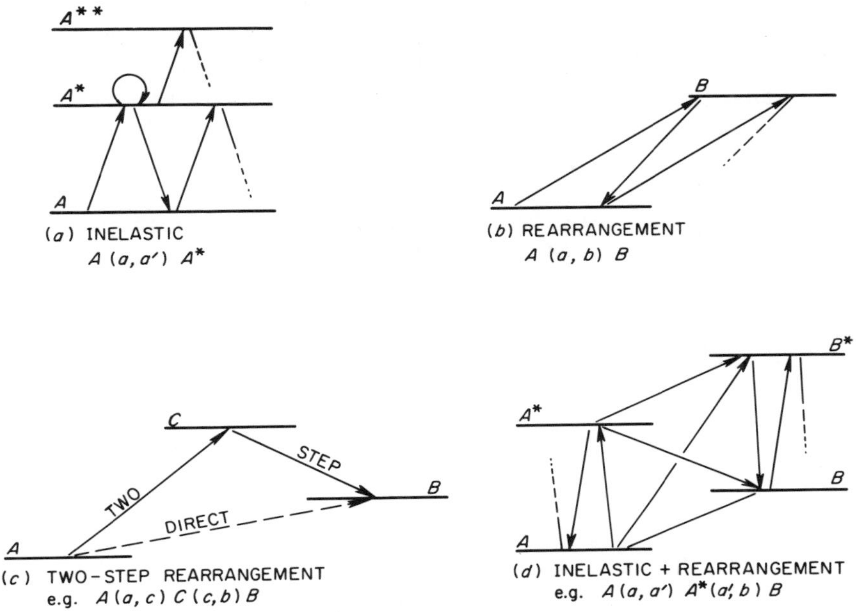

Figure 4.27 Illustrating various types of multi-step processes which may occur in direct reactions. Each arrow indicates one action by the interaction potential V. The Born approximation takes into account only one such arrow

matrix-element $\langle A^* | V | A^* \rangle$ is non-vanishing (the arrow which loops back upon itself in *Figure 4.27a*), this allows a second-order contribution to the inelastic scattering. This latter process is often referred to as *re-orientation* because usually the effect of V in the second step is to change the magnetic substate M_A^* of the spin I_A^* of the excited state, that is to re-orient it. (For the exploitation of this effect in Coulomb excitation measurements, *see* Hausser 1974.)

These extra contributions could be calculated by using a higher-order perturbation theory. However, it is usually more convenient to solve exactly the set of coupled equations (*see* section 3.3.1) corresponding to these few states. In this way, the coupling between them is treated exactly (for a given model), while the effect of the couplings to all the other channels is assumed to be taken care of by the (complex) optical potentials which are used. This procedure is often referred to† as a *coupled-channels* calculation; in atomic physics it is often called the *strong-coupling approximation*. It is widely used to describe the inelastic excitation of collective states, particularly the strong transitions to the lowest 2^+ levels of even nuclei.

†The names in common use to describe the various methods of calculating multi-step processes in nuclear reactions have little to commend them. However, their usage is by now well established and we shall not attempt to advocate more appropriate terms.

The analogous procedure may also be applied to rearrangement reactions, as illustrated schematically in *Figure 4.27b*. It is then often called the *coupled reaction channel* method. This has found somewhat less application because the coupling causing rearrangement transitions is usually weak enough for the Born theory to be used.

The possibility of competition between indirect (two-step) and direct (one-step) contributions may also occur for inelastic scattering. Suppose we add (*see Figure 4.27a*) a second excited state A** which is only weakly coupled to the ground state A but strongly coupled to the excited state A*. It may then be excited via A* as an intermediate state. A typical example would be an even nucleus with a rotational spectrum; A would be the 0^+ ground state, A* the 2^+ state and A** the 4^+ state of the band (*see Figure 1.4*). Generally the direct

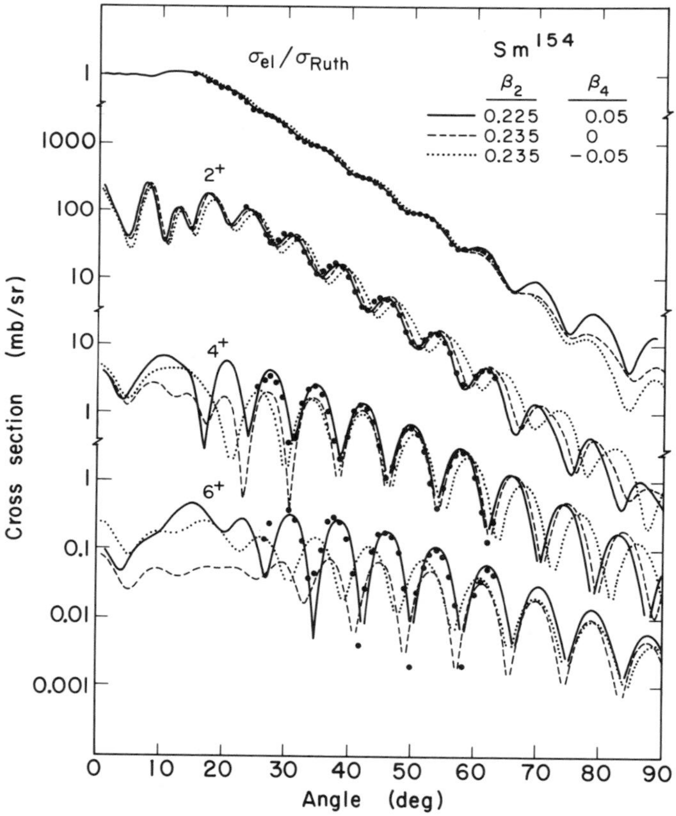

Figure 4.28 Measured and calculated cross-sections for the excitation of several states of a rotational band by the inelastic scattering of α-particles. The cross-sections for the 4^+ and 6^+ states are due largely to multiple $L = 2$ excitations but they also receive very important contributions from the $L = 4$ coupling. (From Glendenning, 1983)

0^+-4^+ coupling is weak but the 0^+-2^+ and 2^+-4^+ couplings are strong. *Figure 4.28* shows an example which also includes the three-step excitation of the 6^+ member of a rotational band. Of course, there is interference between the amplitudes when more than one process contributes to an excitation; observation of the effects of this interference can be a valuable source of information on the characteristics of the various coupling terms.

Again this picture may be extended to rearrangement reactions such as indicated in *Figure 4.27c*. An example of this could be the A(p, n)B reaction, with the indirect process of the A(p, d)C pick-up followed by the C(d, n)B stripping reaction competing with the direct charge-exchange (p, n) process. There are indications that this is an important contribution. The analogous (^{3}He, t) reaction may proceed through the intermediate formation of an α-particle. Another example is that in heavy-ion reactions such as (^{16}O, ^{18}O), the transfer of two nucleons sequentially, such as (^{16}O, ^{17}O) followed by (^{17}O, ^{18}O), may be as probable as their transfer simultaneously. The reader will find no difficulty in constructing many other possibilities of this kind.

Another category of multi-step processes which have been found to be of importance is illustrated by *Figure 4.27d*. This corresponds to the inelastic excitation of the target before a rearrangement or transfer takes place, or the inelastic excitation of the residual nucleus afterwards, or both. It often occurs that the transfer or rearrangement step is sufficiently weak for the Born approximation to be used, but the strong inelastic events have to be calculated by solving the corresponding coupled equations; this procedure is usually called the *coupled-channel Born approximation*. (This can be viewed as a generalisation of the distorted-wave Born approximation, equation 4.57; instead of using the distorted waves of equation 4.56 with only elastic scattering included, we may add the important inelastic waves also, for example as in equation 3.39.) Inelastic excitation of this type will be especially important if the direct route is inhibited; for example, if the A to B* matrix element in *Figure 4.27d* is weak, but the A* to B* one is not, the preliminary excitation of A to A* will be favoured. Such effects have been found to be especially important in some two-nucleon transfer reactions such as (p, t) and (^{16}O, ^{18}O); *see*, for example, *Figure 4.29* and Glendenning (1983).

Yet another example where higher-order effects may be important occurs in deuteron stripping. The deuteron, being loosely bound, is easily broken up by its impact upon a target nucleus. Subsequently the neutron, for example, may be captured and the proton escape, corresponding to a (d, p) reaction. Hence it may be necessary to add at least part of the deuteron break-up wave function (Harvey and Johnson, 1971) to that for simple elastic scattering when using the distorted-wave Born approximation to describe these reactions. A review by Austern *et al.* (1987) describes more recent progress on this problem.

Continuing research has clearly established in many cases the existence and importance of the various higher-order corrections to the simple theories of direct reactions. In other cases we are faced with the danger of merely

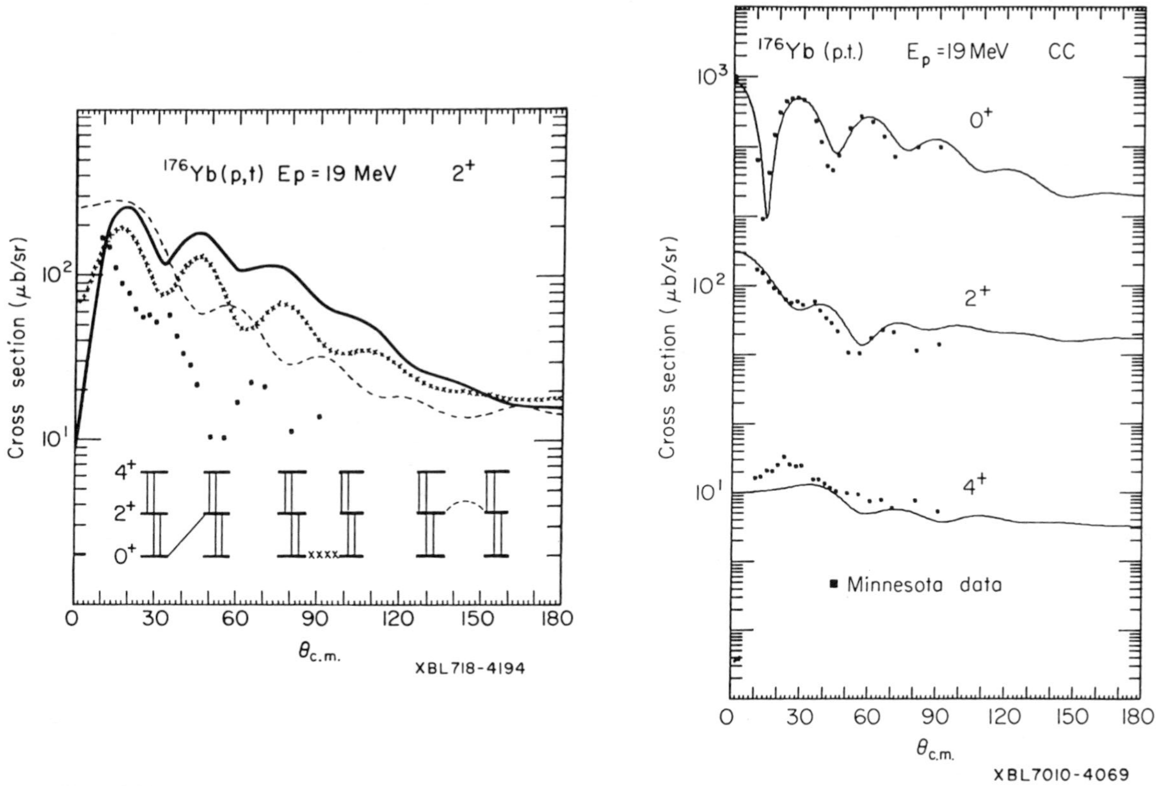

Figure 4.29 The importance of two-step processes involving inelastic excitation before or after the transfer of two neutrons. The separate contributions from the three different possibilities are shown on the left; each is, by itself, larger than the measured cross-section. On the right, their coherent sum (with interferences) is seen to agree with the experimental cross-sections, which are shown as dots. (From Ascuitto *et al*., 1971)

invoking multi-step processes in an *ad hoc* fashion in order to explain away some inconvenient experimental results. Part of the difficulty is that our reaction models themselves are often defined in a somewhat *ad hoc* way, rather than being deduced rigorously from first principles. For example, as discussed in section 4.5.6, the imaginary part of an optical potential is introduced to represent the absorption due to coupling to channels other than the one being studied explicitly. However, a simple potential can only do this in a rough, average, way. Each case must be studied on its own merits to see whether this is adequate or whether attention must be paid explicitly to one or more of these other channels.

4.8 COMPOUND NUCLEUS RESONANCES

At sufficiently low bombarding energies, the interactions between nuclei can exhibit resonance characteristics. These show themselves as sharp peaks in the cross-section as the bombarding-energy is varied (section 2.18.2 and *Figure 2.31*). Typically they may appear for bombarding energies up to several MeV for light nuclei, but may only be seen in the eV region for heavy nuclei (*see Figure 4.30* for some typical excitation functions). In heavy nuclei, because of

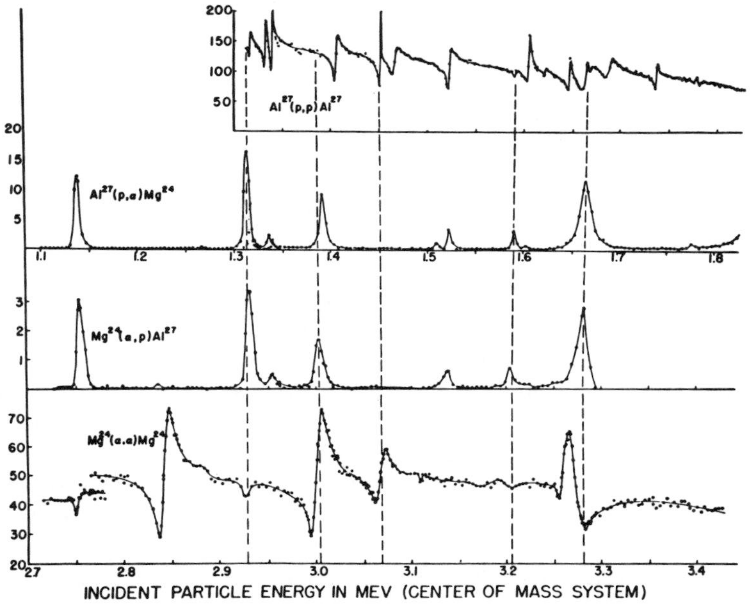

Figure 4.30 Excitation functions for several reactions proceeding through the same compound nucleus ^{28}Si, showing typical resonance structure at low bombarding energies. Note the correspondence in energy of the resonances in different channels; the proton and α-particle energy scales have been adjusted so as to correspond to the same total, or compound nucleus, energy. (From Kaufmann *et al.*, 1952)

the inhibiting effect of the Coulomb barrier for charged particles when these have low energies, the resonances are observed only in neutron-induced reactions. There are some exceptions to these rules, a very important one being the excitation of ***isobaric analogue resonances*** (*see* below).

Compound nucleus resonances are a close analogue of the absorption lines encountered in the phenomenon of the anomalous dispersion of light when passing through some medium (*see*, for example, Jenkins and White, 1957). If the medium contains bound electrons which may oscillate with certain natural frequencies, incident light will undergo resonant scattering and absorption whenever its frequency matches one of these characteristic frequencies. Further, the electron's motion may be affected by a small damping force which dissipates its energy; this damping gives the resonance line a finite width.

Another analogue is the resonant response of an electrical circuit to an oscillatory applied voltage of the appropriate frequency (*see*, for example, Bleaney and Bleaney, 1957). If, for example, a series resonant circuit contains inductance L, capacitance C and resistance R, and an oscillatory voltage $V = V_0 \cos \omega t$ is applied, the circuit responds with the flow of current $I = I_0 \cos(\omega t - \phi)$ where $I_0 = V_0/Z$ and ϕ is a phase shift. Here Z is the impedance of the circuit for the frequency ω

$$Z = \left[R^2 + \left(\omega L - \frac{1}{\omega C}\right)^2\right]^{1/2} \tag{4.69}$$

Energy is dissipated in the circuit at an average rate of $\overline{W} = \frac{1}{2} I_0^2 R$ which is inversely proportional to Z^2. After some rearranging this can be written

$$\overline{W}(\epsilon) = \frac{\frac{1}{8}(V_0^2/R)\,\Gamma(\epsilon)^2}{(\epsilon - \epsilon_0)^2 + \frac{1}{4}\Gamma(\epsilon)^2} \tag{4.70}$$

where we have put $\Gamma = 2\omega R/L$, $\epsilon = \omega^2$ and $\epsilon_0 = \omega_0^2 = (LC)^{-1}$. It may easily be verified that this expression has a maximum for $\epsilon = \epsilon_0$, which corresponds to the well-known resonance condition $\omega = \omega_0 = (LC)^{-1/2}$. That is, as the applied frequency ω is varied, the average current $\overline{I}$ induced and the average energy $\overline{W}$ dissipated (absorbed from the applied signal) reach a maximum at $\epsilon = \epsilon_0$ (*see Figure 4.32* below). The sharpness or width of this resonance as a function of ω or ϵ is given by Γ; since Γ is proportional to the resistance R we see that the width of the resonance is due to the ability of the circuit to dissipate energy.

Such a simple circuit has but a single resonance frequency, but a collection of circuits with coupling between them can exhibit a large number of different resonant frequencies. Nuclei are like this more complicated situation; usually there are very many different possible modes of excitation.

In a nuclear reaction the role of 'applied voltage' is played by a beam of bombarding particles whose energy determines the frequency. A resonance energy is then a characteristic of the compound system, target plus projectile. Although the target has discrete excited states, these only result in the emitted, inelastically scattered, particles having discrete residual energies and hence showing discrete energy losses. An inelastic cross-section does not show a resonance as the bombarding energy crosses one of these excitation energies. However, the energy of the compound system does have a unique energy for a given bombarding energy so that any characteristic or discrete state of this compound nucleus can appear as a resonance in the cross-section for any nuclear reaction as the bombarding energy is varied.

Let the compound system C be formed by bringing together the projectile a and target A at zero bombarding energy. Suppose its binding energy is then S_α; that is, if C were in its ground state, energy S_α would have to be supplied to separate it into a + A. Then with a CM bombarding energy of E_α the compound nucleus C is formed with an excitation energy of $(E_\alpha + S_\alpha)$. If C has a discrete, quasi-stationary, state with excitation energy E_x, this can show as a resonance in the a + A channel at an energy $E_{\alpha r} = E_x - S_\alpha$, provided $E_x > S_\alpha$ (*see* the energy diagram in *Figure 4.31*).

An important feature is that the *same* compound nucleus can be formed from other pairs of nuclei, for example b + B. The same characteristic state of C may then occur as a resonance in the b + B channel at a bombarding energy of

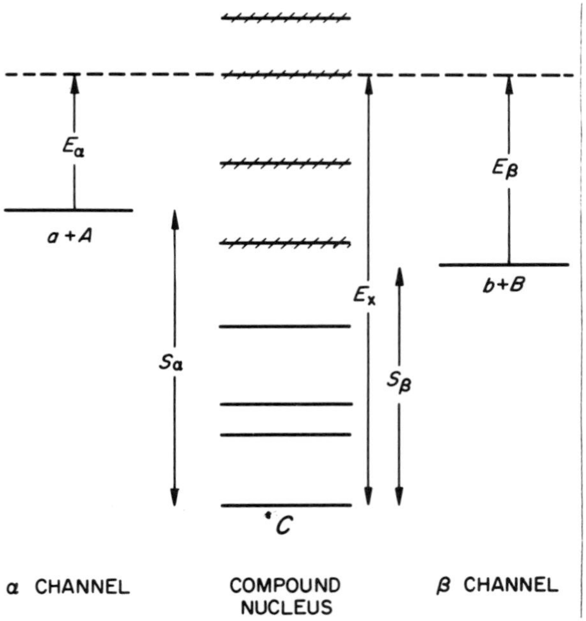

Figure 4.31 Energy-level diagram indicating the energy relationships for a compound nucleus resonance reaction. The incident energy E_α is shown to be in resonance with one of the levels of the compound nucleus at excitation energy E_x. Levels in C with $E_x > S_\beta$ are shown hatched to indicate that they are unstable against particle emission and hence have non-negligible widths

$E_{\beta r} = E_x - S_\beta$ (again provided $E_x > S_\beta$, where S_β is the energy required to separate C into the pair b + B). Observation of this in more than one channel then provides evidence that the resonance is a characteristic of the compound nucleus C and not of the particular pair of nuclei from which it was formed. Consequently we expect that the probability that C decays into a particular channel should be independent of the initial channel through which it was formed; this is the independence hypothesis, equation 2.49 of section 2.18.1.

In the nuclear case, the 'resistive' or dissipative losses which give a resonance a finite width come about because the compound system is unstable and is able to decay (break-up), either back into the initial channel or into any of the other energetically open channels. Thus the nucleus C is radioactive and such a nuclear reaction may be regarded as a kind of artificially induced radioactivity. If C in a particular resonant state has a lifetime τ, the uncertainty principle tells us that this state need not have a unique energy but that it may exist with any energy* within a band of width $\Gamma \approx \hbar/\tau$.

4.8.1 Simple theory of a resonant cross-section

Since a resonant state, if formed at time $t = 0$, will decay with a mean life τ, its wave function ψ must be such that the probability of finding the compound state at time t is

$$|\psi|^2 = |\psi_0|^2 \exp(-t/\tau)$$

This suggests the form

$$\psi = \psi_0 \exp(-i\omega_r t) \exp(-t/2\tau) \tag{4.71}$$

where $\omega_r = E_r/\hbar$ corresponds to the resonance peak energy E_r and ψ_0 is independent of time. The function 4.71 may be expressed as a superposition (wave packet) of stationary states $\phi(\omega)$ with definite energies $E = \hbar\omega$ by means of a Fourier expansion (Messiah, 1962)

$$\psi = \frac{1}{(2\pi)^{1/2}} \int_0^\infty \phi(\omega) \exp(-i\omega t)\, d\omega \tag{4.72}$$

where by inversion

$$\begin{aligned}\phi(\omega) &= \frac{1}{(2\pi)^{1/2}} \int_0^\infty \psi \exp(i\omega t)\, dt \\ &= \frac{1}{(2\pi)^{1/2}} \frac{i\psi_0}{(\omega - \omega_r) + i/2\tau}\end{aligned} \tag{4.73}$$

*Even the so-called bound states (i.e. those stable against emission of one or more nucleons), except for the ground state, are liable to γ-decay, and all may be unstable for β-decay. However the widths Γ due to these processes are so small that for most purposes they may be neglected and the states regarded as having sharp energies. They could show up as resonances in the scattering of γ-radiation by nuclei.

The meaning of equation 4.72 is that the probability of finding the energy $E = \hbar\omega$ in the state ψ is proportional to $|\phi(\omega)|^2$, and from equation 4.73 this is proportional to

$$\frac{1}{(E - E_r)^2 + (\frac{1}{2}\Gamma)^2} \tag{4.74}$$

where $\Gamma = \hbar/\tau$. This is the Lorentzian shape for line broadening. It has a maximum value at $E = E_r$, the resonance energy, and falls to half maximum when $E = E_r \pm \frac{1}{2}\Gamma$ (*see Figure 4.32*).

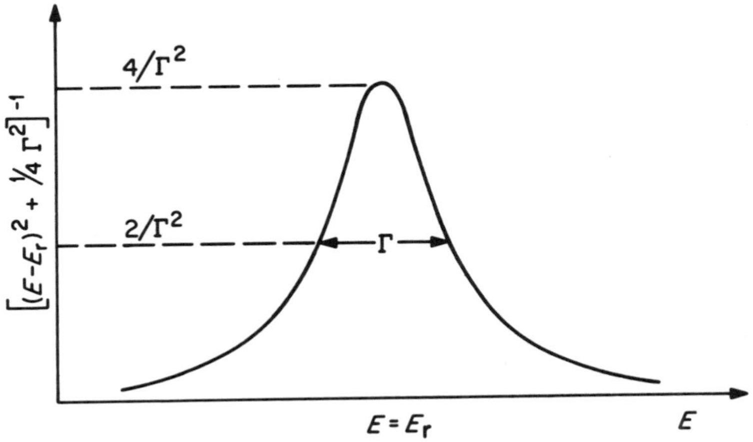

Figure 4.32 Illustrating the Breit–Wigner resonance shape of equation 4.74

The cross-section $\sigma_{\alpha C}$ for forming the compound C in the resonant state from a pair of nuclei $\alpha = a + A$ at an energy E must be proportional to this probability that the resonant state exists in C at that energy. That is, from equation 4.74 it has the form

$$\sigma_{\alpha C} = \frac{F(E)}{(E - E_r)^2 + (\frac{1}{2}\Gamma)^2}$$

where $F(E)$ may be determined by appealing to a simple statistical argument. Consider a system of volume V containing a large number of nuclei a and A in equilibrium with the compound nucleus C. The effective volume swept by a given nucleus a per unit time is $v_\alpha \sigma_{\alpha C}$. The probability of finding a given nucleus A per unit volume is $1/V$. Then the probability per unit time that a given nucleus a will coalesce with a given nucleus A is $v_\alpha \sigma_{\alpha C}/V$ when their relative speed is v_α. When in statistical equilibrium, the number of a + A pairs to be found with relative speeds v_α to $v_\alpha + dv_\alpha$, or relative momenta p_α to $p_\alpha + dp_\alpha$, is proportional to the number of states available in this range (Messiah, 1962)

$$n(p)\,dp = \frac{V p^2\,dp}{2\pi^2 \hbar^3} \tag{4.75}$$

The total probability per unit time of forming C in this resonant state is then obtained by integrating over a range of energies containing $E = E_r$

$$\begin{aligned} P = \int \frac{\sigma_{\alpha C} v_\alpha}{V} \frac{V p_\alpha^2\,dp_\alpha}{2\pi^2 \hbar^3} &= \int \frac{\sigma_{\alpha C}\, p_\alpha^2\,dE}{2\pi^2 \hbar^3} \\ &= \int \frac{\sigma_{\alpha C}\, k_\alpha^2\,dE}{2\pi^2 \hbar} \end{aligned} \tag{4.76}$$

We may assume $k_\alpha^2 F(E)$ varies slowly with energy compared to the Lorentzian 4.74 so that

$$P \approx \frac{k_\alpha^2\ F(E_r)}{2\pi^2 \hbar} \int \frac{dE}{(E - E_r)^2 + (\frac{1}{2}\Gamma)^2} = \frac{k_\alpha^2\,F(E_r)}{\pi \hbar \Gamma} \tag{4.77}$$

If the system is in equilibrium, this is just balanced by the probability per unit time that the compound C has decayed back into a + A. This is defined as $\Gamma_\alpha/\hbar$, where Γ_α is called the *partial width* for decay into the a + A channel. Equating this to P from equation 4.77, we get

$$F(E_r) = \frac{\pi}{k_\alpha^2}\ \Gamma_\alpha \Gamma$$

or

$$\sigma_{\alpha C} = \frac{\pi}{k_\alpha^2} \frac{\Gamma_\alpha \Gamma}{(E - E_r)^2 + (\frac{1}{2}\Gamma)^2} \tag{4.78}$$

Since the *total* rate of decay of C is given by $\Gamma/\hbar$, this must equal the sum of the probabilities for decay into the various open channels, or

$$\Gamma = \sum_\beta \Gamma_\beta \tag{4.79}$$

where the sum includes all the channels β which are energetically available ($\Gamma_\beta = 0$ for a closed channel). This means we could also speak of $\hbar/\Gamma_\beta$ as the partial lifetime for decay into the β channel, so the lifetime of C is the geometric mean of the partial lifetimes, or

$$\frac{1}{\tau} = \sum_\beta \frac{1}{\tau_\beta} \tag{4.80}$$

Further, it follows that the cross-section for the transition from channel α to channel β, i.e. the reaction A(a, b)B, going through this compound nucleus resonance is

$$\sigma_{\alpha\beta} = \frac{\pi}{k_\alpha^2} \frac{\Gamma_\alpha \Gamma_\beta}{(E - E_r)^2 + (\frac{1}{2}\Gamma)^2} \tag{4.81}$$

This has the factored form of equation 2.49, with $G_\beta^C = \Gamma_\beta/\Gamma$, required by the *independence hypothesis* (section 2.18.1) and is the *Breit–Wigner* expression for a resonant cross-section.

We have only given arguments for the *form* of equation 4.81; we do not know *a priori* what values the partial widths Γ_α, Γ_β etc. will assume. It may happen that Γ_α will be very small ($\Gamma_\alpha \ll \Gamma$) or even zero for a particular resonance and a particular channel α. This could occur because the internal structure of that compound state makes it difficult to form it from the a + A pair or because of some angular momentum or parity selection rule acting. In such a case, that resonance will not be observed in any a + A reactions. If instead it is the width Γ_β for the exit channel which is small or zero, the resonance will not be observed in the cross-section for that particular A(a, b)B reaction but may be seen in other a + A reactions.

4.8.2 More formal theory of a resonance

The existence of resonances may be seen in another context. For simplicity we consider *s*-wave neutrons incident on a nucleus. Consider now a surface $r = a$ drawn around the nucleus such that it is just beyond the range of nuclear forces. Outside and on this surface, the elastic *s*-wave for neutrons has the form (section 3.4, equation 3.40)

$$w_0(r) = \exp(-ikr) - \eta_0 \exp(ikr), \quad r \geqslant a \tag{4.82}$$

where $w_0(r)$ is r times the actual wave function $u_0(r)$. The scattering amplitude η_0 may be expressed in terms of the logarithmic derivative f of this wave function at the surface $r = a$, where

$$f = \left(\frac{a}{w}\frac{\mathrm{d}w}{\mathrm{d}r}\right)_{r=a} \tag{4.83}$$

Of course, f depends upon the energy in general, $f = f(E)$, as well as the radius a. With equation 4.82 we get*

$$\eta_0 = \frac{f + ika}{f - ika} \exp(-2ika) \tag{4.84}$$

*Although η_0 appears to depend explicitly upon a, a moment's thought will show that the value of η_0 obtained from equation 4.84 must be independent of the (arbitrary) choice of the value of a provided $r = a$ is beyond the range of interaction. If the value of a is changed, then f must also change in such a way as to exactly compensate in the expression for η_0.

In itself this does not tell us anything new. However the quantity f may be easier to manipulate and to relate to the physics of what is happening inside the 'black box' $r < a$ where the component parts of the compound system are interacting strongly (*see*, for example, Blatt and Weisskopf, 1952). This is one of the basic steps in theories of the R-matrix type (section 3.9). We shall now see how the transformation 4.84 can be used to suggest the existence of resonance phenomena.

Equation 4.84 indicates that if f is real, then $|\eta_0| = 1$, and there is no non-elastic scattering. Further, in order to satisfy unitarity, that is have $|\eta_0| \leqslant 1$, we must have $Im f \leqslant 0$. Now equation 4.84 enables us to write expressions for the s-wave elastic and non-elastic (or absorption) cross-sections, equations 3.62, 3.63 of section 3.4.9, in terms of the real and imaginary parts of $f(E) = x(E) + iy(E)$

$$\sigma_{el} = \frac{\pi}{k^2}\left|1 - \eta_0\right|^2 = \frac{\pi}{k^2}\left|[\exp(2ika) - 1] - \frac{2ika}{x + i(y - ka)}\right|^2 \tag{4.85}$$

$$\sigma_{abs} = \frac{\pi}{k^2}\left[1 - |\eta_0|^2\right] = \frac{\pi}{k^2}\left[\frac{-4kay}{x^2 + (y - ka)^2}\right] \tag{4.86}$$

We return in a moment to our reason for arranging equation 4.85 in that particular way. Equation 4.86 confirms that we must have $Im f\,(= y)$ negative in order that σ_{abs} be positive, as it has to be. We see also that σ_{abs} will be a maximum when $x(E) = 0$; hence if there are certain energies $E = E_r$ for which $x(E_r) = 0$, there will be resonances at those energies.*

4.8.2.1 *Single isolated resonance*

We may expand $x(E)$ in a Taylor series if E is close to one of the E_r for which x has a zero

$$x(E) = (E - E_r)\left(\frac{dx}{dE}\right)_{E=E_r} + \ldots \tag{4.87}$$

Taking just the first term of this expansion, equations 4.85, 4.86 may be written

$$\sigma_{el} = \frac{\pi}{k^2}\left|[\exp(2ika) - 1] + \frac{i\Gamma_\alpha}{(E - E_r) + \frac{1}{2}i\Gamma}\right|^2 \tag{4.88}$$

*We cannot tell from these equations whether such energies E_r exist for a given system or, if they do, what their values are. For that we need to know something about the physical system target-plus-neutron when it is interacting inside the surface $r = a$. However, we *do* know from observation that resonances sometimes occur; then these arguments tell us something about the forms that the scattering amplitudes and cross-sections must assume in such cases.

$$\sigma_{abs} = \frac{\pi}{k^2} \frac{\Gamma_\alpha(\Gamma - \Gamma_\alpha)}{(E - E_r)^2 + (\frac{1}{2}\Gamma)^2} \tag{4.89}$$

where we have defined

$$\Gamma_\alpha = -\frac{2ka}{(\mathrm{d}x/\mathrm{d}E)_{E=E_r}}, \quad \Gamma = -\frac{2ka - 2y}{(\mathrm{d}x/\mathrm{d}E)_{E=E_r}} \tag{4.90}$$

(It may be shown that $\mathrm{d}x/\mathrm{d}E$ is negative, so that Γ_α and Γ are positive quantities. Note also that k, x, y and E all refer to the $\alpha = \mathrm{A} + \mathrm{n}$ entrance channel; we omitted the subscript α for simplicity.)

Comparison with equations 4.78, 4.81 allows us to identify Γ_α and Γ as the neutron width in the entrance channel $\alpha = \mathrm{A} + \mathrm{n}$ and the total width, respectively. By definition the absorption cross-section σ_{abs} does not include elastic scattering, hence the appearance of the difference $(\Gamma - \Gamma_\alpha)$ in equation 4.89; from equation 4.79

$$\Gamma - \Gamma_\alpha = \sum_{\beta \neq \alpha} \Gamma_\beta$$

On the other hand, the compound nucleus formation cross-section, 4.78, includes the possibility of re-emission into the elastic channel. As before, the Breit-Wigner form, 4.81, for decay into a particular final channel β is obtained by taking just the fraction $\Gamma_\beta/(\Gamma - \Gamma_\alpha)$ of the total non-elastic cross-section 4.89, where Γ_β is given by the analogue of the first of equations 4.90 for the β channel at the appropriate energy.

The elastic cross-section, 4.88, shows another term in addition to the resonant one. We may rewrite it as

$$\sigma_{el} = \frac{\pi}{k^2} |A_{pot} + A_{res}|^2 \tag{4.91}$$

where the non-resonant term is called the *potential* or *shape-elastic* scattering amplitude

$$A_{pot} = \exp(2ika) - 1 \tag{4.92}$$

and the other term is the resonance or *compound-elastic* scattering amplitude

$$A_{res} = \frac{i\Gamma_\alpha}{(E - E_r) + \frac{1}{2}i\Gamma} \tag{4.93}$$

The commonly used name 'potential' scattering amplitude for A_{pot} is somewhat misleading because no potential has been introduced. A_{pot} appears because we chose to express the scattering in terms of the boundary conditions on the sphere $r = a$. Indeed A_{pot} is the same as the scattering amplitude for an impenetrable or *hard sphere* of radius a (section 3.4.5). (Such a sphere would force the wave function to vanish at its boundary, $w_0(r = a) = 0$, so that $f = \infty$ and

$\eta_0 = \exp(-2ika)$.) When the bombarding energy is off-resonance so that $|E - E_r| \gg \Gamma$, then $A_{res} \approx 0$ and only A_{pot} remains. The elastic scattering becomes like that from a hard sphere and the absorption cross-section goes to zero. This means that the wave function $w_0(r)$ does not penetrate into the region $r < a$ when the energy is away from the resonance; the compound system A + n cannot exist with an energy such that $|E - E_r| \gg \Gamma$. Formally this separation into two terms A_{pot} and A_{res} appears to be arbitrary because the choice of a is arbitrary. This was true at the stage of equation 4.85. However, in getting to equation 4.91 we have assumed that $x(E)$ has a zero at $E = E_r$ and that there is one and only one such zero that we need to consider for energies $E \approx E_r$. If such a single, isolated resonance corresponds to reality, the widths Γ_α, Γ etc. will have well-defined values corresponding to a definite radius a. This radius thus acquires physical significance; it could be determined empirically from the scattering cross-section off-resonance which is

$$\sigma_{el} \approx \frac{\pi}{k^2} \left|A_{pot}\right|^2 = \frac{4\pi}{k^2} \sin^2 ka \tag{4.94}$$

$$\approx 4\pi a^2$$

if ka is small.

On the resonance the cross-section is given by equation 4.91 and the two amplitudes A_{pot} and A_{res} will interfere; that is, there is a cross-term $2Re(A_{pot}A_{res}{}^*)$. Part of this changes sign in the vicinity of the resonance and

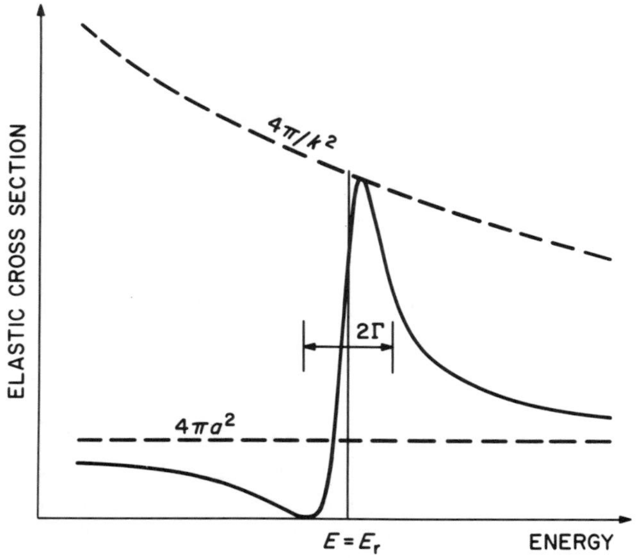

Figure 4.33 Illustrating the interference between resonance and 'potential' amplitudes for elastic scattering at an isolated resonance

may change the shape of the cross-section versus E curve from the Breit–Wigner one (*Figure 4.32*) to one with a characteristic interference dip in it for $E < E_r$ (*see Figure 4.33*). The skew shape of the interference term moves the actual maximum of the resonance slightly above the resonance energy $E = E_r$. An interference dip does not appear in the absorption cross-section at a resonance, and it will be seen in the total cross-section only if the elastic cross-section forms a substantial part of the total. *Figure 4.34* shows some typical measured cross-sections for slow neutrons.

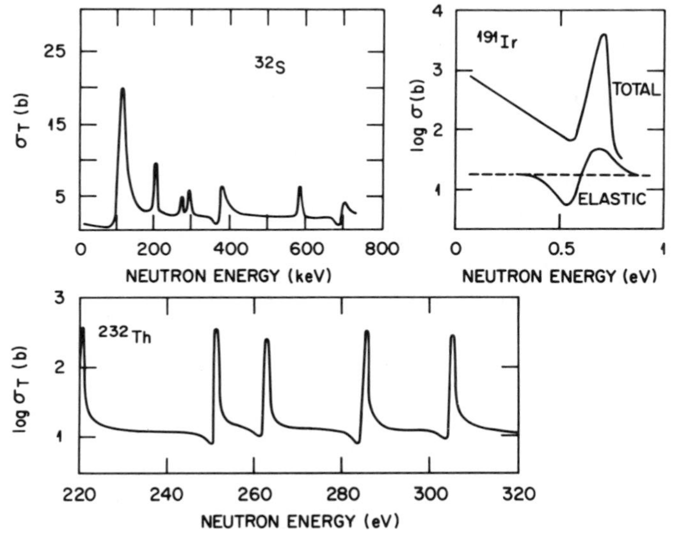

Figure 4.34 Typical cross-sections showing resonances measured for slow neutrons. Note the change in energy scale, especially the change by a factor of 1000 between ^{32}S and ^{232}Th; note also that cross-sections for Ir and Th are plotted on a logarithmic scale. The results for Ir show a pronounced interference minimum in the elastic cross-section near the 0.65 eV resonance, but not in the total. (After Goldberg, 1966)

The potential term A_{pot} is usually much smaller for p-waves, and partial waves with higher angular momentum, at the low energies where isolated resonances are generally seen. Then the cross-section for the elastic scattering of these waves would not show a prominent interference dip preceding each resonance. This fact can often be used to identify easily whether or not an observed resonance was formed by s-waves.

The definition 4.90 of the neutron partial width Γ_α and the total width Γ shows that these widths are not constants but vary with energy; in particular Γ_α for s-wave neutrons is proportional to k or $E^{1/2}$. However, the variation of Γ_α across a resonance may be neglected unless the resonance energy itself is small, $E_r \lesssim \Gamma$. However it is convenient to separate out the energy dependence from the partial width and define an energy-independent *reduced width*, usually denoted γ^2, where

$$\gamma_\alpha^2 = \Gamma_\alpha/2ka = -(\mathrm{d}x/\mathrm{d}E)^{-1}_{E=E_r} \tag{4.95}$$

The widths for other reaction channels and for partial waves with $\ell > 0$ can be treated in a similar fashion (*see* Blatt and Weisskopf, 1952, for example). These reduced widths are closely related to the spectroscopic factors S for bound states that were introduced in section 4.7.4 in connection with direct transfer reactions. The γ_α^2 for a particular resonance is a measure of how much that compound state looks like the target A in its ground state plus a neutron with definite orbital angular momentum (Lane and Thomas, 1958; Lynn, 1968).

4.8.2.2 The wave function at resonance

We can obtain a feeling for the physical significance of the requirement that $Ref = x = 0$ at a resonance by considering the simple case of neutrons incident when only elastic scattering is allowed, the other channels being closed. Then f is real or $y = 0$ and there is no absorption, $\sigma_{\text{abs}} = 0$. If the sphere $r = a$ has been chosen just outside the nuclear surface, we know that once a neutron crosses this sphere (so that $r < a$) it experiences a strong attraction. This attraction means an increase in kinetic energy and hence a decrease in wavelength. The effect is very marked at low bombarding energy because the exterior wavelength is then long but the interior wavelength is short; for example, a neutron entering the nucleus feels on average an attractive potential of strength up to 50 MeV (section 4.5.7.1), so that it has a wavelength Λ of about 4 fm. Outside, a neutron of 1 keV, say, has a wavelength λ of about 900 fm, over 200 times longer. *Figure 4.35* visualises this situation. We must match smoothly at $r = a$ the long wavelength exterior wave and the short wavelength interior wave (both magnitude and derivative must be continuous). When the logarithmic derivative f is large at the boundary, the greater curvature of the wave just inside the nuclear surface forces its amplitude to be much smaller than that of the exterior wave, by a factor of order Λ/λ. The boundary then acts as an efficient reflector, which is why it scatters like a hard sphere. However at resonance, when $f = 0$, the waves match at their crests and they have equal amplitudes. Although this picture is only schematic, the result remains true for the more general situation: a resonance corresponds to a greatly enhanced transmission of the incident particle across the nuclear boundary and into the interior because of good matching conditions which are expressed by $Ref = 0$.

It might be thought that this same argument would also imply that on resonance the particle could easily escape again, resulting in a very short lifetime for the compound system. This is not so, for once the particle has crossed the nuclear surface it is subject to the strong nuclear forces and suffers many collisions with the target nucleons. The chance of it finding its way back out with the same energy is only one of many possibilities and can be very small. The

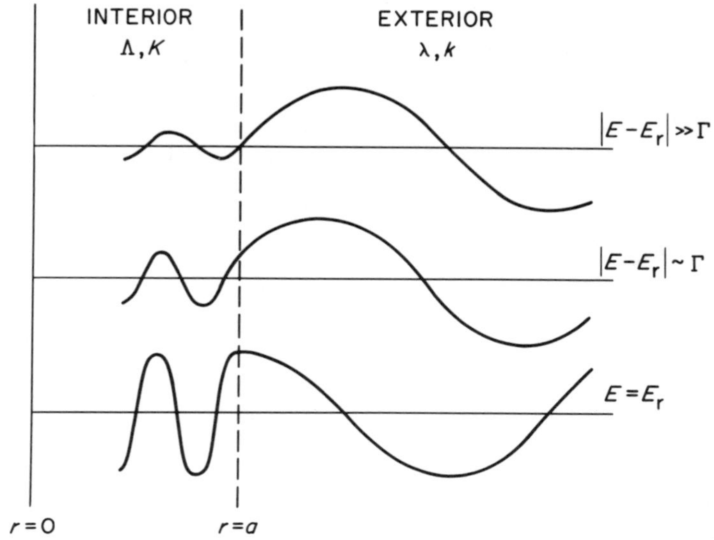

Figure 4.35 Illustrating the wave function for a neutron outside and just inside the surface ($r = a$) of a nucleus on, near and far from a resonance. For clarity the internal and external wavelengths are shown much more nearly equal than is usually the case; in practice, $\lambda \gg \Lambda$ so that the interior amplitude when off-resonance is much smaller than shown here

simple wave function describing the A + n configuration such as pictured in *Figure 4.35* for $r < a$ generally is only a small part of the very complicated compound nucleus wave function which describes the many-nucleon motions in the interior region. It is of importance only because it matches our initial conditions (neutrons bombarding nucleus A).

4.8.2.3 Time delay and interferences

The complicated motions that ensue after a particle has penetrated the nuclear surface and formed a compound nucleus tend to result in a time delay before that particle or another gathers enough energy to be re-emitted. The average of this delay is the compound lifetime, $\tau = \hbar/\Gamma$. The 'potential' scattering amplitude A_{pot} however, which is simply a consequence of the matching conditions at the nuclear surface $r = a$, corresponds to a wave which is reflected immediately and with which no time delay is associated. One may then ask how the potential and resonant scattered waves can interfere, as is implied by equation 4.91, if they are separated in time. This difficulty is resolved when we remember two things. First, because of the uncertainty principle we cannot specify exactly both time and energy simultaneously. When we speak of the time the particle spends interacting with the target, we are describing the behaviour of a small localised wave packet which has to be built up from waves with a range of energies ΔE. (In order to be able to specify time intervals of duration $\Delta t \lesssim \tau = (\hbar/\Gamma)$, we need an energy spread $\Delta E \gtrsim (\hbar/\tau) = \Gamma$. Note that $\hbar \approx 0.66 \times 10^{-21}$ MeV s.) However, equation 4.91 refers to waves of precisely determined energy ($\Delta E = 0$) and wavelength; we then have a continuous wavetrain and the potential and resonant scattered waves cannot be separated in time and space. There is a steady, continuous, flow into and out of the compound state.

Consequently, we integrate the cross-section equation 4.91 over a range ΔE of energies centred on $E = E_r$ with $\Delta E > \Gamma$. However, the reader may determine that the interference term between A_{pot} and A_{res} still does not vanish. Indeed

$$\int_{E-\frac{1}{2}\Delta E}^{E+\frac{1}{2}\Delta E} \sigma_{el}(E)\mathrm{d}E \approx \frac{\pi}{k^2}\left\{4\Delta E \sin^2 ka + \frac{2\pi\Gamma_\alpha^2}{\Gamma} - 4\pi\Gamma_\alpha \sin^2 ka\right\} \qquad (4.96)$$

the three terms being potential, resonant and interference scattering, respectively. We then remember the second fact. When we say that the compound state has a lifetime τ, we do not mean that its decay products are all emitted at a time τ after the formation of the state or even in a small interval around that time. On the contrary, the most probable time delay is zero. Just as in radioactive decay, the decay probability decreases exponentially from the time of formation, falling by e^{-1} in time τ, as the wave function 4.71 shows explicitly. Then the profile of the wave at time t after the centre of the incident wave packet hits the target is like that shown schematically in *Figure 4.36*. The precise form of the leading edge of the resonantly scattered wave depends upon the profile of the incident wave packet, which is also the profile of the 'potential' scattered wave. Hence there is always some overlap between the two scattered waves and therefore some interference. Since the width of the wave packet is $\sim \hbar v/\Delta E$ while the decay length of the resonant wave is $\sim \hbar v/\Gamma$, where v is the speed of the neu-

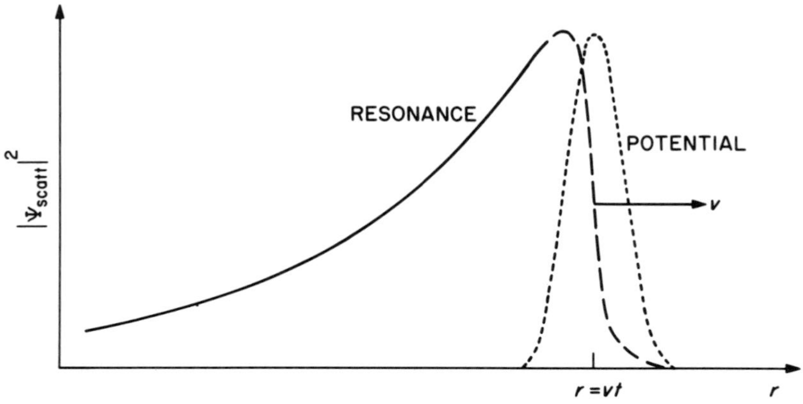

Figure 4.36 Profile of the intensity of a wave packet, moving with velocity v, at time t after its centre has struck a target nucleus. The shape-elastic or 'potential' wave and the resonantly scattered wave are shown separately. The trailing edge of the resonant wave intensity is proportional to $r^{-2} \exp[(r - vt)/v\tau]$ where $\tau = \hbar/\Gamma$ is the lifetime. In general the two waves will have different amplitudes.

trons, this overlap will be of order $\Gamma/\Delta E$. This is what we find for the interference term in equation 4.96 relative to the potential term.

The time delay between the absorption and emission of charged particles at a compound nucleus resonance can be observed by measuring the bremsstrahlung emitted during the reaction. A time delay τ leads to a phase difference $\exp(iE_\gamma \tau/\hbar)$ between the amplitudes of bremsstrahlung radiation (of energy E_γ) emitted before the formation and after the decay of the compound state. By observing the effects of the interference of these two amplitudes it is possible to measure the time delay. Such measurements have been made (Maroni *et al.*, 1976).

A simple account of the time dependence of the scattering of wave packets is given by Taylor (1972). *See* also Austern (1970), Yoshida (1974).

4.8.3 Resonances with charged particles

The cross-sections for scattering and reactions involving charged particles may also exhibit resonances. The repulsive Coulomb potential affects the values of the various partial widths Γ_α, but otherwise the single-level resonance formulae remain valid for non-elastic reactions when isolated resonances are observed. The Rutherford scattering amplitude must be added in for the elastic scattering and the appropriate changes made to the shape elastic term A_{pot}. Further, the *phase* of the resonance amplitude A_{res} is modified. Full details are given by Blatt and Weisskopf (1952), Lane and Thomas (1958).

4.8.4 Angular momentum and spin

Up to this point we have ignored angular momentum and its effects. In particular, a compound nucleus resonant state will be characterised by a definite spin J and parity π. This will impose constraints upon the partial waves that may contribute to its formation. For example, if the nuclei a and A do not have spins, the resonance can only occur in that partial wave with $\ell = J$. This implies a parity $\pi = \pi_a \pi_A (-)^J$ where π_a and π_A are the parities of a and A. If the compound state has the opposite parity, it cannot be formed at all in the a + A channel.

In addition to these selection rules, angular momentum introduces a *statistical weight* factor $g_\alpha(J)$ into the expression for the cross-section for formation of the compound state C. Further, if both particles a and A have spin, i_a and I_A say, more than one *channel spin* S and more than one partial wave ℓ may feed the resonance. For example, equation 4.78 then becomes

$$\sigma_{\alpha C}^J = \frac{\pi}{k_\alpha^2} \sum_{S\ell} \frac{g_\alpha(J) \Gamma_{\alpha S \ell}^J \Gamma^J}{(E - E_r)^2 + (\frac{1}{2}\Gamma^J)^2}, \quad g_\alpha(J) = \frac{2J+1}{(2i_a+1)(2I_A+1)} \tag{4.97}$$

where we have noted explicitly that the partial widths Γ_α will in general depend upon S and ℓ. The derivation of $g(J)$, the definition of channel spin and angular momentum considerations generally were discussed in detail in section 3.8.

4.8.5 Limits on the cross-sections

If the $\alpha = A + a$ channel is the only one open for the decay of the compound state C, then it accounts for the whole width of the state and $\Gamma^J = \sum_{S\ell} \Gamma^J_{\alpha S\ell}$. In this case, the maximum cross-section (at $E = E_r$) is

$$\sigma^J_{\alpha C}(\max) = \frac{4\pi}{k_\alpha^2}\, g(J) \tag{4.98}$$

From equation 4.97, $g(J)$ is a maximum for $i_a = I_A = 0$, so that $\ell = J$; then

$$\sigma^J_{\alpha C}(\max) = \frac{4\pi}{k^2}\,(2\ell + 1) \tag{4.99}$$

The non-elastic or absorption cross-section 4.89 is largest when the entrance channel α contributes only one-half of the total width, or $\Gamma_\alpha = \sum_{S\ell} \Gamma^J_{\alpha S\ell} = \frac{1}{2}\Gamma^J$. Then at the maximum ($E = E_r$) we have

$$\sigma_{abs}(\max) = \frac{\pi}{k^2}\, g(J) \tag{4.100}$$

or for spinless particles

$$\sigma_{abs}(\max) = \frac{\pi}{k^2}\,(2\ell + 1) \tag{4.101}$$

Because of the $(\pi/k^2) = \pi\lambda\!\!\!^{-2}$ factor, these cross-sections may be very large at low energy because $\lambda\!\!\!^{-}$ itself is large. If the energy is measured in eV, we have $\pi\lambda\!\!\!^{-2} \approx 6.5 \times 10^5/E$ barns.

4.8.6 Overlapping resonances

The discussion so far has been about a single, isolated resonance. However in general $Ref = x(E)$ (section 4.8.2) may have many zeros and hence the cross-section will exhibit many resonances. Even at low energies it often happens that the spacing in energy between two or more resonance levels is comparable to or smaller than their widths. In that case there can be interference between the corresponding resonance amplitudes. The effects can be seen by simply taking a sum of the single-level Breit–Wigner amplitudes such as in equation 4.93, one for each level. More sophisticated treatments are available however, such as use of the R-matrix theory (Lane and Thomas, 1958). General resonance theories have been reviewed by Robson (1974).

When the density of resonances is high and their overlap is considerable, we must resort to studying their average behaviour. This is discussed below in section 4.9.

4.8.7 Resonances as poles in the scattering matrix

Earlier we drew an analogy between a resonance seen in a nuclear reaction and the resonant response of an electric circuit to an applied voltage. We saw that resonant response corresponds to a minimum in the impedance Z or a maximum in the *admittance,* Z^{-1}. Now it is common in circuit theory to introduce complex numbers; instead of equation 4.69, the impedance is replaced by

$$Z(\omega) = R + i\left(\omega L - \frac{1}{\omega C}\right)$$

which we may rewrite

$$Z(\epsilon) = \frac{iL}{\omega}\left[(\epsilon - \epsilon_0) - \tfrac{1}{2}i\Gamma(\epsilon)\right]$$

using the same notation as in equation 4.70. In practice we encounter real values of ϵ, but $Z(\epsilon)$ may be generalised for complex ϵ by analytic continuation. Then we see that we would get infinite response ($Z \to 0$) when $\epsilon = \epsilon_0 + \frac{1}{2}i\Gamma$. More complicated circuits will exhibit a number of such poles in the complex-ϵ plane, each corresponding to a resonance. In a similar way, the scattering matrix $\eta(E)$ or $S(E)$ for a nuclear collision may possess poles in the complex-energy (E) plane corresponding to resonances; *see* equation 4.84 for example. There are extensive theoretical developments along these lines which we shall not pursue here (*see* the many texts on formal scattering theory for further details, for example Sitenko, 1971). It may be shown, for example, that *bound* states of the system correspond to poles in this generalised S-matrix at negative energies and these are closely related to the positive energy resonance poles. This reinforces our intuitive identification of a resonance as a quasi-stationary state of the compound system.

4.8.8 Isobaric analogue resonances

In general we only expect to see individual compound nucleus resonances at low bombarding energies where their widths are less than their spacings. Exceptions to this rule are the isobaric analogue resonances (IAR) which may occur at high excitation energy (10–20 MeV), yet still be relatively narrow ($\Gamma \sim 100$ keV), even in heavy nuclei. Such a resonance is distinguished from a continuous background of states by its isospin quantum number (*see*, for example, Wilkinson, 1969).

If we add a proton to a nucleus (N, Z) which has isospin $T_0 = \frac{1}{2}(N - Z)$, we can form a nucleus $(N, Z + 1)$ in states with isospin $T = T_0 + \frac{1}{2}$ or $T_0 - \frac{1}{2}$. (Here N and Z refer to the number of protons and neutrons, respectively.) The states with lowest energy will have $T = T_0 - \frac{1}{2}$. The lowest $T = T_0 + \frac{1}{2}$ state will

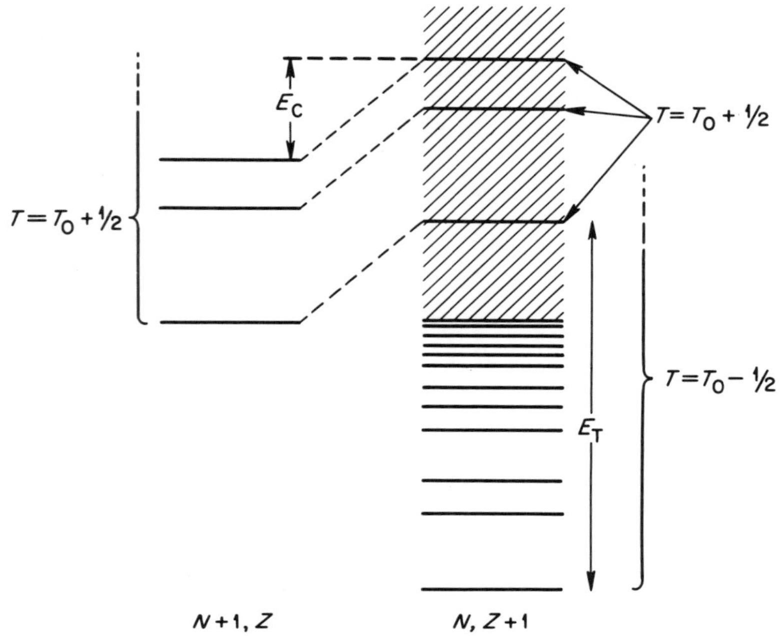

Figure 4.37 Energy-level diagram illustrating the energy relationships for isobaric analogue resonances. E_C is the shift due to the Coulomb potential of the extra proton and the neutron–proton mass difference. E_T is the excitation energy of the lowest state with $T = T_0 + \frac{1}{2}$ in the $(N, Z + 1)$ nucleus. The hatched area represents many close levels

be the analogue of the ground state of the adjacent $(N + 1, Z)$ nucleus which also has $T = T_0 + \frac{1}{2}$; that is to say, it will have the same structure except for the conversion of a neutron into a proton. (It will have a higher energy because of the additional Coulomb energy Δ_C experienced by the extra proton, less the neutron–proton mass difference. The energy shift is then $E_C = \Delta_C - (m_n - m_p)c^2$. The ground state of the $(N, Z + 1)$ nucleus, with $T = T_0 - \frac{1}{2}$, will be lower in energy because the added proton can be rearranged into a more tightly bound configuration. Consequently (*see Figure 4.37*) the IAR may have quite a high excitation energy in the $(N, Z + 1)$ compound nucleus; in ^{209}Bi, for example, the analogue of the ^{209}Pb ground state appears with an excitation of 18.7 MeV and is formed by 15-MeV protons incident on ^{208}Pb.)

There will be in the $(N, Z + 1)$ nucleus analogues of all the low-lying states in the $(N + 1, Z)$ nucleus (*see Figure 4.37*). The spacing between these $T = T_0 + \frac{1}{2}$ states in the $(N, Z + 1)$ nucleus will be very much larger than that between the dense $T = T_0 - \frac{1}{2}$ states at the same energy; consequently the former may show up as discrete resonances on a smooth background composed of the latter when we scatter protons from the (N, Z) nucleus.

When an IAR is formed by scattering protons from the (N, Z) nucleus, it appears as a resonance in the $(N, Z+1)$ compound nucleus. The various modes of decay of this IAR then can give information about its structure and, by implication, about the structure of the state in the $(N+1, Z)$ nucleus of which it is the analogue. Because it is formed by the addition of a nucleon and decays by the emission of a nucleon, the kind of information we gain is like that obtained from direct stripping and pick-up reactions (section 4.7.4). Alternatively, a charge exchange reaction, like (p, n) or (^{3}He, t), with the $(N+1, Z)$ nucleus as target can excite in the residual $(N, Z+1)$ nucleus the state which is the analogue of the target ground state simply by changing a target neutron into a proton. The Q-value of the (p, n) reaction is simply $Q = -\Delta C$ (*see Figure 4.37*).

An IAR has a finite width Γ for two reasons. The first, which we may denote $\Gamma\uparrow$, occurs because the state is unstable and may decay back into the entrance channel or into any other channels that are open to it. Secondly, the $T_0 + \frac{1}{2}$ state may mix with the $T_0 - \frac{1}{2}$ background states. If there were no Coulomb forces, the charge-independence of nuclear forces would ensure that isospin was

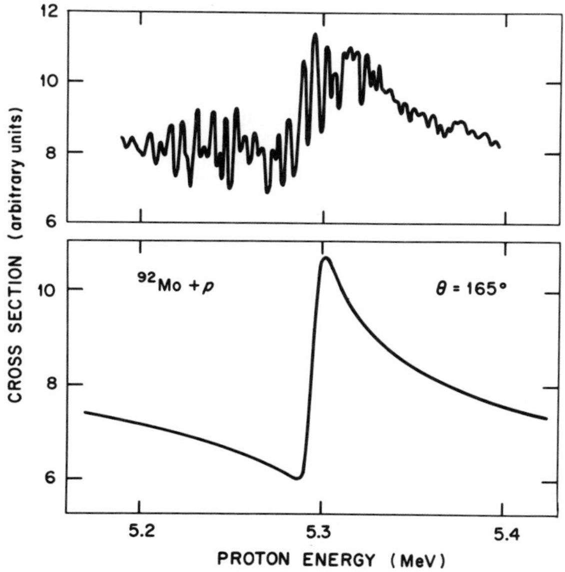

Figure 4.38 Comparison of the results of measurements on the excitation function for elastic scattering across an isobaric analogue resonance using good (above) and poor (below) energy resolution. (Note the depressed zero for the cross-section scale.) Poor resolution shows a simple resonance; better resolution shows that it is composed of many fine-structure components. The measurements shown were made at an angle of 165° for protons scattering from a ^{92}Mo target. Hence the resonance is in the compound nucleus ^{93}Tc and is the analogue of a state in the nucleus ^{93}Mo. (After Richard *et al.*, 1964)

a good quantum number and such mixing would not occur. However the presence of the Coulomb interaction means that isospin is not a precise quantum number and it causes some mixing. This broadens the IAR further; it contributes a width which we may denote $\Gamma\downarrow$, so that $\Gamma = \Gamma\uparrow + \Gamma\downarrow$. Indeed, under favourable circumstances the IAR can be resolved experimentally into a number of fragments corresponding to the $T = T_0 - \frac{1}{2}$ states into which some $T = T_0 + \frac{1}{2}$ character has been mixed. *Figure 4.38* illustrates this by showing the results of measurements with both poor and good energy resolution.

This kind of fragmentation of a simple state by mixing into more complicated background states will be met again when we discuss single-particle potential resonances and strength functions (section 4.9.5).

4.9 CONTINUUM OR STATISTICAL THEORY OF THE COMPOUND NUCLEUS

Even if isolated resonances are observed for low bombarding energies, as the energy is increased they tend to occur closer together and their widths tend to increase until finally we have the complete overlapping of many resonances. On average the cross-sections are then smoothly varying functions of the energy, although in some cases measurements with good energy resolution may still detect rapid and apparently random variations (or 'fluctuations') with energy. These are discussed later in section 4.9.4. First, however, we shall discuss the average behaviour of cross-sections in this so-called *continuum* region.

The formation of a compound nucleus through an isolated resonance leads to cross-sections of the Breit–Wigner type, equation 4.81, which exhibit the independence of formation and decay described in section 2.18.1. For the reasons discussed there, we expect to see the independence hypothesis, equation 2.49, satisfied in the continuum region also. This hypothesis may be written

$$\sigma_{\alpha\beta} = \sigma_{\alpha \mathrm{C}}\, G_\beta^{\mathrm{C}} = \sigma_{\alpha \mathrm{C}}\, \Gamma_\beta^{\mathrm{C}}/\Gamma^{\mathrm{C}} \qquad (4.102)$$

where we have expressed the relative probability G_β^{C} of decay into the β channel as the ratio of a partial 'width' Γ_β to a total 'width' Γ. Here a width is a convenient device for expressing a decay rate (or reciprocal of a lifetime) in energy units, through analogy with the uncertainty relation

$$\Gamma_\beta = \hbar/\tau_\beta, \quad \Gamma = \sum_\beta \Gamma_\beta = \hbar/\tau \qquad (4.103)$$

etc. These quantities are not widths in the sense of being the widths of definite energy levels, although they can be interpreted in terms of the average widths of the underlying compound nucleus resonance levels (*see* below, section 4.9.4).

The reciprocity relation, equation 2.47, gives another expression for G_β^{C}. Let

$\sigma_{\beta\alpha}$ be the cross-section for the inverse reaction, measured at the same total energy, then (ignoring spins)*

$$k_\alpha^2 \, \sigma_{\alpha\beta} = k_\beta^2 \, \sigma_{\beta\alpha}$$

Using the form 4.102, we get immediately

$$G_\beta^C / G_\alpha^C = k_\beta^2 \, \sigma_{\beta C} / k_\alpha^2 \, \sigma_{\alpha C}$$

or, since $\sum\limits_\beta G_\beta^C = 1$

$$G_\beta^C = \frac{k_\beta^2 \, \sigma_{\beta C}}{\sum\limits_\gamma k_\gamma^2 \, \sigma_{\gamma C}} \tag{4.104}$$

where the sum runs over all the open channels γ by which the compound nucleus C can be formed at that energy or into which it can decay. (Although all these channels are open, the corresponding formation cross-sections $\sigma_{\gamma C}$ are seldom susceptible to direct measurements because they mostly would involve targets in excited states.)

If there are very many channels available for the decay of C, then any individual G_β^C may be quite small. In particular this means that the contribution to elastic scattering from compound formation and decay will be small and the scattering will be dominated by the shape-elastic or potential term (*see* section 4.9.5). Also, the emission of a charged particle is strongly inhibited by the Coulomb barrier when one is not very far above its threshold energy. Hence, provided it is energetically allowed, neutron emission will tend to be the dominant decay mode (*see* section 6.8). If that is the case, the compound formation cross-section may be directly measured by measuring the cross-section for neutron production.

The hypothesis of the independence of formation and decay is subject to the constraints of conservation laws. Besides the total energy, the total angular momentum J, its projection J_z on the z-axis and the parity π of the system have to be conserved also. Various amounts of orbital angular momentum of the incident particle relative to the target will contribute, and this means that in general various values of the total J and π will participate. These J and π are also the spin and parity of the intermediate compound nucleus. Conservation of J

*If the nuclei have spins, we must include the corresponding statistical weight factors g (*see* section 3.8). Then we would have

$$\frac{k_\alpha^2 \sigma_{\alpha\beta}}{g_\alpha} = \frac{k_\beta^2 \sigma_{\beta\alpha}}{g_\beta}\,; \quad G_\beta^C = \frac{k_\beta^2 \, \sigma_{\beta C} / g_\beta}{\sum\limits_\gamma k_\gamma^2 \, \sigma_{\gamma C} / g_\gamma}$$

and π means that the independence hypothesis, equation 2.49, will only hold in general for the contribution from each pair of J and π values

$$\sigma_{\alpha\beta}(E, J, \pi) = \sigma_{\alpha}^{\mathrm{C}}(E, J, \pi)\, G_{\beta}^{\mathrm{C}}(E, J, \pi) \tag{4.105}$$

Since different modes of forming the compound system may result in different distributions of J and π values, they will sample somewhat different sets of compound states. As a consequence, the cross-section, when summed over all possible J and π

$$\sigma(E) = \sum_{J,\pi} \sigma(E, J, \pi) \tag{4.106}$$

may show departures from the simple overall independence (*see*, for example, Fluss *et al.*, 1969). This is especially true for heavy-ion collisions which involve very large amounts of angular momentum compared to formation of the same compound system by, say, neutron capture which brings in only small amounts of angular momentum. Similarly, neutron emission from the compound nucleus will only carry away a small amount of angular momentum. Consequently (HI, xn) reactions (where HI means heavy ion and xn denotes emission of one or more neutrons) tend to decay to states in the residual nucleus with high spins; indeed they are an invaluable tool for studying such states. At the same time, these reactions will only exhibit independence of formation and decay in the generalised sense of equations 4.105, 4.106.

4.9.1 Statistical model for formation and decay of the compound nucleus

A 'black nucleus' or strong absorption model illustrates the main features. We assume that the incident particle is absorbed once it has penetrated into the nuclear interior. Hence the cross-section $\sigma_{\alpha\mathrm{C}}$ is proportional to the transmission coefficient T_α, the probability that the particle a will be able to cross the surface of the nucleus A

$$\sigma_{\alpha\mathrm{C}} = \frac{\pi}{k_\alpha^2}\, g_\alpha T_\alpha \tag{4.107}$$

(where g_α is the statistical weight due to angular momentum—*see* section 3.8.1). Now T_α may have a magnitude as large as unity but it is generally less than unity because the potential of mutual interaction of the pair (including the Coulomb barrier if the projectile a is not a neutron) causes the waves to be partially reflected at the surface. If partial waves with $\ell > 0$ are considered, the centrifugal barrier may also result in $T_\alpha < 1$. This just means that the larger the value of ℓ, the less chance that the particle will hit the nucleus. The simplest assumption would be the sharp cut-off model of section 4.1; this means $T_\alpha = 1$ for $\ell \leqslant kR$ and $T_\alpha = 0$ for $\ell > kR$, where R is the nuclear radius. This ignores the reflections that may occur at the surface.

The simple example of S-wave neutrons was studied in section 3.6.1, where it was assumed that there were only ingoing waves just inside the nuclear surface. (This is often called the black nucleus model.) This gave a reflection coefficient

$$R = \left|\eta_0\right|^2 = \left|\frac{K - k}{K + k}\right|^2$$

and a transmission coefficient

$$T = 1 - R = \frac{4kK_0}{|K + k|^2} \tag{4.108}$$

where $K = K_0 + ib$ is the complex wave number in the absorptive region just inside the nucleus. If $K \gg k$, $T \approx 4kK_0/(K_0^2 + b^2)$. This is the case for low-energy neutrons and with equation 4.107 this gives again the $1/v$ law for formation of the compound nucleus by neutron capture

$$\sigma_{\alpha C} \approx \frac{4\pi K_0}{k_\alpha(K_0^2 + b^2)}\, g_\alpha = \frac{4\pi K_0 \hbar}{\mu_\alpha(K_0^2 + b^2)}\, g_\alpha\left(\frac{1}{v_\alpha}\right) \tag{4.109}$$

A more sophisticated treatment of the transmission across the nuclear surface uses the optical model (section 4.5). This model also allows us to include the cases where absorption is not complete when the surface is crossed but there remains a finite chance of escape. However, the cross-sections are still written in the form 4.107 and the factor T_α is still referred to as a transmission coefficient.

Using equations 4.102, 4.104 and 4.107, the statistical model expression for the cross-section for the A(a, b)B reaction becomes

$$\sigma_{\alpha\beta} = \frac{\pi}{k_\alpha^2}\, \frac{g_\alpha\, T_\alpha\, T_\beta}{\sum_\gamma T_\gamma} \tag{4.110}$$

As it stands, this expression is somewhat schematic. When using it, we must take proper account of the conservation of angular momentum and parity. There is a contribution like 4.110 for each value of the total angular momentum and parity and each partial wave and channel spin.

This formalism is often referred to as Hauser–Feshbach theory; it has been reviewed in detail by Vogt (1968). It can be extended to predict differential cross-sections also. One characteristic of the angular distributions of the reaction products resulting from the statistical model is that they are symmetric about $\theta = 90°$ (in the CMS), in contrast to those from direct reactions which tend to be peaked in the forward direction (*see Figure 2.33*). Compound reactions which do not involve large amounts of angular momentum (such as initiated by nucleons and other light ions at low and medium energies) have angular distributions which generally do not vary rapidly with angle (that shown schematically in *Figure 2.33* is typical). However, if there is a large amount of angular momentum (as can be the case in heavy-ion reactions), the angular distribution tends to be more strongly peaked in the forward ($\theta = 0°$) and backward ($\theta = 180°$)

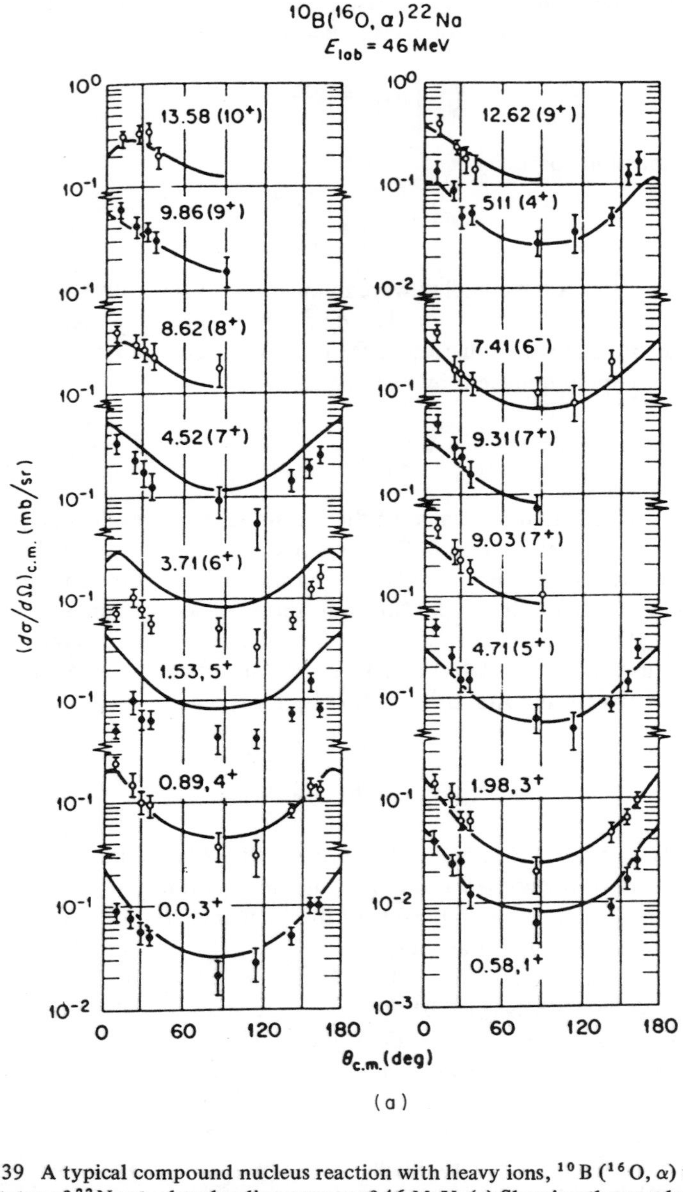

Figure 4.39 A typical compound nucleus reaction with heavy ions, ^{10}B (^{16}O, α) to various excited states of ^{22}Na at a bombarding energy of 46 MeV. (a) Showing the angular distributions to be strongly peaked in the forward and backward directions and symmetric about 90°. (Note that the cross-section scale is logarithmic.) The curves were calculated using the Hauser–Feshbach theory described in section 4.9. (b) Showing that the sum of the differential cross-sections to various excited states has the $1/\sin\theta$ form predicted by simple classical considerations. (From Gomez del Campo *et al.*, 1974)

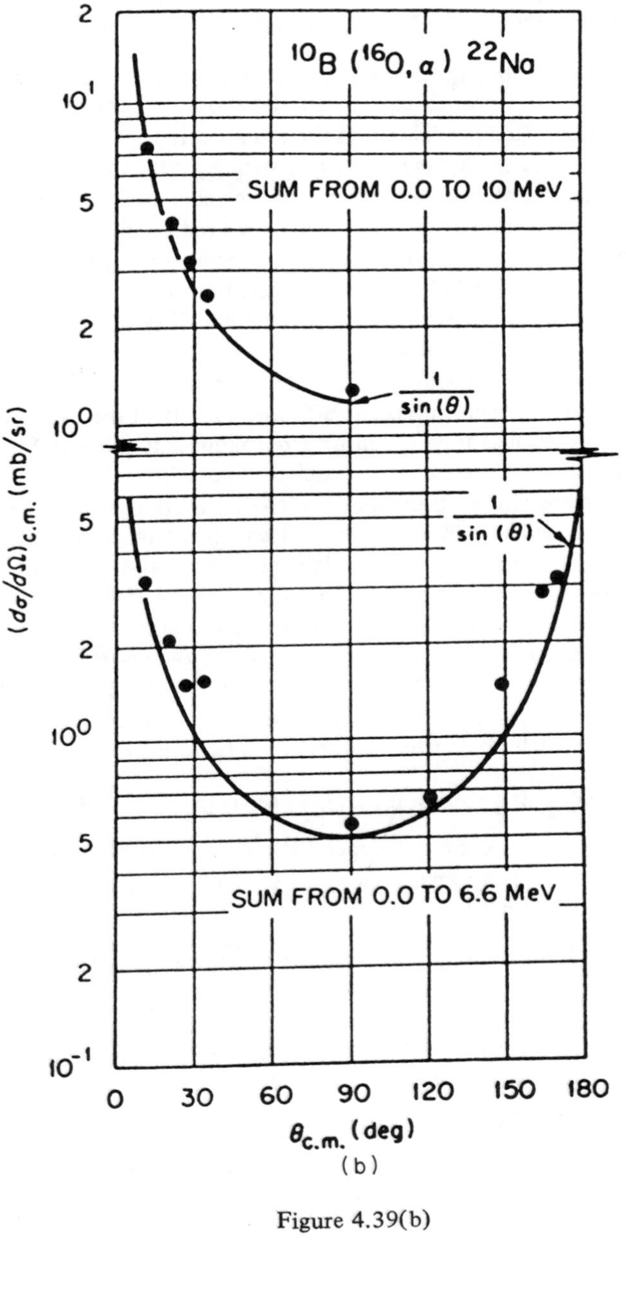

$^{10}B(^{16}O,\alpha)^{22}Na$
SUM FROM 0.0 TO 10 MeV
1/sin(θ)
SUM FROM 0.0 TO 6.6 MeV
$(d\sigma/d\Omega)_{c.m.}$ (mb/sr)
$\theta_{c.m.}$ (deg)
(b)

Figure 4.39(b)

directions (*see Figure 4.39* for example); in the limit this tends to $1/\sin\theta$ (section 2.18.7).

4.9.2 *The evaporation model for decay of the compound nucleus*

The branching probability for decay to a particular state of a particular final nucleus is given by G_β^C, which equation 4.104 shows to be proportional to the energy E_β of the emitted particle times the cross-section for the inverse capture process. When E_β is large, G_β^C is not varying strongly with the emitted energy. However, the number of available levels in the residual nucleus increases rapidly (exponentially) as its excitation energy increases (and hence as the emitted particle energy decreases). But then as the emitted energy E_β becomes small, the transmission or penetrability for the particle becomes more important and G_β^C decreases rapidly. For example, for neutron emission the $1/v$ law shows that G_β^C is proportional to $E_\beta^{1/2}$ for small E_β. These two facts explain the main features of the neutron spectrum illustrated in *Figure 2.36*. When the emitted particle is charged, the spectrum is cut off more sharply at the low energy end, as in *Figure 2.36*, because the Coulomb barrier reduces the penetrability more rapidly as the energy of the emitted particle is reduced.

These spectra are similar to the Maxwellian distribution in energy of particles emitted from a small hole in a container of gas in thermal equilibrium. This suggests that the compound nucleus might be treated as a thermodynamic system. First we note from equation 4.104 that the probability of decay to a particular level of the residual nucleus B is proportional to the energy E_β of the emitted particle b times the cross-section $\sigma_\beta(E_\beta)$ for the inverse capture. Next, let the number of levels in B with excitations from ϵ to $\epsilon + d\epsilon$ be $\rho(\epsilon)d\epsilon$. If the maximum kinetic energy available in the channel β is $E_{\beta 0}$, then $\epsilon = E_{\beta 0} - E_\beta$. The number of particles b emitted with energies E_β to $E_\beta + dE_\beta$ is then

$$n(E_\beta)dE_\beta = \text{constant} \times E_\beta\, \sigma_\beta(E_\beta)\rho(E_{\beta 0} - E_\beta)dE_\beta \tag{4.111}$$

The level density $\rho(\epsilon)$ is known to depend strongly on the excitation energy so, rather than make a linear expansion of ρ itself, let us expand its logarithm, $\mathcal{S}(\epsilon) = \log \rho(\epsilon)$, about $\epsilon = E_{\beta 0}$

$$\mathcal{S}(E_{\beta 0} - E_\beta) \approx \mathcal{S}(E_{\beta 0}) - E_\beta(d\,\mathcal{S}(E)/dE)_{E=E_{\beta 0}} \tag{4.112}$$

This immediately gives

$$\rho(E_{\beta 0} - E_\beta) = \rho_0 \exp(-E_\beta/T) \tag{4.113}$$

where

$$\rho_0 = \exp(\mathcal{S}(E_{\beta 0})) \quad \text{and} \quad T^{-1} = (d\,\mathcal{S}/dE)_{E=E_{\beta 0}} \tag{4.114}$$

(Thus we see equation 4.112 is equivalent to approximating ρ by an exponential, rather than a linear, dependence on energy in the vicinity of $\epsilon = E_{\beta 0}$.) Then equation 4.111 becomes

$$n(E_\beta)dE_\beta = \text{constant}' \times E_\beta\, \sigma_{\beta C}(E_\beta)\exp(-E_\beta/T) \tag{4.115}$$

If the energy dependence of $\sigma_{\beta C}$ can be ignored, this is just the Maxwellian distribution for a temperature T (if we measure T in units of energy k_B where here k_B means Boltzmann's constant.) This interpretation is valid because $\rho(\epsilon)\mathrm{d}\epsilon$ is the number of states available to nucleus B in the range $\mathrm{d}\epsilon$ at excitation ϵ, hence $\mathcal{S}(\epsilon)$ can be considered to be its entropy at that excitation. Then equation 4.114 is the usual thermodynamic relation between temperature and entropy. We note that it is the temperature of the *residual* nucleus, not the temperature of the compound nucleus before emission.

The Maxwellian spectrum is distorted by the energy dependence of the absorption cross-section $\sigma_{\beta C}$. For neutrons this dependence is weak except for small E_β. The Coulomb barrier causes a large distortion for charged particles, as shown in *Figure 2.36*. Thus we see the overall features of the spectrum are due to the temperature of the system, which determines the availability at the surface of particles with various energies, while $\sigma_{\beta C}$ describes the probability that they will escape across the surface.

The maximum of the distribution 4.115 (if we ignore the energy dependence of $\sigma_{\beta C}$) occurs at $E_\beta = T$ and the average energy is $2T$. The peak energy $E_\beta = T$ should be small compared to the maximum available, $E_{\beta 0}$, in order that the expansion 4.111 should be valid. Typical nuclear temperatures T are of the order of an MeV for excitations ϵ of 10 MeV or so. For a Fermi gas, T can be expressed in terms of ϵ as

$$\epsilon = aT^2 \tag{4.116}$$

where a is a constant for a given nucleus, but which varies with the mass A of the nucleus. We find empirically that $a \sim (A/10)\ \mathrm{MeV}^{-1}$. The relation 4.115 also can be used to give an expression for the entropy of the system

$$\mathcal{S}(\epsilon) = \int \frac{\mathrm{d}\epsilon}{T} = 2\,(a\epsilon)^{1/2} + \text{constant}$$

and hence for the level density

$$\rho(\epsilon) = \exp \mathcal{S}(\epsilon) = \text{constant} \times \mathrm{e}^{2(a\epsilon)^{1/2}} \tag{4.117}$$

From measured spectra we obtain a rough estimate of the value of the constant in equation 4.117 of about $(10/A)\ \mathrm{MeV}^{-1}$ for even-even nuclei, about $(20/A)$ MeV^{-1} for even-odd nuclei and about $(100/A)\ \mathrm{MeV}^{-1}$ for odd-odd nuclei.* (This behaviour reflects the greater stability of even-even nuclei as compared to odd-odd, with even-odd falling between.)

Since the most probable energy loss upon evaporation is quite small, the residual nucleus may be left with sufficient excitation to enable it to emit

*It must be emphasised that the considerations of this section are very schematic. The numerical values suggested for the constants are only given as a very rough guide. Indeed the form 4.117 for the level density is itself only a rough approximation. Certainly it is not applicable at low excitation energies where there are relatively few levels and even at higher excitations the more sophisticated theories give much more complicated expressions for

another neutron, and so on. Successive application of the above arguments then yields expressions for reactions like (a, 2n), (a, 3n), etc.

4.9.3 Pre-equilibrium decays

Our discussions so far have emphasised two extreme views of the progress of a nuclear reaction. A direct reaction is one that is completed after the initial collision of the projectile with, say, a target nucleon. The compound nucleus is viewed as being formed after the projectile is captured, initiating a number of such collisions which distribute the incident energy over the whole nucleus so that a kind of thermal equilibrium is attained. It is not clear how many collisions are necessary for this to occur. In any case, we can expect that there are events intermediate between these extremes, consisting of the escape of an appreciable part of the incident energy after a relatively few collisions. Indeed, the energy spectra of particles emitted from many reactions, particularly those induced by projectiles with relatively high energies, are found to deviate from the simple shapes shown in *Figure 2.36*. They show an excess of high-energy particles over those expected for evaporation from a 'hot' system in thermal equilibrium. *Figure 4.40* shows, as an example, the inelastic protons emitted after bombardment of ^{56}Fe by protons of 29, 39 and 62 MeV. The spectrum for the lowest incident energy appears to consist of an evaporation peak plus the high-energy proton groups from direct excitation of the low-lying states of ^{56}Fe. At the higher energies, however, it becomes clear that there is a flat continuum of protons with energies between these two extremes.

This emission, before full thermal or statistical equilibrium of the compound system is reached, is called *pre-equilibrium* or *precompound* decay. Such emission will contribute mainly to the higher-energy part of the spectrum of emitted particles, corresponding to energies $E_\beta \gg T$. For example, one description of these pre-equilibrium decays (Griffin, 1966) resulted in a term in the energy distribution which is only decreasing linearly with the energy E_β of the emitted

---

$\rho(\epsilon)$ (*see*, for example, Lynn, 1968; Huizenga and Moretto, 1972; Richter, 1974). In particular it is shown that level density depends upon the spin I of the levels

$$\rho(\epsilon, I) = \rho(\epsilon, 0)\,(2I + 1)\exp\left[-\frac{I(I+1)}{2\sigma^2}\right]$$

where σ is known as the *spin cut-off parameter*. The value of σ^2 varies with A but is of the order of $\sigma^2 = 10$. The factor $(2I + 1)$ arises simply because a 'level' of spin I actually constitutes $(2I + 1)$ degenerate magnetic substates. The exponential cut-off reflects the fact that, as I increases, more of the excitation energy ϵ is taken up in the rotational motion and less is available for thermal motion so that the effective temperature is reduced.

The more detailed considerations also lead to some additional dependence of $\rho(\epsilon)$ on ϵ over the exponential in equation 4.117 as well as taking into account odd–even differences and the effects of shell structure.

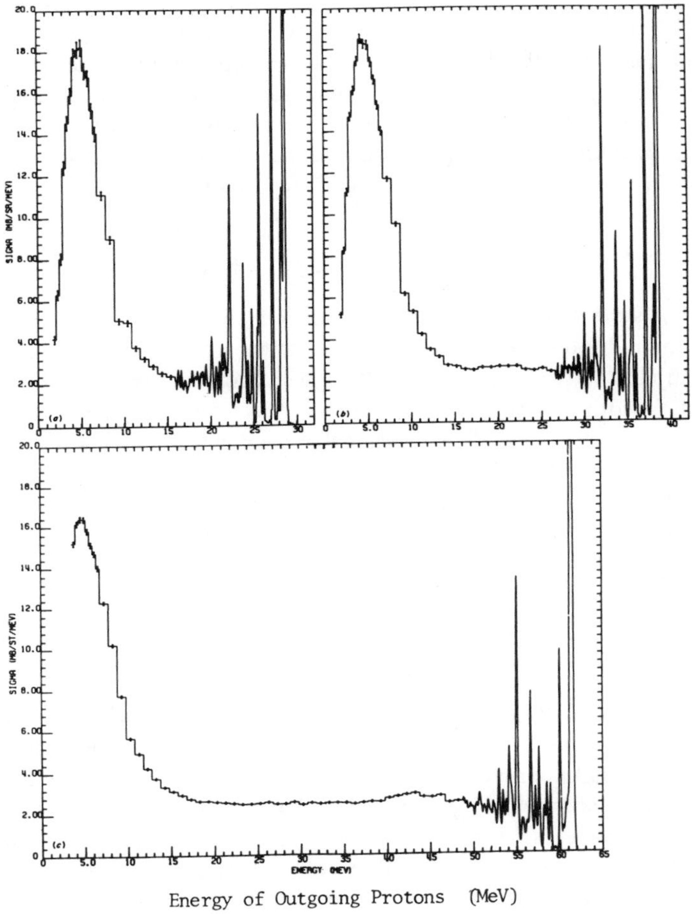

Figure 4.40 The energy spectra of inelastic protons emitted after bombardment of ^{55}Fe by protons of 29 (upper left), 39 (upper right) and 62 MeV (lower). The highest-energy protons emitted (right-hand end of spectra) correspond to the excitation (mainly by direct reactions) of discrete, low-lying, excited states in ^{56}Fe; the peaks to the left correspond to evaporation from the compound nucleus ^{57}Co. (From Bertrand and Peele, 1973, where examples of the spectra for other emitted particles and other targets are also given)

particle when E_β is large. The evaporation spectrum 4.114 is decreasing exponentially in this energy region so that any appreciable amount of precompound emission would dominate.

The evidence for these pre-equilibrium processes and the models used to describe them have been reviewed by Blann (1975) and Hodgson (1987).

4.9.4 Fluctuation phenomena

When we reach bombarding energies at which the compound resonances are dense and strongly overlapping, it still may not be true that the cross-sections vary smoothly with energy. One way of seeing this is to postulate that the transition *amplitude* can be written as a sum of single-level resonance terms (like that for elastic scattering in equation 4.93, for example)

$$A_{\beta\alpha} = \sum_j \frac{(\Gamma_\alpha^j \Gamma_\beta^j)^{1/2} \exp(i\,\delta_{\alpha\beta}^j)}{(E - E_r^j) + \frac{1}{2} i\,\Gamma^j} \tag{4.118}$$

where $\delta_{\alpha\beta}^j$ is the phase of the contribution from the jth level, at energy E_r^j. Now the compound states with which these resonance amplitudes are associated have very complicated structures at these high excitation energies and we can expect the partial width amplitudes $(\Gamma_\alpha)^{1/2}$, $(\Gamma_\beta)^{1/2}$, etc. and the phases $\delta_{\alpha\beta}$ to vary wildly from level to level. This suggests that we regard these quantities as random variables and treat them statistically. This enables us to express the average cross-section in terms of average values of the partial widths. But this view also implies *statistical fluctuations* about these average cross-sections. These fluctuations can be large. For example, in a reaction where there is a single entrance channel α and a single exit channel β, this approach predicts a root mean square deviation equal to the average itself, or

$$\overline{\sigma_{\alpha\beta}^2} - \overline{\sigma}_{\alpha\beta}^2 = \overline{\sigma}_{\alpha\beta}^2 \tag{4.119}$$

In practice, there is usually more than one channel contributing (several partial waves, channel spins, etc.); if these behave as statistically independent, they will reduce the overall fluctuations. If there are N channels contributing equally, the right side of equation 4.119 will be reduced by a factor N^{-1}. As the bombarding energy increases, so does the number of channels, and the fluctuations are eventually damped out and a smooth energy variation results. The averaging over a spread of energies owing to the finite energy resolution of an actual measurement has a similar effect. The existence of these fluctuations has been verified experimentally in a large variety of situations (*see Figure 4.41* for example).

We may ask how the values of the cross-sections are distributed. Let the distribution be $P(x)$, where $x = \sigma/\overline{\sigma}$ and $\overline{\sigma}$ is the average value, so that the probability of finding a cross-section ratio between x and $x + \delta x$ is $P(x)\delta x$. *Figure 4.41* includes histograms showing the measured distributions. If there were a single reaction amplitude with a random distribution of values, we would expect

$$P_1(x) = e^{-x}$$

This could occur if the nuclei in both exit and entrance channels all had zero spin, as for the transition to the ground state of ^{24}Mg in the $^{12}C(^{16}O, \alpha_0)$ reaction, and *Figure 4.41* indicates agreement with the data in this case. An interesting consequence of the distribution P_1 is that the most probable value of the cross-section is zero!

If there were N statistically independent amplitudes with equal variances, the distribution would be

$$P_N(x) = \frac{N}{(N-1)!} (Nx)^{N-1} e^{-x}$$

There can be more than one amplitude when any of the nuclei has non-zero spin, corresponding to transitions between different magnetic substates. *Figure 4.4!* shows that the data for transitions to the excited states of ^{24}Mg can be fitted with $N > 1$.

We may also study the *energy correlation* of such a fluctuating cross-section. A measure of the degree to which the average cross-section at energy E is correlated statistically with that at energy $E + \epsilon$ is given by

$$C(\epsilon) = \frac{\overline{\sigma(E+\epsilon)\sigma(E)} - \overline{\sigma(E)}^2}{\overline{\sigma(E)}^2} = \frac{\overline{(\sigma(E+\epsilon) - \overline{\sigma(E)})(\sigma(E) - \overline{\sigma(E)})}}{\overline{\sigma(E)}^2} \tag{4.120}$$

When the two bombarding energies E and $E + \epsilon$ are close together (i.e. when ϵ is small) many of the same compound states are excited and contribute to both cross-sections $\sigma(E)$ and $\sigma(E + \epsilon)$ in the same way. As ϵ increases, however, entirely different sets of compound levels are being sampled and the two cross-sections become unrelated; $C \to 0$ as ϵ increases. We may write this as

$$C(\epsilon) = \frac{\overline{\Gamma}^2}{\epsilon^2 + \overline{\Gamma}^2} \tag{4.121}$$

and $\overline{\Gamma}$ may be interpreted as the mean width of the underlying compound states. C becomes small when $\epsilon \gg \overline{\Gamma}$.

In general, more than one channel will contribute to a reaction; if the particles have spins, transitions between different pairs of magnetic substates are possible. Then, for example, equation 4.121 becomes

$$C(\epsilon) = \frac{1}{N} \frac{\overline{\Gamma}^2}{\epsilon^2 + \overline{\Gamma}^2} \tag{4.122}$$

if there are N independent channels.

The existence of direct reactions indicates that the amplitudes from different compound levels are not completely random but that there is some correlation between them. One way of seeing this is based upon the complementarity between time and energy. If the reactions are associated with a long time delay $\tau = \hbar/\overline{\Gamma}$ (on average) because of formation and decay of the compound system, a description of the process only needs knowledge about the nuclear states in a narrow energy interval $\Delta E \sim \overline{\Gamma}$. However, the existence of a rapid direct reaction,

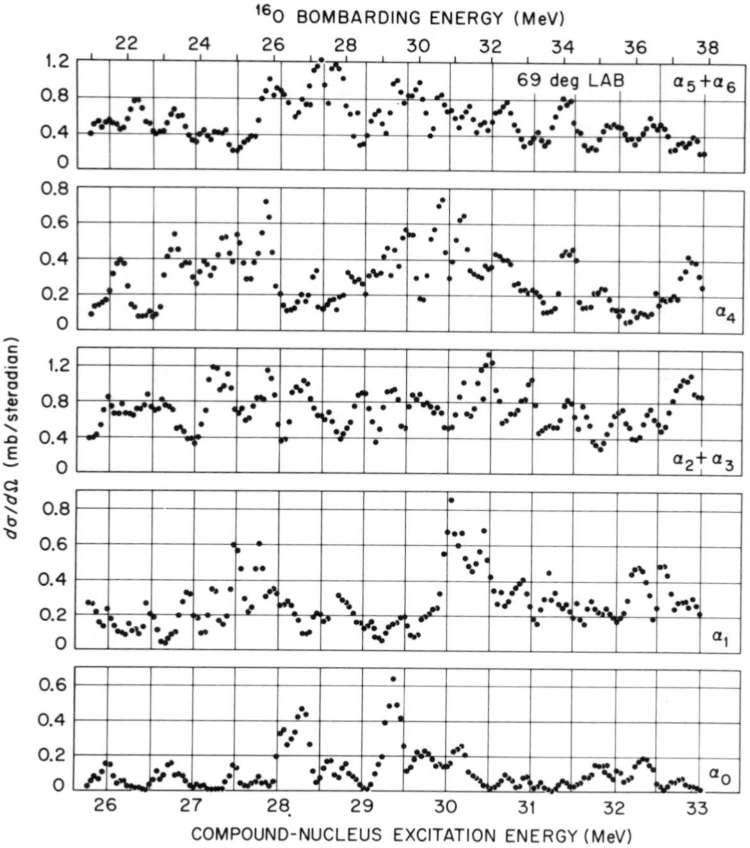

(a)

Figure 4.41 (a) Typical examples of fluctuations in the cross-section at $\theta_L = 69°$ for the reaction $^{12}C(^{16}O, \alpha)^{24}Mg$ to various excited states as the bombarding energy is varied. (b) The probability distributions $P(\sigma/\bar{\sigma})$ where $\bar{\sigma}$ is the average cross-section; the histograms are the experimental values, the curves are the statistical distribution functions P_N for N independent random variables. (From Halbert *et al.*, 1967)

with a short time delay $\ll \tau$, implies the coherent participation of nuclear states over a much wider range of energies $\Delta E \gg \bar{\Gamma}$. The presence of these correlations between the phases of the contributions to 4.118 from different resonances means that the average of the amplitude $A_{\beta\alpha}$ does not vanish. We may identify this average with the direct reaction amplitude which varies slowly with energy. The total amplitude then fluctuates about this average; the mean *square* deviation does not vanish and corresponds to the compound nucleus cross-section.

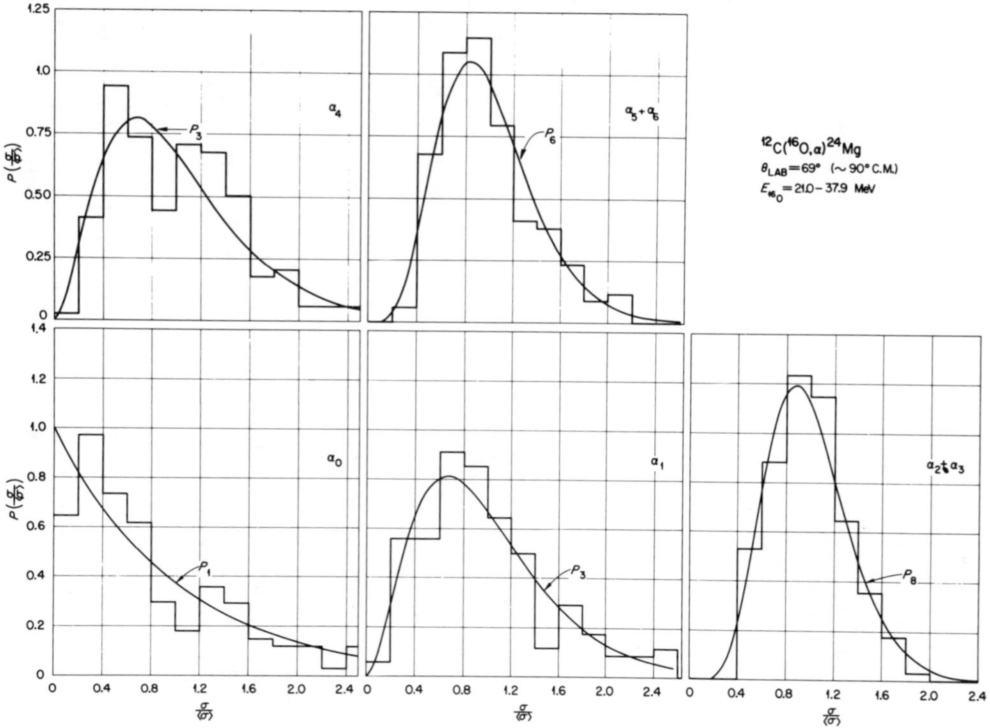

(b)

Figure 4.41(b)

In addition to these long-range correlations which correspond to direct reactions, there may be relatively strong *local* correlations among levels in a small energy region (over some hundreds of keV, for example, for medium-weight nuclei) which will lead to an *intermediate structure* in the cross-section versus energy curves even after they have been averaged over an interval much larger than the average width $\overline{\Gamma}$ of individual levels (*see*, for example, *Figure 2.27*). These structures may be associated with the effects of relatively simple excitations of the compound system, the so-called *doorway states* (*see*, for example, Feshbach, 1965, 1975; Kennedy and Schrils, 1968).

The existence of direct reactions and their associated correlations will also modify the energy correlation function 4.121 or 4.122. The direct-reaction contributions vary smoothly with energy and imply a correlation between the amplitudes for energies which differ by much more than $\overline{\Gamma}$. For the simple case of $N = 1$, equation 4.121 becomes

$$C(\epsilon) = (1 - y^2)\,\frac{\overline{\Gamma}^2}{\epsilon^2 + \overline{\Gamma}^2} \qquad (4.123)$$

where y is a measure of σ_{D}, the direct part of the average cross-section $\overline{\sigma}$

$$y = \frac{\sigma_{\mathrm{D}}}{\overline{\sigma}} \qquad (4.124)$$

There is an extensive theory of statistical fluctuations in nuclear reactions, initiated by Brink and also by Ericson, among others. They are often referred to as 'Ericson fluctuations'. (*See*, for example, reviews by Ericson, 1966; Ericson and Mayer-Kuckuk, 1966.)

4.9.5 Direct reactions

Although direct and compound nuclear reactions represent opposite extremes in reaction processes, they are not mutually exclusive. Both may contribute to a given reaction, especially at low bombarding energies, so that in the present context the total reaction amplitude may be written

$$A_{\beta\alpha} = A^{\mathrm{D}}_{\beta\alpha} + A^{\mathrm{C}}_{\beta\alpha} \qquad (4.125)$$

Following the discussion in section 4.9.4, we can identify* the direct amplitude

*The reader may be confused because the present discussion, especially equation 4.125, implies a clean distinction between direct reactions and compound nuclear reactions, whereas earlier discussions (e.g. sections 4.7.6 and 4.9.3) had indicated that direct and compound represent the extremes of a whole spectrum of possibilities of varying complexity. This is because the definition 4.126 of direct reactions, while unambiguous, is not identical to a definition in terms of, for example, the Born series of perturbation theory (sections 3.3.3 and 4.7.2). However, the underlying physical ideas are similar in both cases and, to some degree, it is a matter of convention (and convenience) which definition we adopt (*see*, for example, the discussion by Austern, 1970).

$A^{\mathrm{D}}_{\beta\alpha}$ with the average of the amplitude 4.118. The compound amplitude $A^{\mathrm{C}}_{\beta\alpha}$ can still be written in the form of equation 4.118 but with modified numerators such that the average of $A^{\mathrm{C}}_{\beta\alpha}$ is zero. Consequently $A^{\mathrm{D}}_{\beta\alpha}$ will vary smoothly with energy while $A^{\mathrm{C}}_{\beta\alpha}$ fluctuates rapidly about zero

$$\overline{A_{\beta\alpha}} = A^{\mathrm{D}}_{\beta\alpha}, \quad \overline{A^{\mathrm{C}}_{\beta\alpha}} = 0 \tag{4.126}$$

At a given energy the two amplitudes $A^{\mathrm{D}}_{\beta\alpha}$ and $A^{\mathrm{C}}_{\beta\alpha}$ are coherent and interfere. However if we average over a range of energies the fluctuating nature of $A^{\mathrm{C}}_{\beta\alpha}$ relative to $A^{\mathrm{D}}_{\beta\alpha}$ ensures that the interference terms will average to zero.* Then we may simply add the cross-sections calculated for the two types of process. Typically we could use the distorted-wave theory (section 4.7.2) for the direct reaction part and the Hauser–Feshbach theory (section 4.9.1) for the compound part. *Figure 4.42* shows examples of this kind. As the bombarding energy is increased, the number of channels which are open for the decay of the compound nucleus becomes large. The probability that it will decay into those channels leading to low excited states of the residual nucleus (hence those associated with a high energy for the emitted particle) becomes correspondingly small. However these are the channels fed preferentially by direct reactions so that only at low energies are both processes likely to be of comparable importance. Exceptions may occur if the direct process is inhibited by some selection rule; even then it is possible that a semi-direct process (section 4.7.6) may be as important as a true compound nucleus process.

4.10 THE OPTICAL MODEL AT LOW ENERGIES AND THE NEUTRON STRENGTH FUNCTION

One of the earliest applications of the optical model (section 4.5) was to the scattering and absorption of low-energy neutrons, including the energy regions where discrete resonances are observed. Clearly the model can only be used to interpret energy-averaged quantities in these circumstances.

Let us consider the relationship between the energy-averaged elastic scattering amplitudes $\overline{\eta}$ and the energy-averaged cross-sections $\overline{\sigma}$. For simplicity, we restrict

*Again the taking of an average over energy may be thought of as a way of specifying times more precisely by building localised wave packets. Then the vanishing of the interferences after averaging corresponds to the appearance of a time delay between the emission of the products from the fast direct reaction and those from the slower compound decay. Such time delays have been measured (Kanter *et al.*, 1975) by using the crystal blocking technique. The target used is in crystalline form. The yield of charged particles from such nuclei emitted along a crystallographic plane or axis is reduced by scattering from the other atoms; the row of atoms casts a 'shadow'. If the emission following a nuclear reaction is prompt, it will originate from a lattice site and suffer this blocking. However, if there is a time delay, the recoiling nucleus will move out of the lattice site before emission and the blocking will be reduced.

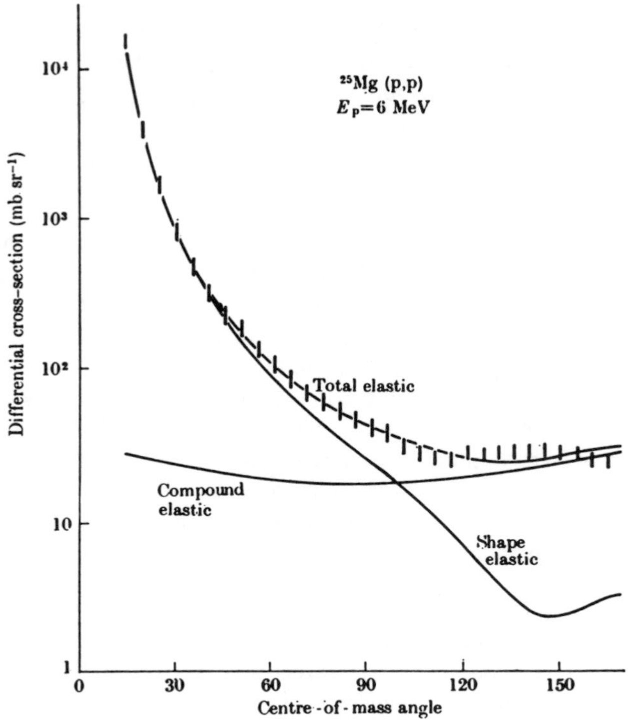

Figure 4.42 Examples of reactions where both the formation and decay of a compound nucleus and a direct reaction contribute. Left, an example of elastic proton scattering at a fairly low energy (from Gallmann *et al.*, 1966). Right, a deuteron stripping reaction (from Porto *et al.*, 1969). The compound nucleus cross-sections (dotted curves) were calculated using the Hauser–Feshbach theory described in section 4.9. According to the statistical assumption, there is no interference between the compound and direct contributions (dashed curves) so that these cross-sections simply add to give the resultant cross-section (full curve)

ourselves to *s*-waves. One simple way of defining the average over energy of a quantity $f(E)$ is to take

$$\overline{f(E)} = \frac{1}{\Delta} \int_{E-\frac{1}{2}\Delta}^{E+\frac{1}{2}\Delta} f(E)\mathrm{d}E \tag{4.127}$$

where Δ is large enough to include many resonances so that we have $\Delta \gg D$, where D is the average spacing between resonances, as well as $\Delta \gg \overline{\Gamma}$. From sections 3.4 and 3.5, the averaged total, elastic and non-elastic cross-sections may be given in terms of the elastic scattering amplitude η. Ignoring any nuclear spins

$$\overline{\sigma_{\text{tot}}} = \frac{2\pi}{k^2} (1 - Re\,\overline{\eta}) \tag{4.128}$$

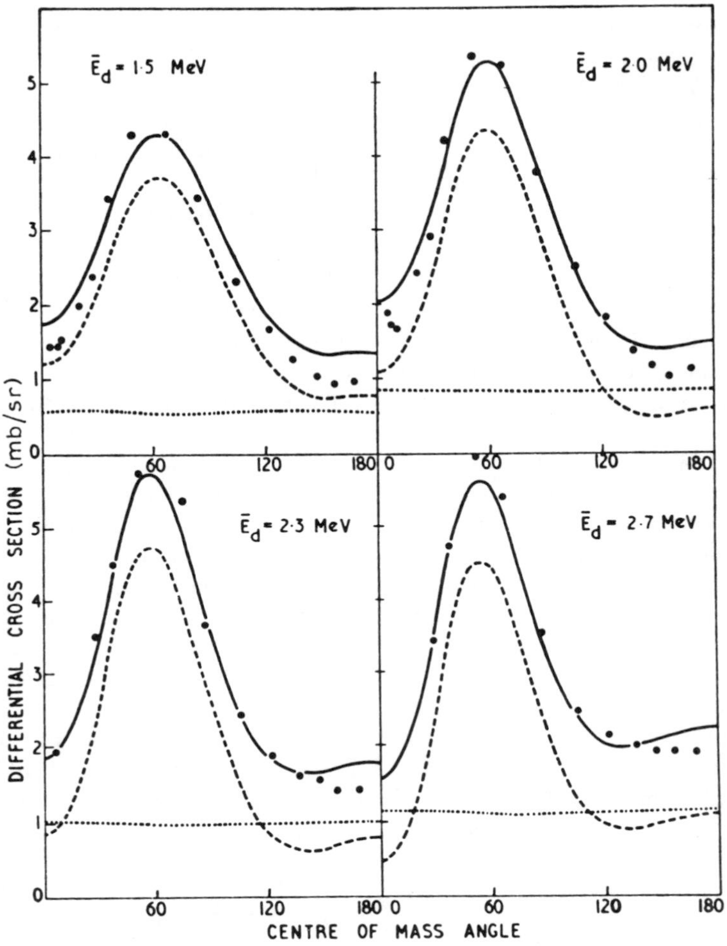

$$\overline{\sigma_{el}} = \frac{\pi}{k^2}\ \overline{|1-\eta|^2} = \frac{\pi}{k^2}\ |1-\overline{\eta}|^2 + \sigma_{fluc} \tag{4.129}$$

$$\overline{\sigma_{ne}} = \frac{\pi}{k^2}\ (1 - \overline{|\eta|^2}) = \frac{\pi}{k^2}(1 - |\overline{\eta}|^2) - \sigma_{fluc} \tag{4.130}$$

where we have assumed that k^2 does not vary significantly over the range Δ. (Alternatively we could discuss the averages of $k^2\sigma$ rather than σ itself.) We have been able to write the last two cross-sections in terms of the average scattering amplitude $\overline{\eta}$ by introducing the quantity

$$\sigma_{fluc} = \frac{\pi}{k^2}\ (\overline{|\eta|^2} - |\overline{\eta}|^2) \tag{4.131}$$

This is proportional to the mean square deviation of η from its average value $\overline{\eta}$, which is why it is called the fluctuation cross-section.

The equation 4.128 suggests that we require the optical model to yield the average scattering amplitude $\overline{\eta}$, and hence to give the correct $\overline{\sigma_{tot}}$, rather than require it to give the average elastic cross-section $\overline{\sigma_{el}}$. Let us denote the elastic and absorption cross-sections obtained from this optical model (i.e. from $\overline{\eta}$) by σ_{se} and σ_a, respectively. Then equations 4.127, 4.128 read

$$\overline{\sigma_{el}} = \sigma_{se} + \sigma_{fluc}, \quad \overline{\sigma_{ne}} = \sigma_a - \sigma_{fluc} \tag{4.132}$$

This suggests an interpretation of σ_{fluc}. In the optical model as defined here, σ_a corresponds to the flux removed from the entrance channel either to form the compound nucleus or to go into direct reactions. The difference σ_{fluc} between this and the true non-elastic cross-section σ_{ne} must then correspond to re-emission of part of this flux from the compound nucleus back into the elastic channel.* We may call this *compound elastic* scattering, or $\sigma_{fluc} = \sigma_{ce}$. The potential scattering σ_{se} is then called *shape elastic*. (This name arises because this contribution to the scattering is due to a nucleus having a quite well-defined surface or 'shape' and to the matching conditions at that surface.) We can simply add cross-sections to obtain σ_{el} rather than having to add amplitudes because, as discussed in the preceding sections, the interference terms average to zero when we average over energy. As mentioned earlier, the corresponding time delay between the compound elastic and shape elastic emissions has actually been observed by crystal blocking techniques (Kanter *et al.*, 1975).

We may average over energy explicitly if we express the amplitude as a sum of resonance terms, as in equation 4.118. For elastic scattering we must add the 'potential' term (as in equation 4.88) so that the scattering matrix element η for neutron elastic scattering becomes

$$\eta = \exp(-2i\delta)\left[1 - \sum_j \frac{i\Gamma_n^j}{(E - E_j) + \frac{1}{2}i\Gamma^j}\right] \tag{4.133}$$

*In section 3.4, the non-elastic cross-section σ_{ne} was denoted σ_{abs}, or absorption cross-section. One must be careful to distinguish between that absorption cross-section and the one σ_a defined by equations 4.130, 4.132. The $\sigma_{ne} = \sigma_{abs}$ is a measurable reaction cross-section. However, the σ_a includes that fraction of the compound nucleus formation which decays back into the elastic channel, albeit with a time delay, so that $\sigma_a = \sigma_{ne} + \sigma_{ce}$. This feedback into the elastic channel decreases as the bombarding energy is raised, so that σ_{ce} tends to zero and σ_a approaches $\sigma_{ne} = \sigma_{abs}$. At the opposite extreme, if the bombarding energy is below the threshold for any non-elastic reaction then $\sigma_{ne} = 0$ but σ_a need not be zero; however any compound formation then has to return to the elastic channel, $\sigma_a = \sigma_{ce}$.

The name 'absorption cross-section' is often used for both quantities σ_{abs} or σ_a, although the context usually makes clear which is meant.

where $\delta = ka$ is the hard sphere phase shift. The average of this in an energy region Δ, as defined by equation 4.127, is

$$\bar{\eta} = \exp(-2i\delta)\left(1 - \sum_j \frac{\pi \Gamma_n^j}{\Delta}\right) \approx \exp(-2i\delta)\left(1 - \frac{\pi \bar{\Gamma}_n}{D}\right) \qquad (4.134)$$

since the number of resonances in the interval Δ is Δ/D and $\bar{\Gamma}_n$ is the average neutron width in this interval. If the resonances are well separated ($D \gg \bar{\Gamma}$), the absorption cross-section obtained from the optical model is

$$\sigma_a \equiv \frac{\pi}{k^2}(1 - |\bar{\eta}|^2) \approx \frac{2\pi^2}{k^2}\frac{\bar{\Gamma}_n}{D} \qquad (4.135)$$

The fluctuation cross-section is obtained from equation 4.131; if $D \gg \bar{\Gamma}$

$$\sigma_{\text{fluc}} = \frac{2\pi^2}{k^2 D}\overline{\left(\frac{\Gamma_n^2}{\Gamma}\right)} \approx \frac{2\pi^2}{k^2}\frac{\bar{\Gamma}_n}{D}\left(\frac{\bar{\Gamma}_n}{\bar{\Gamma}}\right) \qquad (4.136)$$

The last step consists of approximating the average of a product by the product of the averages. The physical meaning of σ_{fluc} now becomes clear by comparison with the equation 4.135; σ_{fluc} is the absorption cross-section σ_a times the average branching ratio $G_n^C = \bar{\Gamma}_n/\bar{\Gamma}$ for decay of the compound nucleus back into the entrance channel; hence it describes the compound elastic scattering.

Similarly, we can obtain the average of the actual non-elastic cross-section from equations 4.127 and 4.129.

$$\sigma_{ne} = \frac{\pi}{k^2}(1 - \overline{|\eta|^2}) \approx \frac{2\pi^2}{k^2}\frac{\bar{\Gamma}_n}{D}\frac{\bar{\Gamma} - \bar{\Gamma}_n}{\bar{\Gamma}} \qquad (4.137)$$

We see that $(\bar{\Gamma} - \bar{\Gamma}_n)/\bar{\Gamma}$ is the branching into all channels other than the elastic; if there were no other channels open, we would have $\bar{\Gamma} = \bar{\Gamma}_n$ so that $\sigma_{ne} = 0$ and $\sigma_a = \sigma_{ce}$.

The quantity $\bar{\Gamma}_n/D$ plays a central role in all considerations of this type. It is called a strength function because it tells us how the neutron width, hence the strength of neutron capture (compare equation 4.135) is distributed per unit energy interval. Following equation 4.95, the uninteresting energy-dependent factor $2ka$ is often removed and the strength function $s(E)$ is defined* as the

*The term 'strength function' is often used in different ways. Theorists like to work with quantities like $\overline{\gamma_n^2}$. Experimentalists, however, measure directly quantities like $\bar{\Gamma}_n$; consequently an observable strength function is $\bar{\Gamma}_n/D$. This is often reduced to correspond to a standard energy E_0 (E_0 = 1 eV usually); since Γ_n is proportional to k, this means that the quantity $(E_0/E)^{1/2}$ $(\bar{\Gamma}_n/D)$ is defined as an operational strength function.

(dimensionless) quantity

$$s(E) = \frac{\overline{\Gamma}_n}{2kaD} = \frac{\overline{\gamma_n^2}}{D} \tag{4.138}$$

The absorption cross-section σ_a (the cross-section for formation of the compound nucleus if direct reactions are negligible) was written in terms of a transmission coefficient T_α in equation 4.107. We see that for slow neutrons this transmission coefficient is proportional to the strength function

$$T_n = 2\pi \frac{\overline{\Gamma}_n}{D} = 4\pi ka\, s(E) \tag{4.139}$$

(Since at the moment we are ignoring any nuclear spin, $g_\alpha = 1$ in equation 4.107.) Using this identification, the compound elastic cross-section 4.136 can be re-written in the Hauser–Feshbach form 4.110

$$\sigma_{ce} = \sigma_{fluc} = \frac{\pi}{k^2} \frac{T_n\, T_n}{\sum_\gamma T_\gamma} \tag{4.140}$$

We can obtain a measure of the strength function for a 'black' nucleus from section 4.9.1. Equation 4.108 with equation 4.139 results in

$$\frac{\overline{\Gamma}_n}{D} \approx \frac{2k}{\pi K_0}, \quad s(E) \approx \frac{1}{\pi K_0 a}, \quad \text{if } K \approx K_0 \gg k \tag{4.141}$$

If the wave number K_0 just inside the nuclear surface corresponds to motion in an attractive potential of about 40 MeV, we have $s(E) \approx 1/4a$ if a is measured in fm. The corresponding value of $(E_0/E)^{1/2}\ \overline{\Gamma}_n/D$ is $\approx 1 \times 10^{-4}$ for the standard energy of $E_0 = 1$ eV.

Experimentally, the strength function may be extracted from total cross-section measurements. From equations 4.128 and 4.134 we have

$$\sigma_{tot} \approx \frac{2\pi^2}{k^2} \frac{\overline{\Gamma}_n}{D} + 4\pi a^2 = \frac{4\pi^2 a}{k} s(E) + 4\pi a^2 \tag{4.142}$$

where we have approximated $\sin\delta \approx \delta = ka$ and assumed $D \gg \overline{\Gamma}_n$. *Figure 4.43* shows the results for slow neutrons scattered from nuclei with a wide range of masses, together with the predictions (solid curve) for a typical optical model potential. The black nucleus prediction of equation 4.139 is constant, independent of mass (if the interior wave number K is assumed constant), with a value of about 1×10^{-4}. However, the measured values show large but systematic variations about this value. In particular, there are two 'giant resonances', one peaking at $A \approx 50$ and one at $A \approx 160$. The nucleus is more transparent than

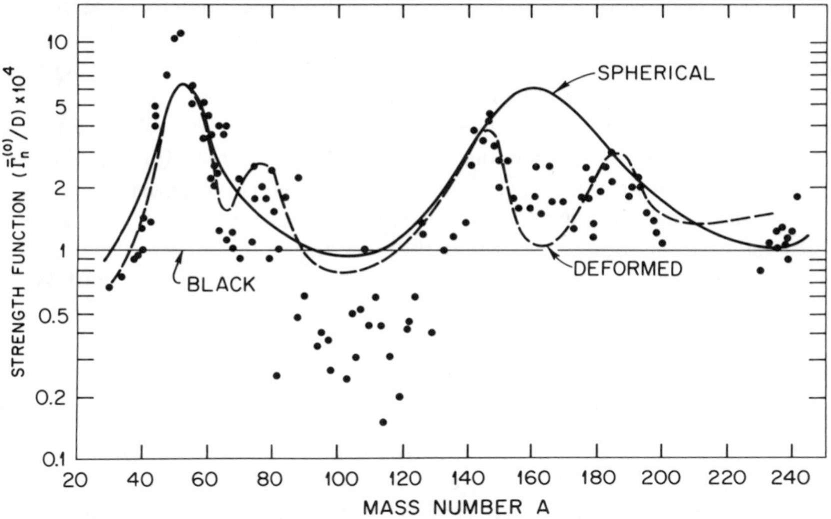

Figure 4.43 The *s*-wave strength function for low-energy neutrons scattered by target nuclei with various mass numbers A. Plotted are measured values of $\bar{\Gamma}_n^{(0)}/D$ where $\bar{\Gamma}_n^{(0)} = (E_0/E)^{1/2} \times \bar{\Gamma}_n$ and $E_0 = 1$ eV. Note the logarithmic scale. Large uncertainties are associated with most of these values; for clarity these uncertainties have not been indicated in the figure. The solid curve corresponds to the predictions for a semi-transparent spherical nucleus, represented by an optical model potential. The dashed line is similar but takes into account the possibility that the nuclei may be deformed away from the spherical shape. The 'black nucleus' model is represented by the straight line at 1×10^{-4}

equation 4.139 implies and we need the more general optical model (section 4.5) to describe the scattering. As we increase the mass A of the target, we also increase the radius $R = r_0 A^{1/3}$ of the optical model potential well. As we pass the positions of the two peaks in *Figure 4.43*, we reach the radii at which first the $3s_{1/2}$ and then the $4s_{1/2}$ neutron shell model orbits just become bound, so we can identify the peaks in the strength function with the occurrence of these single-particle potential resonances at very low energies. On resonance, the neutron wave function is large inside the potential well and hence inside the nucleus (compare *Figure 4.35*); there is enhanced opportunity for capture of the neutron to form a compound nucleus.

As we have discussed elsewhere, this picture does not imply that there is a single excited state in any of these nuclei which can be identified with a single neutron bound in, or scattering resonantly from, the potential due to the other nucleons. We know that at these excitation energies (corresponding to the separation energy of a neutron or $E_x \sim 8$ MeV) there is a very high density of levels; hence the giant resonance is only representing some average property of these levels. Indeed, the strength function is the average of the probability of finding in the wave function of an actual level or resonance in nucleus A a piece corresponding to the nucleus (A – 1) in its ground state with the extra neutron in an s-wave potential-scattering state.* If the nucleons in the target nucleus (A – 1) could not be excited, slow neutron scattering would only show this broad ($\Gamma \sim 1$ MeV) potential resonance. However the incident neutron can excite individual target nucleons in many differing ways, thereby forming the many narrow resonant states actually observed. The original potential resonance strength is fragmented among all these states. We can say this more formally; consider the wave function ψ_i describing the structure of the ith resonance level in nucleus A. This may be expanded on any complete set of wave functions ϕ_λ with coefficients a_λ^i

$$\psi_i = \sum_\lambda a_\lambda^i \, \phi_\lambda$$

Most of the ϕ_λ correspond to complicated many-nucleon excitations. However one term, say ϕ_0, will correspond to a neutron in an s-state with the (A – 1) nucleus in its ground state. Then if we average $|a_0^i|^2$ over the resonances in an energy interval $\Delta E \gg D$, the average spacing between resonances, this quantity is proportional to the neutron s-wave strength function. If the interaction between the incident neutron and the target nucleons were very strong, this 'single particle' strength would be distributed over a very wide range of excitation energies; the strength function would be flat, corresponding to the black nucleus picture, and we should see no giant resonance. The fact that we do indeed see such giant resonances means that the interaction is of intermediate strength; the nucleus is partially transparent. This is precisely the same situation that allows the independent particle, or shell, model of nuclear structure to have any validity. It is meaningful to think of nucleons moving freely in independent orbits; they do collide with one another (causing configuration mixing) but not strongly enough to completely mix up the underlying single particle orbits.

The reader will observe from *Figure 4.43* that the giant resonance near $A = 160$ is not very well defined. Indeed, the region around $A \approx 160$ is one where strongly deformed nuclei occur (*see* section 1.7.3) and this results in the $3s_{1/2}$ single-particle resonance being split into two fragments. This can be understood crudely if we think of the single-particle resonance as occurring when an integral number of half-wavelengths just fit into the potential well. When the nucleus is deformed, for example into an ellipsoidal shape, it has *two* characteristic radii, corresponding to its major and minor axes, consequently a resonance along the major axis occurs for somewhat lighter masses, and along

*Occasionally a single resonance may be found in *light* nuclei which has a large part of the single-particle strength. The classic example (Johnson, 1973) is the 1.00-MeV resonance in $^{16}O + n$ scattering which corresponds to the $1d_{3/2}$ shell model orbit. This resonance is relatively narrow ($\Gamma \sim 94$ keV) because its energy is less than the centrifugal barrier for $\ell = 2$.

the minor axis for somewhat heavier masses, than one would have expected from the formula $R = r_0 A^{1/3}$ for the average radius. These two fragments of the giant resonance can be discerned in *Figure 4.43*. The dashed curve shows the predictions for a non-spherical optical model potential; these were calculated using the coupled-equations technique of section 3.3.1 and reproduce the main features of the data.

The discussion of this section can be extended to other partial waves with $\ell > 0$, and to proton scattering. Information has been obtained on the strength function for *p*-wave neutrons, one peak of which occurs for masses with $A \approx 100$. Because of the Coulomb barrier, proton strength functions are difficult to obtain, but the available evidence indicates peaks in the strength function for *s*-wave protons at $A \approx 70$ and 230.

A very detailed discussion of low-energy neutron scattering is given by Lynn (1968).

4.11 NUCLEAR REACTIONS WITH LIGHT IONS OF HIGH ENERGIES

'High' is a relative term. In the context of nuclear physics, high energy may be loosely defined as meaning projectiles with bombarding energies greater than about 100 MeV per nucleon. As the bombarding energy becomes high, it is no longer appropriate to think of the formation of a compound nucleus. The initial collision is so fast and violent that the first few events are almost always direct reactions. In addition to the simple direct reactions described in section 4.7, a series of individual nucleons (or small groups of nucleons) may be ejected (called 'spallation'), or the target may break into several large fragments (called 'fragmentation'). If the projectile is composite, it may be disrupted itself. The nuclei remaining after this phase of the reaction are likely to be left with internal excitation which may then be lost by the evaporation-type processes described in the previous sections.

In a reaction induced by a very high energy nucleon, the projectile has a sufficiently short wavelength that it can be said to 'see' individual nucleons, that is, it becomes sensitive to the granular structure of a nucleus. (At 1 GeV, a proton has $\lambda\!\!\!^{-} \simeq 0.12$ fm.) Its history can then be described as a series of successive scatterings from individual nucleons (*see* section 3.11 and *Figure 3.16*). Analysis of measurements of this type can then, in principle, give detailed information on the distribution of nucleons in the target and their correlation—for example, how probable it is to find two target nucleons within a certain distance of each other.

A very high energy particle passing through a nucleus is hardly deflected at all; it is so very much more energetic than the target nucleons that when it collides with one, the others are simply spectators. These conditions make it convenient to use the methods of optical diffraction theory to describe the scattering; that is, they favour the use of approximations like the eikonal and

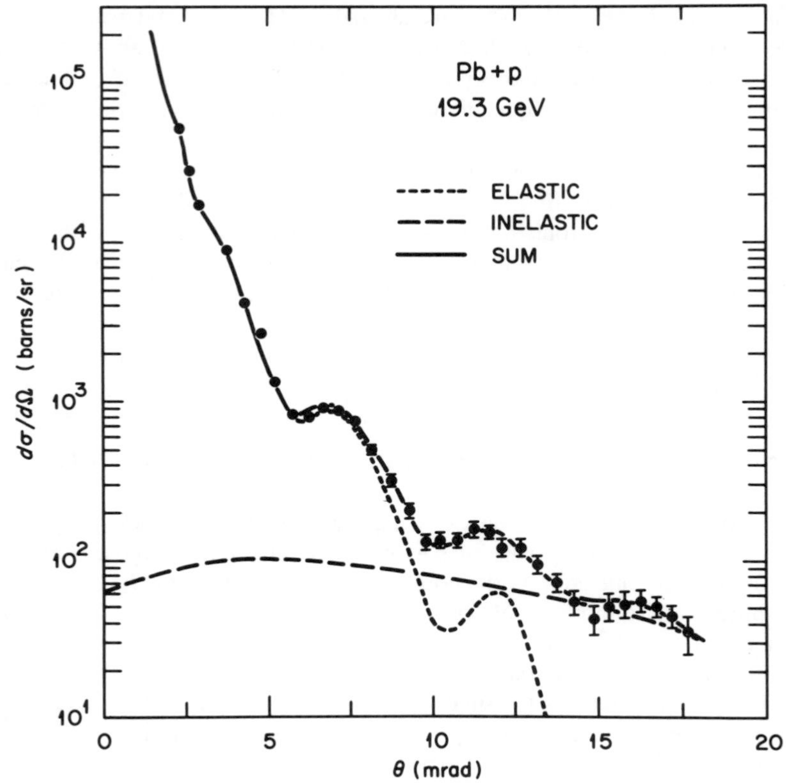

Figure 4.44 Cross-sections measured for the scattering of 19.3-GeV protons by ^{208}Pb nuclei. The measurements include both elastic and inelastic events (from Bellettini *et al.*, 1966). The curves correspond to calculations using the theory of Glauber (1970) (*see* section 3.10.2.2). Note the angular scale; 1 mrad = 0.0573°

impulse approximations (*see* sections 3.10, 3.11). However, some attention must be paid to the use of relativistic kinematics. In this sense, the description of nuclear reactions at very high energies becomes much simpler and more transparent than it is at low energies.

The incident particle tends on average to lose only a small part of its energy and momentum in each collision with a target nucleon and a single scattering is the most likely event. Consequently elastic and almost-elastic scattering are the most probable and their differential cross-sections are very strongly peaked in the forward direction. (*Figure 4.44* shows the results of measurements of the scattering of 20-GeV protons from Pb. Note the angle scale; 1 mrad ≈ 0.057°.) Of course, good energy resolution becomes increasingly difficult to achieve as measurements are made at increasingly higher energies. A 'typical' energy of the first excited state of a nucleus is ~ 1 MeV, much smaller than the kinetic energy of a proton with a bombarding energy of ~ 1 GeV, say. As a result, many high-energy measurements (as in *Figure 4.44*) do not distinguish between elastic and inelastic events.

High energies are to be preferred for the knock-out reactions such as (p, 2p) which were described in section 4.7.5. Then it is more reasonable to neglect refraction and absorption effects (or at least it is possible to make simple corrections for them) and the interpretation of the measurements is much more direct.

A new feature which enters when the bombarding energy becomes sufficiently high is the possibility of producing new particles such as pi-mesons or pions; this can occur through elementary events like

$$p + n \rightarrow \pi^0 + p + n$$

and

$$p + n \rightarrow \pi^- + p + p$$

etc. The first anti-protons were observed following the bombardment of copper nuclei by 6.3-GeV protons.

The theoretical description of the scattering of high-energy protons and pions from nuclei has been reviewed recently by Glauber (1970); *see* also Jackson (1970) and, more recently, Chaumeaux *et al.* (1978).

4.12 REACTIONS BETWEEN HEAVY IONS

Heavy-ion physics may be said to be the study of nuclear reactions induced when projectiles with mass $A > 4$ are accelerated and strike a target. Some general characteristics of these reactions and the possibilities of learning new things about nuclei with them were discussed in section 2.18.12. This is a rapidly developing subject so that it is inevitable that some of the things we say now will become obsolete with the growth of our understanding of the processes involved. The interested reader is urged to consult current review articles and the proceedings of conferences on the subject (*see*, for example, Bromley, 1984; Bass, 1980).

Heavy-ion reactions allow us to bring together two relatively large pieces of nuclear matter in which the kinetic energy of relative motion is already almost equally shared among all the nucleons. In itself this favours the formation of a compound nucleus. However the two heavy ions can also carry large amounts of angular momentum and it may not be possible to form a stable compound system with such a high spin. This has led to the picture of intermediate types of process such as deep inelastic scattering in which the two ions make sufficiently strong contact that they are able to share a large fraction of the incident energy (and convert it into energy of internal excitation—they 'heat up'), and also perhaps transfer a few nucleons, yet they do not stay together long enough to

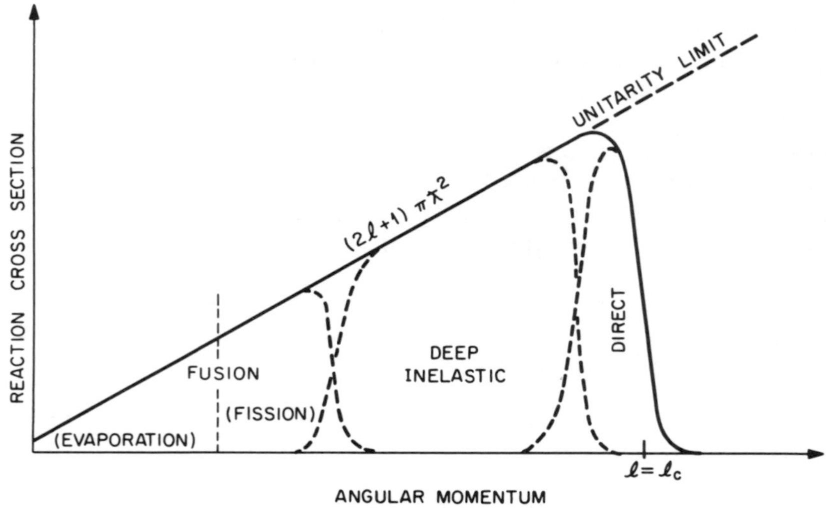

Figure 4.45 Schematic picture of the contributions from different partial waves to the reaction cross-section for a collision between two heavy ions. The relative proportions of the various non-elastic events varies with energy and the masses of the ions

be considered a compound nucleus in the usual sense. Of course the extreme peripheral collisions are direct reactions of the usual type (as discussed earlier for strongly absorbed systems). The angular momentum structure of the reaction or absorption cross-section for these collisions may be visualised schematically as in *Figure 4.45* which indicates how the various contributions to the absorption cross-section vary with the angular momentum in the entrance channel. (The word 'schematic' should be emphasised; the proportions of the various processes depend upon the energy and the nature of the two ions. Also, the divisions between them are unlikely to be as sharp as the figure implies.) *Figure 4.45* also indicates in a qualitative way that, of the systems that do truly fuse, those with the larger angular momenta may be more likely to decay by fissioning into two large fragments while those with smaller spins will probably decay by the evaporation of low-energy particles (section 4.9.2), especially neutrons.

When we examine the distribution of masses and of kinetic energies of the products of a collision of two heavy ions, we see a strong peak, which we call *quasi-elastic*, corresponding to only small losses of kinetic energy and in which there are two fragments with masses and charges very close to those of the projectile and the target. These are the direct reactions and include the true elastic scattering. In addition, we may see a broad and sometimes strong group of outgoing nuclei, in which an appreciable fraction of the incident kinetic energy has been converted into internal excitation energy; these we call the products of *deep inelastic* (or *damped*) reactions. These products are also found to be primarily binary and to have masses and charges that differ by only a few units from those of the projectile and target, so that although a considerable energy transfer has taken place, only a few nucleons have been exchanged. (Complete fusion into a compound nucleus and its consequent evaporation

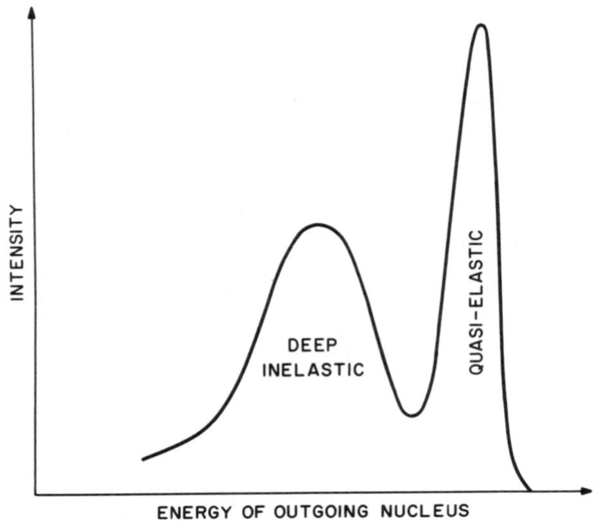

Figure 4.46 Schematic behaviour of the energy spectrum of particles resulting from the collision of two heavy ions (such as $^{84}Kr + ^{209}Bi$ at 600 MeV). The detailed shape and proportions depends upon the two ions, their relative energy and the angle of observation (*see*, for example, Schroder and Huizenga, 1984)

would be extremely unlikely to result in these products.) Often the energy spectrum of the projectile-like fragments will appear as in *Figure 4.46*. These strongly damped collisions may account for a large fraction of the reaction cross-section for heavier systems (such as $^{136}Xe + ^{209}Bi$), with complete fusion being a much less likely event.

The angular distributions of these deep-inelastic products are strongly correlated with the amount of kinetic energy lost. If the energy loss is not too large, the ejectiles may be concentrated in a small range of angles in the forward direction, such as in *Figure 4.47*. This peak appears in the vicinity of the scattering angle for the Rutherford trajectory which corresponds to a peripheral collision. This suggests a glancing collision in which close enough contact is made for the transfer of energy and a few nucleons (corresponding to trajectory 2 in *Figure 3.11*), followed quickly by separation again. On this picture direct reactions result from less close collisions (for example, trajectory 1 in *Figure 3.11*), while greater energy losses would imply more intimate contact (for example, trajectory 3 in *Figure 3.11*) in which the two ions rotate about one another for a longer time, perhaps even orbiting. During this time they retain their dinuclear configuration without complete fusion occurring. After break-up, the nuclei resulting from these events would exhibit broader angular distributions.

Our abbreviated discussion and *Figures 4.46* and *4.47* can scarcely do justice to the rich variety of phenomena encountered in these damped reactions. The reader is directed to Schroder and Huizenga (1984) and Bass (1980) for more detailed reviews.

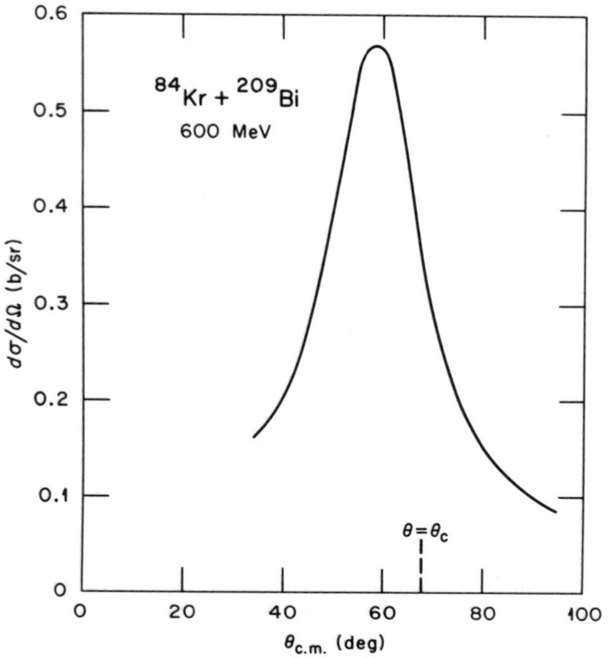

Figure 4.47 Schematic angular distribution of the deep-inelastic reaction products following the collision of two heavy ions (such as ^{84}Kr + ^{209}Bi at 600 MeV). The angle $\theta = \theta_c$ corresponds to the classical scattering angle for a grazing collision. Again, the precise shape depends upon the two ions and their relative energy (*see*, for example, Schroder and Huizenga, 1984)

Whatever the final outcome of these investigations, we see here an example of the unique opportunities that heavy-ion reactions provide for the study of the bulk (or 'macroscopic') dynamical properties of nuclear matter when infused with large amounts of energy, opportunities scarcely afforded by reactions with light ions. (The latter are complementary in the sense that they are particularly suitable for studies of 'microscopic' properties, such as the motions of individual nucleons and the properties of individual quantum states.)

There are events in which the two ions do fuse but, because of high angular momentum, the compound nucleus is unstable against fission. Spontaneous fission, or that induced by light ions (such as in neutron capture) is only seen for systems with low spin; consequently, heavy-ion reactions allow us to extend our studies of this phenomenon to systems with large spins. Further, because of the greater variety of projectiles available, these studies can be extended to many new compound nuclei.

Those compound nuclei which do not fission may be said to decay by evaporation in the way discussed in section 4.9.2; in particular, the Hauser-Feshbach theory of the statistical model (section 4.9.1) appears to be very successful in describing these events. With light ions the characteristic angular distribution of these evaporation products is not very anisotropic (*see Figure 2.33* for example); however, the large angular momenta involved in heavy-ion reactions can result in very anisotropic distributions (although they remain symmetric about $\theta = 90°$) as shown in *Figure 4.39* for example.

At the other extreme, the very peripheral collisions result in direct reactions which are analogues of those already discussed for light ions, although again a number of new features enter. One is that much larger amounts of angular momentum are involved; this allows more angular momentum to be transferred during the reaction and also helps to make more valid a semi-classical description of the reaction. Another feature is that the Coulomb repulsion between the ions often plays a more important role than it does with light ions. When the bombarding energy is below or close to the Coulomb barrier, the reaction products show an angular distribution which is a simple 'bell-shaped' peak centred near the deflection angle for the glancing Rutherford orbit. As the energy is increased above the Coulomb barrier, the angular distribution begins to show structure and eventually assumes a diffraction pattern like those discussed in section 4.4. *Figure 4.48* shows examples for one- and two-nucleon transfers, analogues of the stripping and pick-up reactions with light ions discussed in section 4.7.4. Inelastic scattering shows similar effects with the exception that for the smaller angular momentum transfers (low multipoles, especially $L = 2$ and 3) the Coulomb field results in explicit Coulomb excitation contributions also (section 4.4.4). Below the Coulomb barrier this Coulomb excitation dominates and is an important tool for measuring transition probabilities; because the Coulomb interaction is strong for heavy ions, multiple inelastic excitation (section 4.7.6) becomes important and enables us to Coulomb excite states with quite high spins, especially in nuclei exhibiting rotational spectra, by repeated $E2$ transitions up the band.

Another important feature of direct reactions with heavy ions is the possibility of transferring groups of four or more nucleons. Of particular interest is the transfer of an α-particle-like cluster of two neutrons and two protons. This enables us to probe the validity of α-particle models of the lighter nuclei and the possibility of α-clustering in heavier nuclei. The reactions (^{6}Li, d), (^{16}O, ^{12}C) and (^{20}Ne, ^{16}O) and their inverses are amongst those that have been used for this purpose. (We note that the pick-up of an α-particle from a target nucleus is essentially the artificially induced radioactive α-decay of that nucleus.)

The study of heavy-ion reactions at high energies ($>$ 100 MeV per nucleon) is still in its early, though vigorous, infancy (Nagamiya *et al.*, 1984). There is the possibility of exciting shock waves in the nuclei during collisions at supersonic energies. At more extreme, 'ultra-relativistic', energies ($\gtrsim$ 100 GeV per nucleon) there is the hope that a collision between two heavy ions may result in the momentary formation of a very hot and highly condensed state of nuclear matter known as a 'quark–gluon plasma'. In this state, the nucleons are brought so close together that they lose their identities and dissolve into a plasma of their constituent particles (Shuryak, 1988).

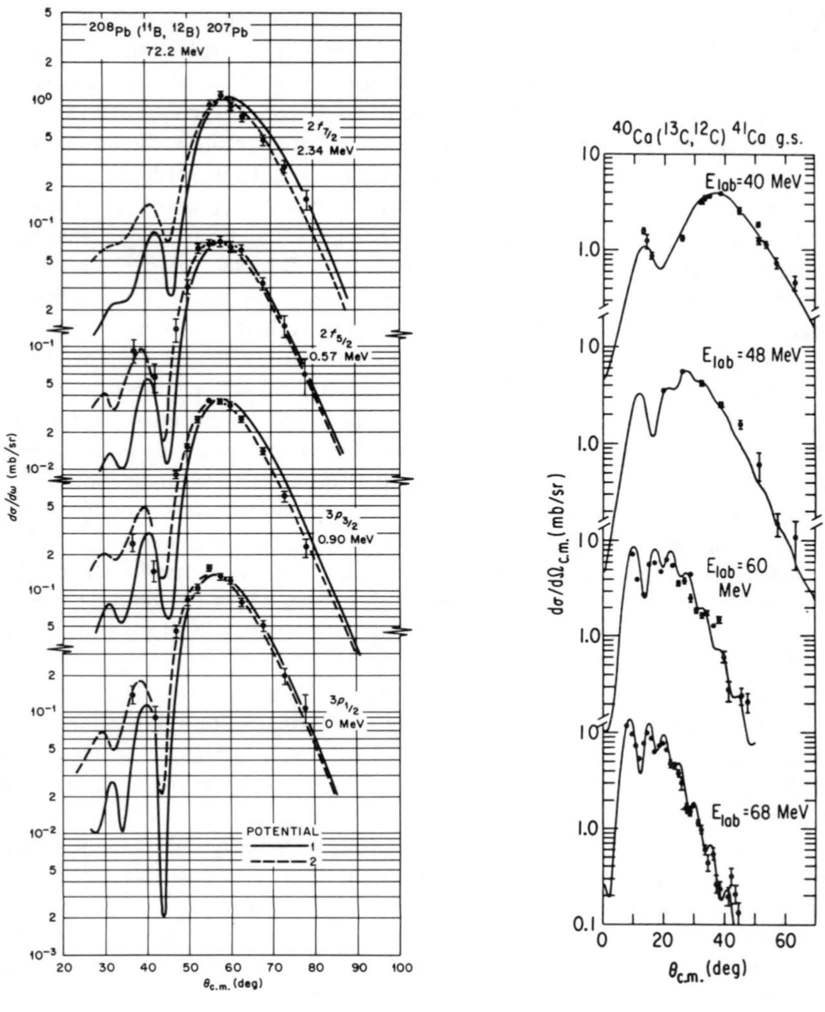

(a) (b)

Figure 4.48 Typical angular distributions for peripheral or direct reactions between heavy ions in which one or two nucleons is transferred and the residual nuclei are left with little or no excitation energy. (a) 'Bell-shaped' curves typical of reactions initiated by bombarding energies close to the Coulomb barrier (from Toth *et al.*, 1976). (b) Transition from a bell-shape to a diffraction pattern as the bombarding energy is increased (from Bond *et al.*, 1973). (c) An example of an oscillating angular distribution when two neutrons are transferred (from LeVine *et al.*, 1974). In each case, the curves represent theoretical calculations using the distorted-wave Born approximation. The notation g.s. means a transition to the ground state of the final nucleus

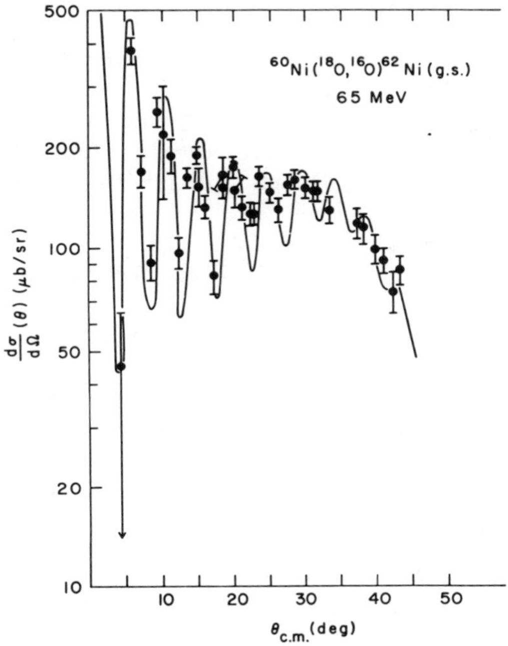

(c)

REFERENCES

Abramowitz, M. and Stegun, I. A. (1970). *Handbook of Mathematical Functions.* New York; Dover Publications

Alder, K. and Winther, A. (1975). *Electromagnetic Excitation: Theory of Coulomb Excitation with Heavy Ions.* Amsterdam; North-Holland

Ascuitto, R. J., Glendenning, N. K. and Sorensen, B. (1971). *Phys. Lett.* Vol. 34B, 17

Austern, N. (1970). *Direct Nuclear Reactions.* New York; Wiley

Austern, N. and Blair, J. S. (1965). *Annals of Physics.* Vol. 33, 15

Barschall, H. H. and W. Haeberli (1971). *Polarization Phenomena in Nuclear Reactions.* Madison; University of Wisconsin Press

Bass, R. (1980). *Nuclear Reactions with Heavy Ions.* Berlin; Springer-Verlag

Bassel, R. H., Satchler, G. R., Drisko, R. M. and Rost, E. (1962). *Phys. Rev.* Vol. 128, 2693

Belletini, G., Cocconi, G., Diddens, A. N., Lillethun, E., Matthiae, G., Scanlon, J. P. and Wetherell, A. M. (1966). *Nucl. Phys.* Vol. 79, 609

Bertrand, F. E. and Peele, R. W. (1973). *Phys. Rev.* Vol. C8, 1045

Bethe, H. A. (1971). *Ann. Rev. Nucl. Sci.* Vol. 21, 93
Biedenharn, L. C. and Brussaard, P. J. (1965). *Coulomb Excitation*. Oxford; Oxford University Press
Blair, J. S. (1966). In *Lectures in Theoretical Physics, VIIIC*. Eds. Kunz, P. D., Lind, D. A. and Brittin, W. E. Boulder; University of Colorado Press
Blair, J. S., Farwell, G. W. and McDaniels, D. K. (1960). *Nucl. Phys.* Vol. 17, 641
Blann, M. (1975). *Ann. Rev. Nucl. Sci.* Vol. 25, 123
Blatt, J. M. and Weisskopf, V. F. (1952). *Theoretical Nuclear Physics*. New York; Wiley
Bleaney, B. I. and Bleaney, B. (1957). *Electricity and Magnetism*. Oxford; Oxford University Press
Blin-Stoyle, R. J. (1955). *Phil. Mag.* Vol. 46, 973
Bond, P. D., Garrett, J. D., Hansen, O., Kahana, S., LeVine, M. J. and Schwarzschild, A. Z. (1973). *Phys. Lett.* Vol. 47B, 231
Broek, H. W., Yntema, J. L., Buck, B. and Satchler, G. R. (1964). *Nucl. Phys.* Vol. 64, 259
Bromley, D. A. (1984). *Treatise on Heavy Ion Science*. New York; Plenum Press
Butler, S. T. (1957). *Nuclear Stripping Reactions*. Sydney; Horwitz Publications.
Butler, S. T., Austern, N. and Pearson, C. (1958). *Phys. Rev.* Vol. 112, 1227
Chasman, C., Kahana, S. and Schneider, M. J. (1973). *Phys. Rev. Lett.* Vol. 31, 1074
Chaumeaux, A., Layly, V. and Schaeffer, R. (1978). *Annals of Physics*. Vol. 116, 247
Cohen, B. L., Fulmer, R. H. and McCarthy, A. L. (1962). *Phys. Rev.* Vol. 126, 698
Ericson, T. (1966). In *Lectures in Theoretical Physics, VIIIC*. Eds. Kunz, P. D., Lind, D. A. and Brittin, W. E., Boulder; University of Colorado Press
Ericson, T. and Mayer-Kuckuk, T. (1966). *Ann. Rev. Nucl. Sci.* Vol. 16, 183
Fernandez, B. and Blair, J. S. (1970). *Phys. Rev.* Vol. C1, 523
Feshbach, H. (1965). In *Nuclear Structure Study with Neutrons*. Eds. Neve de Mevergnies, M., Van Assche, P. and Vervier, J. Amsterdam; North-Holland
Fluss, M. J., Miller, J. M., D'Auria, J. M., Dudey, N., Foreman, B. M., Kowalski, L. and Reedy, R. C. (1969). *Phys. Rev*. Vol. 187, 1449
Frahn, W. E. (1984). In *Treatise on Heavy Ion Science*, Vol. 1. Ed. D. A. Bromley. New York; Plenum Press
Frahn, W. E. (1985). *Diffractive Processes in Nuclear Physics*. Oxford; Oxford University Press
Frahn, W. E. and Gross, D. H. E. (1976). *Annals of Physics*. Vol. 101, 520
Frahn, W. E. and Rehm, K. E. (1978). *Phys. Reps.* 37C, 1
Gallmann, A., Wagner, P., Franck, G., Wilmore, D. and Hodgson, P. E. (1966). *Nucl. Phys*. Vol. 88, 654
Glauber, R. J. (1970). In *High Energy Physics and Nuclear Structure*. Ed. Devons, S. New York; Plenum Press
Glendenning, N. K. (1983). *Direct Nuclear Reactions*. New York; Academic Press

Goldberg, D. A., Smith, S. M. and Burdzik, G. F. (1974). *Phys. Rev.* Vol. C10, 1362
Goldberg, M. D. (1966). Neutron Cross Sections (Supplements to Brookhaven National Laboratory Report BNL-325). U.S. Government Printing Office, Washington, D.C.
Gomez del Campo, J., Ford, J. L. C., Robinson, R. L., Stelson, P. H. and Thornton, S. T. (1974). *Phys. Rev.* Vol. C9, 1258
Greenlees, G. W., Pyle, G. J. and Tang, Y. C. (1968). *Phys. Rev.* Vol. 171, 1115
Griffin, J. J. (1966). *Phys. Rev. Lett.* Vol. 17, 478
Halbert, M. L., Durham, F. E. and van der Woude, A. (1967). *Phys. Rev.* Vol. 162, 899
Harvey, J. D. and Johnson, R. C. (1971). *Phys. Rev.* Vol. C3, 636
Hausser, O. (1974). In *Nuclear Spectroscopy and Reactions, Part D.* Ed. Cerny, J. New York; Academic Press
Henley, E. M. and Yu, D. V. L. (1964). *Phys. Rev.* Vol. B135, 1152
Hillis, D. L., Gross, E. E., Hensley, D. C., Bingham, C. R., Baker, F. T. and Scott, A. (1977). *Phys. Rev.* Vol. C16, 1467
Hintenberger, F., Mairle, G., Schmidt-Rohr, U., Wagner, G. J. and Turek, P. (1968). *Nucl. Phys.* Vol. A115, 570
Hodgson, P. E. (1971). *Nuclear Reactions and Nuclear Structure.* Oxford; Oxford University Press
Hodgson, P. E. (1987). *Reps. Prog. Phys.* Vol. 50, No. 9
Huizenga, J. R. and Moretto, L. G. (1972). *Ann. Rev. Nucl. Sci.* Vol. 22, 427
Jackson, D. F. (1970). *Nuclear Reactions.* London; Methuen
Jackson, D. F. (1971). *Advances in Nuclear Physics.* Vol. 4, 1
Jacob, G. and Maris, Th. A. J. (1966). *Rev. Mod. Phys.* Vol. 38, 121
Jacob, G. and Maris, Th. A. J. (1973). *Rev. Mod. Phys.* Vol. 45, 6
Jenkins, F. A. and White, H. E. (1957). *Fundamentals of Optics.* New York; McGraw-Hill
Jeukenne, J. P., Lejeune, A. and Mahaux, C. (1977). *Phys. Rev.* Vol. C16, 80
Johnson, C. H. (1973). *Phys. Rev.* Vol. C7, 561
Kanter, E. P., Hashimoto, Y., Leuca, I., Temmer, G. M. and Alvar, K. R. (1975). *Phys. Rev. Lett.* Vol. 35, 1326
Kaufmann, S. G., Goldberg, E., Koester, L. J. and Mooring, F. P. (1952). *Phys. Rev.* Vol. 88, 673
Kennedy, H. P. and Schrils, R. (1968). *Intermediate Structure in Nuclear Reactions.* Lexington; University of Kentucky Press
Kerlee, D. D., Blair, J. S. and Farwell, G. W. (1957). *Phys. Rev.* Vol. 107, 1343
Lane, A. M. and Thomas, R. G. (1958). *Rev. Mod. Phys.* Vol. 30, 257
Lee, L. L., Schiffer, J. P., Zeidman, B., Satchler, G. R., Drisko, R. M. and Bassel, R. H. (1964). *Phys. Rev.* Vol. 136, B971
LeVine, M. J., Baltz, A. J., Bond, P. D., Garrett, J. D., Kahana, S. and Thorne, C. E. (1974). *Phys. Rev.* Vol. C10, 1602
Lynn, J. E. (1968). *The Theory of Neutron Resonance Reactions.* Oxford;

Oxford University Press
Macfarlane, M. H. and Schiffer, J. P. (1974). In *Nuclear Spectroscopy and Reactions, Part B*. Ed. Cerny, J. New York; Academic Press
McGowan, F. K. and Stelson, P. H. (1974). In *Nuclear Spectroscopy and Reactions, Part C.* Ed. Cerny, J. New York; Academic Press
McVoy, K. W. (1967). *Annals of Physics*. Vol. 43, 91
McVoy, K. W., Heller, L. and Bolsterli, M. (1967). *Rev. Mod. Phys*. Vol. 39, 245
Madsen, V. A. (1974). In *Nuclear Spectroscopy and Reactions, Part D*. Ed. Cerny, J. New York; Academic Press
Maroni, C., Massa, I. and Vannini, G. (1976). *Nucl. Phys*. Vol. A273, 429
Marshak, H., Langsford, A., Wong, C. Y. and Tamura, T. (1968). *Phys. Rev. Lett.* Vol. 20, 554
Messiah, A. M. (1962). *Quantum Mechanics, I and II*. Amsterdam; North-Holland
Nadasen, A., Schwandt, P., Singh, P. P., Jacobs, W. W., Bacher, A. D., Debevec, P. T., Kaitchuk, M. D. and Meek, J. T. (1981). *Phys. Rev*. Vol. C23, 1023
Nagamiya, S., Randrup, J. and Symons, T. J. M. (1984). *Ann. Rev. Nucl. Part. Sci*. Vol. 34, 155
Park, J. Y. and Satchler, G. R. (1971). *Particles and Nuclei*. Vol. 1, 233
Peterson, J. M. (1962). *Phys. Rev.* Vol. 125, 955
Porto, V. G., Ueta, N., Douglas, R. A., Sala, O., Wilmore, D., Robson, B. A. and Hodgson, P. E. (1969). *Nucl. Phys*. Vol. A136, 385
Preston, M. A. and Bhaduri, R. K. (1975). *Structure of the Nucleus*. Reading, Mass.; Addison-Wesley
Rapaport, J. (1982). *Phys. Reps*. Vol. 87, 25
Richard, P., Moore, C. F., Robson, D. and Fox, J. D. (1964). *Phys. Rev. Lett.* Vol. 13, 343
Richter, A. (1974). In *Nuclear Spectroscopy and Reactions, Part B*, Ed. Cerny, J. New York; Academic Press
Robson, D. (1974). In *Nuclear Spectroscopy and Reactions, Part D*. Ed. Cerny, J. New York; Academic Press
Rost, E. (1962). *Phys. Rev.* Vol. 128, 2708
Satchler, G. R. (1969). In *Isospin in Nuclear Physics*. Ed. Wilkinson, D. H. Amsterdam; North-Holland
Satchler, G. R. (1983). *Direct Nuclear Reactions*. Oxford; Oxford University Press
Satchler, G. R., Halbert, M. L., Clarke, N. M., Gross, E. E., Fulmer, C. B., Scott, A., Martin, D., Cohler, M. D., Hensley, D. C., Ludemann, C. A., Cramer, J. G., Zisman, M. S. and DeVries, R. M. (1978). *Nucl. Phys*. Vol. A298, 313
Schroder W. U. and Huizenga, J. R. (1984). In *Treatise on Heavy Ion Science*, Vol. 2. Ed. D. A. Bromley. New York; Plenum Press
Shuryak, E. V. (1988). *The QCD Vacuum, Hadrons and the Superdense Matter*. Singapore; World Scientific
Silva, R. J. and Gordon, G. E. (1964). *Phys. Rev.* Vol. 136, B618
Sitenko, A. G. (1971). *Lectures in Scattering Theory*. Oxford; Pergamon

Taylor, J. R. (1972). *Scattering Theory*. New York; Wiley
Toth, K. S., Ford, J. L. C., Satchler, G. R., Gross, E. E., Hensley, D. C., Thornton, S. T. and Schweizer, T. C. (1976). *Phys. Rev.* Vol. C14, 1471
Towner, I. S. and Hardy, J. C. (1969). *Advances in Physics*. Vol. 18, 401
Uberall, H. (1971). *Electron Scattering from Complex Nuclei*. New York; Academic Press
Vogt, E. (1968). *Advances in Nuclear Physics*. Vol. 1, 261
von Oertzen, W. (1974). In *Nuclear Spectroscopy and Reactions, Part B*. Ed. Cerny, J. New York; Academic Press
Weisskopf, V. F. (1957). *Nucl. Phys.* Vol. 3, 423
Wilkinson, D. H. (1969). *Isospin in Nuclear Physics*. Amsterdam; North-Holland
Yoshida, S. (1974). *Ann. Rev. Nucl. Sci.* Vol. 24, 1

EXERCISES FOR CHAPTER 4

4.1 (i) Using the Fraunhöfer diffraction model described in section 4.3.1, deduce interaction radii R for ^{A}Ca + α systems from the measurements shown in *Figure 4.6*. (Assume that the minima correspond to the zero in $J_1(x)$ at $x = 10.17$.)

How large is the correction for deflection by the Coulomb field in these cases? (*See* equation 4.12.) Express the radii in the form 4.6; what is the value of r_0?

(ii) Deduce an interaction radius R for ^{58}Ni + α from *Figure 4.20*, assuming that the minimum near $\theta = 43°$ corresponds to $x = 13.32$. What is the significance of this radius in terms of *Figure 4.21*? What classical orbital angular momentum ℓ would give a Rutherford orbit whose distance of closest approach was R? Use *Figure 4.1a* to estimate the corresponding transmission coefficient T_ℓ.

4.2 An ion incident upon a target nucleus will react with it if their distance of closest approach is equal to or less than an interaction radius R. If the incident ion is neutral, this leads classically to a reaction cross-section of πR^2.

Suppose the ion is charged. Then it has to surmount a Coulomb barrier $V_B = Z_1 Z_2 e^2/R$ in order to react with the target (*see Figures 2.35, 3.7*). Use the classical relations of section 2.10 (*see* equation 4.7 also) to show that the reaction cross-section at a bombarding energy E becomes

$$\sigma(E) = \pi R^2 \left[1 - \frac{V_B}{E}\right], \quad E \geqslant V_B$$

$$= 0, \qquad E < V_B$$

Interpret this result in terms of *Figure 4.3*. Draw graphs of $\sigma(E)$ versus E and $1/E$.

4.3 Consider a bundle of parallel rays incident upon a sphere of radius R and refractive index μ. Derive an expression for the average path length within the sphere. Show that this reduces to the result quoted in section 4.5.2 when $\mu = 1$.

4.4 Verify that equation 4.21 reduces to that given by the eikonal approximation of equation 3.117 if $V \ll E$.

4.5 Assume that the range of v in the folding integral 4.22 is much smaller than the radius of the density distribution ρ. Use a Taylor series to expand $\rho(\mathbf{r})$ about the point $\mathbf{r}_p$. Give an approximate expression for $U(\mathbf{r}_p)$ in terms of $\rho, \nabla^2 \rho$ and the volume integral J_0 and mean square radius $\langle r^2 \rangle$ of the interaction v.

Use the result to demonstrate that U falls to one-half of its central value at a radius close to that where the density has reached one-half of its central value but that U has a more diffuse surface.

4.6 Generalise equation 4.22 for the nucleon potential $U(r)$ to allow the proton–neutron interaction to be more attractive than the proton–proton interaction, $v_{np} < v_{pp}$. Assume that the neutron and proton distributions in the target have the same shape, so that $\rho_n = (N/Z)\rho_p$. Express the result in the form

$$U(r) = U_0(r) \pm \frac{N - Z}{A} U_1(r)$$

What value of the ratio v_{np}/v_{pp} is required to give $U_1 = \frac{1}{2} U_0$? What difference would this imply between the depths of the potentials for neutrons and protons incident upon ^{208}Pb if U_0 is a potential well of depth 50 MeV? Which potential is deeper?

4.7 Show that the loss of flux from a beam of particles moving through a region where they experience a complex potential is given by equation 4.30.

Relate this result to the mean free path given by equation 4.29.

4.8 Estimate the strength of the imaginary potential for a proton with an incident energy of 20 MeV moving into a gas of other nucleons if the total proton–nucleon scattering cross-sections can be represented by

$$\sigma_{pi} \approx A_i/(B_i + E_0)$$

where E_0 is the energy in the CMS of the colliding pair. Use the values A_n = 4.32 MeV b, B_n = 0.54 MeV, A_p = 1.55 MeV b and B_p = 0.85 MeV. Assume also that the incident nucleon experiences an attractive real potential of about −50 MeV and that the target nucleons may be treated as at rest. How does the result depend upon the ratio N/Z of the number of neutrons to the number of protons in the target gas? How does the calculated imaginary potential vary as the bombarding proton energy is varied?

Discuss the changes that would be introduced by (i) the motion of the target nucleons and (ii) the Pauli Exclusion Principle that does not allow a nucleon to scatter into states below the Fermi level because these are already occupied. (*See*, for example, Greenlees *et al.*, 1968.)

4.9 From one analysis of measurements on the elastic scattering of protons, α-particles and ^{16}O ions, with bombarding energies of 10 MeV per nucleon, the depths used for the real and imaginary parts of the optical potentials were $(V, W) = (55, 8)$, $(200, 45)$ and $(750, 525)$ MeV, respectively. Calculate the corresponding mean free paths and compare these to typical nuclear dimensions.

How valid is the approximation, equation 4.29, in these cases?

4.10 The three potentials illustrated in *Figure 4.20*, which give very similar scattering for 43-MeV α-particles, have the Woods–Saxon form of equation 4.33 with the same shape for both real and imaginary parts (i.e. $R' = R$ and $a' = a$). The values of the parameters are

V (MeV)	W (MeV)	R (fm)	a (fm)
69	16	5.810	0.556
105	19	5.593	0.554
146	21	5.404	0.557

Show that asymptotically these potentials are proportional to $\exp(-r/a)$ with strengths $V\exp(R/a)$ (real parts) and $W\exp(R/a)$ (imaginary parts). Study their numerical values for large r values ($r \gtrsim 7$ fm); how similar are the various potentials in this region?

4.11 Match an interior wave $A\sin(Kr + \Delta)$ on to an exterior wave $\sin(kr + \delta)$ at the surface $r = a$. Obtain expressions for the amplitude A when $(ka + \delta) = n\pi$ and $(n + \frac{1}{2})\pi$, where n is an integer. Which condition corresponds to a resonance? (Assume $K \gg k$). Compare your results with *Figure 4.35*.

4.12 Consider the s-wave scattering of neutrons by a real potential of square well form

$$U(r) = -V, \quad r \leqslant R$$
$$= 0, \quad r > R$$

Show that $\Delta = 0$ in the preceding exercise. Derive an expression for the s-wave phase shift δ at a bombarding energy E. Obtain the limit for the elastic cross-section σ_{el} as the bombarding energy goes to zero. Show that σ_{el} vanishes in this limit if $\tan K_0R = K_0R$ (where $K_0 = (2mV/\hbar^2)^{1/2}$) but becomes infinite if an odd number of quarter-wavelengths fit into the well.

Show that the absorption cross-section is a maximum in the latter case if the potential also contains a small imaginary part. [Hint: use the results of the preceding exercise and equation 4.30.]

Note that the resonance condition for which σ_{el} is a maximum depends only upon the product VR^2 and not V and R separately.

4.13 Using the results of the preceding exercise, find the potential depth V for which the neutron strength function has maxima at $A = 60$ and 164 corresponding respectively to 3s and 4s scattering resonances at zero energy. Assume that the potential radius is $R = 1.3 \times A^{1/3}$ fm.

Protons entering a nucleus of charge Ze and radius R experience a repulsive Coulomb potential whose average value is $\bar{V}_C = \frac{4}{3}(Ze^2/R)$. Use this result to predict approximately at what mass numbers A the 3s and 4s resonances would occur for protons of low energy if their nuclear potential is the same as that just deduced for neutrons. [Hint: expand to lowest order in $\bar{V}_C/V$ and use typical values of Z from Appendix C5.]

There are experimental indications that proton strength function maxima occur for $A \sim 70$ and 230. What does this imply for the depth of the nuclear potential for protons compared to neutrons?

4.14 Give the complete Born approximation (or Butler) expression for the amplitude $T_{p,d}$ for the pick-up reaction A(p, d)B without making the zero-range approximation of equation 4.64. Evaluate the deuteron transform of equation 4.66 by using the Hulthen wave function

$$\psi_d(s) = C\,\frac{e^{-\alpha s} - e^{-\beta s}}{s}$$

You may obtain C from the normalisation condition for ψ_d. The constant α is given by the binding energy of the deuteron

$$\epsilon_d = \alpha^2\hbar^2/2\mu_d$$

It is found that $\beta \approx 7\alpha$ gives a reasonable description of the physical deuteron. What happens in the limit $\beta \to \infty$? Identify the constant D_0 of the zero-range approximation.

Give an alternative derivation of D_0 by using the Schrödinger equation to transform equation 4.66 and applying Green's theorem.

Obtain an explicit expression for the integral over r_n by restricting it to $r_n \geqslant R$ and using the asymptotic form for the neutron radial wave function

$$u_{nL}(r_n) = N_{nL}\, h_L^{(1)}(iK_n r_n), \quad r_n \geqslant R$$

Here K_n is given by the neutron binding energy, $\epsilon_n = K_n^2\hbar^2/2\mu_n$, while $h_L^{(1)}$ is a spherical Hankel function of the first kind (Abramowitz and Stegun, 1970). [Hint: choose the z-axis along $\mathbf{Q} = \mathbf{k}_d - (M_B/M_A)\mathbf{k}_p$ and use equations 3.35, A39.]

4.15 The differential cross-sections shown in *Figure 4.25* for the ^{56}Fe(d, p) ^{57}Fe reactions induced by 15-MeV deuterons show major peaks at angles of approximately 0°, 10°, 17° and 27°. Use the semi-classical model of section 4.7.1 to estimate the angular momentum transfer L if the reaction radius is $R = 1.6 \times A^{1/3}$ fm.

What values of the interaction radius R would be required to match these peaks with the simple Born expression of equation 4.54?

Would the use of the more complete Butler expression (*see* the preceding exercise) lead to improved agreement with the data? Discuss the effects due to the distortion of the deuteron and proton waves.

4.16 Confirm explicitly that each of equations 4.70, 4.73 and 4.84–4.86 may be obtained from the equations which precede them.

Evaluate numerically the elastic cross-section 4.88 or 4.91 for energies in the vicinity of a resonance, using a variety of choices of values for E_r, Γ and a, and plot the results as shown in *Figure 4.33*. For simplicity, assume that $\Gamma_\alpha = \Gamma$, i.e. that only elastic scattering can occur. What is the effect of having $\Gamma_\alpha < \Gamma$? What is the maximum value of the cross-section in that case?

In particular, study a case at low energy where Γ is of the same order as E_r. (Use the result of equations 4.90, 4.95 that $\Gamma \propto E^{1/2}$.)

Verify that the interference minimum always occurs for $E < E_r$.

4.17 The $^{115}In(n, \gamma)$ reaction exhibits a very strong *s*-wave resonance at a neutron bombarding energy of $E_r = 1.457$ eV (Goldberg, 1966). The peak cross-section is 38100 barns. The ground state of ^{115}In has a spin of $I_A = \frac{9}{2}$ and the resonance has a spin $J = 5$. Given that $\Gamma = \Gamma_n + \Gamma_\gamma$ and that $\Gamma_\gamma > \Gamma_n$, estimate the value of Γ_n/Γ_γ at this resonance.

Given that $\Gamma = 0.075$ eV and that the potential scattering is due to a hard sphere of radius $a = 6.5$ fm, use equation 4.91 and the considerations of section 4.8.4 to calculate the elastic cross-section for neutron energies from zero to $E \approx 2E_r$. (Remember that $\Gamma_n \propto E^{1/2}$ and that only the $S = J = 5$ part of the potential scattering amplitude will interfere with the resonance term.)

Compare the variation with energy of the elastic cross-section with that for the capture reaction. Is there a noticeable interference dip in the curve of *total* cross-section versus energy?

4.18 Verify the relation 4.96. Using the uncertainty principle, give reasons for the incident wave packet having a spatial dimension of order $\hbar v/\Delta E$ if it contains energies in a range ΔE. Indicate why the trailing edge of the resonantly scattered wave packet in *Figure 4.36* falls exponentially with a $1/e$ decay length of $\hbar v/\Gamma$.

4.19 Derive equation 4.97 using the considerations of section 3.8.

How must equations 4.88, 4.91 for the elastic cross-section be generalised when more than one channel spin S and total angular momentum J are allowed?

4.20 Deduce the results 4.119 and 4.121 from the postulated form of the amplitude, equation 4.118.

4.21 Verify that in a (p, 2p) reaction the two protons will be emitted in directions perpendicular to one another but coplanar with the incident proton if the scattering is quasi-free. Show how this relation changes if the emitted particles have different masses, such as in a (p, pα) reaction.

Consider the changes that occur if the target nucleon is also in motion but with a velocity which is randomly directed.

4.22 Two particles with spins **I** and **i** interact via a potential which is proportional to $\mathbf{l}\cdot\mathbf{i}$, where **l** is their relative orbital angular momentum. Show that $\mathbf{I}^2$, $\mathbf{i}^2$ and $\mathbf{l}^2$ remain constants of the motion. Which of the quantities $(\mathbf{I}+\mathbf{i})^2$, $(\mathbf{i}+\mathbf{l})^2$ and $(\mathbf{I}+\mathbf{l})^2$ also remains a constant of the motion? Derive an expression for the expectation value of $\mathbf{l}\cdot\mathbf{i}$.

4.23 A compound nucleus with spin $J = 2$ is formed with a z-component M. It decays by the emission of p-wave α-particles, leaving a residual nucleus with spin 1. Using the results of Appendix A, calculate the angular distribution of the emitted α-particles when (i) $M = 2$, and (ii) $M = 0$.

Appendix A. Angular Momentum and Spherical Harmonics

The total angular momentum of an isolated system is conserved; it is a constant of the motion. If we change to a new, perhaps more convenient, coordinate system related to the old by a rotation, the wave function for the system transforms in a definite way depending upon the value of the angular momentum. If the system is made up of two or more parts, each of them may have angular momentum, and we need to know the quantum rules for combining these to a resultant. We cannot do more here than introduce and summarise a few salient points about these matters. More detailed discussions of the quantum theory of angular momentum are available in several treatises (for example, Brink and Satchler, 1968; *see* also Messiah, 1962).

A general word of caution is in order. The literature on this subject presents a wide variety of conventions for notation, normalisation and phase of the various quantities which occur. Sometimes these conventions represent largely arbitrary choices, sometimes they emphasise different aspects of the quantities involved. If the reader is aware of this possibility, and if he adopts a consistent set of formulae, little difficulty should be encountered.

A1 ANGULAR MOMENTUM IN QUANTUM THEORY

An isolated system will have a total angular momentum whose square has a discrete value given by $j(j+1)\hbar^2$, where $\hbar$ is Planck's constant divided by 2π and where j is an integer or integer-plus-one-half according to whether the system as a whole is a boson or a fermion.* That is to say, the wave function of the system will be an eigenfunction of $\mathbf{J}^2$, the operator for the square of the angular momentum

$$\mathbf{J}^2 |j\rangle = j(j+1)|j\rangle \tag{A1}$$

(Henceforth, we shall use natural units so that $\hbar = 1$.) Note that we often speak of 'the angular momentum j' when strictly we mean that the *square* of the angular momentum is $j(j+1)$. The state $|j\rangle$ cannot be an eigenfunction of the operator $\mathbf{J}$ because the various components of the vector $\mathbf{J}$ do not commute with each other. Instead their commutation relations can be summarised by the formula

$$\mathbf{J} \times \mathbf{J} = i\mathbf{J} \tag{A2}$$

Hence a system cannot simultaneously be an eigenfunction of the operators for more than one component of $\mathbf{J}$. Each component of $\mathbf{J}$ commutes with $\mathbf{J}^2$, however, so we are free to choose $\mathbf{J}^2$ and one component to generate angular momentum eigenfunctions. It is conventional to choose the z-component J_z and we label its eigenvalues by m

$$J_z |jm\rangle = m|jm\rangle \tag{A3}$$

The lack of commutativity between the components J_i can then be pictured by using the vector model, Figure A1. In this model, the vector $\mathbf{J}$ has a definite magnitude $[j(j+1)]^{1/2}$ and a definite projection m on the z-axis, but its direction is uncertain because of the uncertainty in J_x and J_y. This uncertainty is represented by imagining the vector to be precessing around the z-axis so that its x and y components fluctuate.

The prototypical angular momentum operators are those corresponding to the orbital angular momentum $\mathbf{L}$ of a particle with momentum $\mathbf{p}$. Explicitly

$$\mathbf{L} = \mathbf{r} \times \mathbf{p}, \quad \text{with} \quad p_x = -i\,\partial/\partial x, \text{ etc.}$$

or in polar coordinates

$$\begin{aligned} L_\pm &= L_x \pm iL_y \\ &= \pm\, e^{\pm i\phi} \left\{ \frac{\partial}{\partial\theta} \pm i \cot\theta \frac{\partial}{\partial\phi} \right\} \\ L_z &= -i \frac{\partial}{\partial\phi} \end{aligned} \tag{A4}$$

*A nucleus is made up of A nucleons, each of which is a fermion and has an intrinsic spin of $\frac{1}{2}\hbar$. Their relative orbital motions can have angular momenta which are integral multiples of $\hbar$. The rules of angular momentum addition (*see* below) then ensure that the total angular momentum or 'nuclear spin' is an integral multiple of $\hbar$ if A is even or an integer-plus-one-half times $\hbar$ if A is odd. Therefore nuclei with A even are bosons, while those with A odd are fermions.

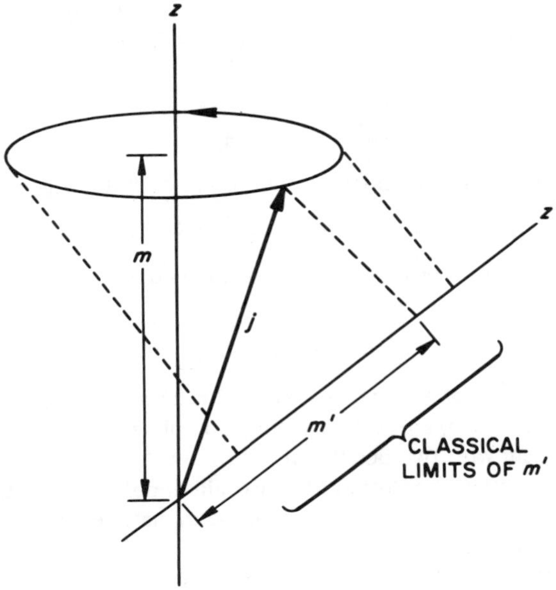

Figure A1 Vector model of an angular momentum vector J with a magnitude $\{[j(j+1)]\}^{1/2}$ and a z-projection equal to m. It precesses about the z-axis so that its x- and y-components are variable. Its projection m' upon another z'-axis also fluctuates; classically only the range of m' values shown is allowed but quantum indeterminacy allows $-j \leqslant m' \leqslant j$

and

$$\mathbf{L}^2 = -\left\{\frac{1}{\sin\theta}\frac{\partial}{\partial\theta}\left(\sin\theta\frac{\partial}{\partial\theta}\right) + \frac{1}{\sin^2\theta}\frac{\partial^2}{\partial\phi^2}\right\} \tag{A5}$$

The expression A5 for $\mathbf{L}^2$ is one which occurs when the Schrödinger equation for particle moving in a central field is separated into radial and angular equations (Messiah, 1962). The angular part of the wave function is an eigenfunction of L_z and $\mathbf{L}^2$; these eigenfunctions are called spherical harmonics and denoted $Y_\ell^m(\theta,\phi)$. They obey the eigenvalue equations

$$\mathbf{L}^2\, Y_\ell^m(\theta,\phi) = \ell(\ell+1)\, Y_\ell^m(\theta,\phi) \tag{A6}$$

$$L_z\, Y_\ell^m(\theta,\phi) = m\, Y_\ell^m(\theta,\phi) \tag{A7}$$

Spherical harmonics and their properties are discussed in more detail in section A3.

Another important set of angular momentum operators are those for spin-$\frac{1}{2}$ particles. For these we have $j = \frac{1}{2}$ and $\mathbf{J} = \frac{1}{2}\boldsymbol{\sigma}$, with the vector σ representing the three 2×2 matrices of Pauli. With J_z diagonal these are

$$\sigma_x = \begin{pmatrix} 0 & 1 \\ 1 & 0 \end{pmatrix}, \quad \sigma_y = \begin{pmatrix} 0 & -i \\ i & 0 \end{pmatrix}, \quad \sigma_z = \begin{pmatrix} 1 & 0 \\ 0 & -1 \end{pmatrix} \tag{A8}$$

with

$$\boldsymbol{\sigma}^2 = \sigma_x^2 + \sigma_y^2 + \sigma_z^2 \tag{A9}$$

Then the eigenfunctions for an intrinsic spin of 1/2 obey the eigenvalue equations

$$\boldsymbol{\sigma}^2 \,|\tfrac{1}{2}, m\rangle = 3\,|\tfrac{1}{2}, m\rangle, \quad \sigma_z\,|\tfrac{1}{2}, m\rangle = 2m\,|\tfrac{1}{2}, m\rangle \tag{A10}$$

Besides the commutation properties embodied in equation A2, the σ matrices obey the anti-commutation relations

$$\sigma_i\sigma_j + \sigma_j\sigma_i = 2\delta_{ij} \tag{A11}$$

in particular, $\sigma_x^2 = \sigma_y^2 = \sigma_z^2 = \mathbf{1}$, where $\mathbf{1}$ is the 2×2 unit matrix.

Together with $\mathbf{1}$, the σ_i are sufficient for a complete description of a spin-$\frac{1}{2}$ system, i.e. a system with two possible states ($m = \pm\frac{1}{2}$). For this reason they can also be used to represent the isospin of an isospin doublet such as the neutron–proton pair. When used to represent isospin, the σ_i matrices are usually denoted τ_i (*see* Wilkinson, 1969).

It is often convenient to use two different coordinate systems which are oriented in space in different directions. [One example is in the description of a nucleus with a permanently deformed (non-spherical) shape. It is helpful to use a set of axes fixed along the principal axes of the nucleus (the *body-fixed axes*) as well as a set independently fixed in space (the *space-fixed axes*). If the nucleus is rotating, the body-fixed axes will be rotating with respect to the space-fixed ones.] Then a state with angular momentum j which has a definite projection m on one z-axis will have a distribution of projections m' with respect to the other z'-axis. This distribution comes from the quantum uncertainty in the x and y components and can also be visualised using the vector model as being due to the precession of $\mathbf{J}$ about the z-axis (*Figure A1*). The rotational transformation which determines this distribution in m' is described in standard texts (Brink and Satchler, 1968; Messiah, 1962).

A2 ANGULAR MOMENTUM COUPLING AND SYSTEMS COMPOSED OF TWO OR MORE PARTS

Often we have systems made up of two parts, each with angular momentum. These parts may be two different particles, or perhaps the spin and orbital properties of a single particle. Each part has associated with it an angular momentum operator and its z-component. Let these be $\mathbf{J}_1$, $\mathbf{J}_{1z}$, $\mathbf{J}_2$, and $\mathbf{J}_{2z}$. Then we have two choices for a set of four commuting operators for the combined system. One choice is

$$\mathbf{J}_1^2, \mathbf{J}_{1z}, \mathbf{J}_2^2, \mathbf{J}_{2z} \tag{A12}$$

The corresponding eigenfunctions are products of the eigenfunctions for each part and obey the eigenvalue equations

$$\mathbf{J}_i^2 \,|j_1 m_1\rangle\,|j_2 m_2\rangle = j_i(j_i+1)\,|j_1 m_1\rangle\,|j_2 m_2\rangle \tag{A13}$$

$$J_{iz}\,|j_1 m_1\rangle\,|j_2 m_2\rangle = m_i\,|j_1 m_1\rangle\,|j_2 m_2\rangle \tag{A14}$$

where $i = 1$ or 2.

The other choice of a set of commuting operators is in terms of the total angular momentum of the combined system, $\mathbf{J} = \mathbf{J}_1 + \mathbf{J}_2$. It is

$$\mathbf{J}_1^2,\ \mathbf{J}_2^2,\ \mathbf{J}^2 = (\mathbf{J}_1 + \mathbf{J}_2)^2,\ J_z = (J_{1z} + J_{2z}) \tag{A15}$$

The eigenfunctions of the coupled system we will write as $|j_1 j_2 JM\rangle$. They obey the eigenvalue equations

$$\mathbf{J}_i^2\,|j_1 j_2 JM\rangle = j_i(j_i+1)\,|j_1 j_2 JM\rangle,\quad i = 1, 2 \tag{A16}$$

$$\mathbf{J}^2\,|j_1 j_2 JM\rangle = J(J+1)\,|j_1 j_2 JM\rangle \tag{A17}$$

$$J_z\,|j_1 j_2 JM\rangle = M\,|j_1 j_2 JM\rangle \tag{A18}$$

For given j_1 and j_2, the values of J are restricted by *the triangular condition* of vector addition

$$j_1 + j_2 \geqslant J \geqslant |j_1 - j_2| \tag{A19}$$

and the allowed J ranges between these limits in integer steps.

The choice A12 would be especially useful if the system were isolated and if there was no interaction between the two parts. Then the angular momentum and its orientation for each part would remain constants of the motion, as is expressed by equations A13, A14. However, if there is an interaction between the two parts, it is likely that the individual components m_i will not remain constant even if the magnitudes j_i do: the two vectors $\mathbf{j}_i$ will tend to precess around their resultant $\mathbf{J}$ instead of each precessing independently about the z-axis. If the system as a whole is isolated, $\mathbf{J}^2$ and J_z will remain constants. These two extremes are pictured using the vector model in *Figure A2*.

These two descriptions are not independent but their eigenfunctions are related by a unitary transformation. This may be written explicitly as

$$|j_1 j_2 JM\rangle = \sum_{m_1 m_2} |j_1 m_1\rangle |j_2 m_2\rangle \langle j_1 j_2 m_1 m_2 | j_1 j_2 JM\rangle \tag{A20}$$

or the inverse

$$|j_1 m_1\rangle |j_2 m_2\rangle = \sum_J |j_1 j_2 JM\rangle \langle j_1 j_2 JM | j_1 j_2 m_1 m_2\rangle \tag{A21}$$

These equations define the (real and symmetric) *Wigner* or *Clebsch–Gordan coefficient*

$$\langle j_1 j_2 m_1 m_2 | j_1 j_2 JM\rangle = \langle j_1 j_2 JM | j_1 j_2 m_1 m_2\rangle \tag{A22}$$

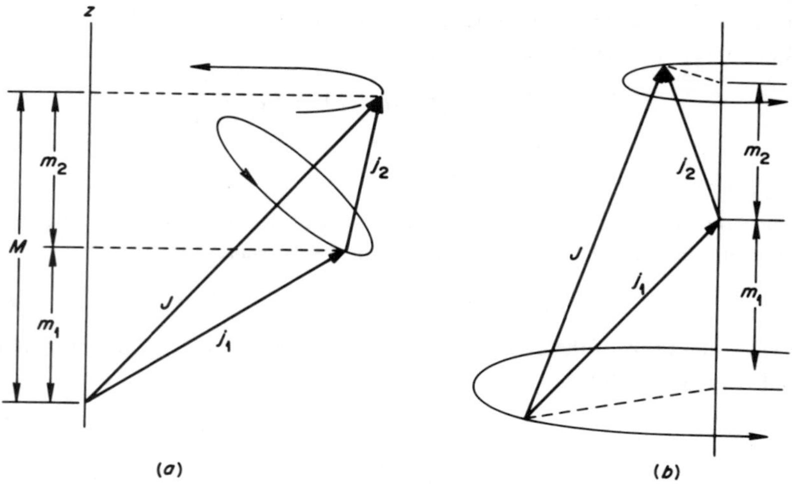

Figure A2 Vector picture of two angular momenta j_1 and j_2. (a) In the coupled (J, M) representation, j_1 and j_2 have a resultant J with projection M. J precesses about the z-axis, while j_1 and j_2 precess about J; m_1 and m_2 are indeterminate. (b) In the uncoupled $(j_1 m_1 j_2 m_2)$ representation j_1 and j_2 precess independently about the z-axis; their resultant J fluctuates in direction and magnitude

For brevity, this is often written $\langle j_1 j_2 m_1 m_2 | JM \rangle$ and we shall follow this convention. The coefficient vanishes unless J satisfies the condition A19; further, since $J_z = J_{1z} + J_{2z}$, we must have $M = m_1 + m_2$.

The vector model gives a physical meaning to the transformations A20 and A21. The left side of *Figure A2* represents a system with definite J and M; the individual $\mathbf{j}_i$ then precess around $\mathbf{J}$ in a correlated fashion and their corresponding projections m_i fluctuate. Equation A20 expresses the distribution of the m_i; $\langle j_1 j_2 m_1 m_2 | JM \rangle^2$ is the probability that observation would yield the values m_1 and m_2. Classically, the range of m_i allowed for a given J, M would be determined geometrically; quantum uncertainties allow all values of m_i from $-j_i$ to j_i, provided $m_1 + m_2 = M$.

The right side of *Figure A2* pictures the state $|j_1 m_1\rangle |j_2 m_2\rangle$ where the two j_i vectors precess around the z-axis in an uncorrelated way. Their resultant $\mathbf{J}$ then fluctuates, and equation A21 gives the distribution of J values. The square of the coefficient is now the probability of finding a particular value of J.

Equations A20 and A21 represent a unitary transformation so the coefficients must satisfy orthonormality relations

$$\sum_{m_1 m_2} \langle JM | j_1 j_2 m_1 m_2 \rangle \langle j_1 j_2 m_1 m_2 | J'M' \rangle = \delta_{JJ'} \delta_{MM'} \tag{A23}$$

and

$$\sum_{JM} \langle j_1 j_2 m_1 m_2 | JM \rangle \langle JM | j_1 j_2 m_1' m_2' \rangle = \delta_{m_1 m_1'} \delta_{m_2 m_2'} \tag{A24}$$

Hence no given coefficient may exceed unity. An explicit general expression can be given for the coefficients, as well as simple formulae for special cases (Brink and Satchler, 1968). Numerical tables are also available. Physical arguments can be used in some special cases. For example, suppose we have the system $|j_1 j_1\rangle |j_2 j_2\rangle$ in which each part has the maximum allowed z-component. Then the resultant $M = j_1 + j_2$ is also a maximum and the rule A19 shows that only $J = j_1 + j_2$ is allowed. Consequently, because of the normalisation condition A24, $\langle j_1 j_2 j_1 j_2 | J = j_1 + j_2, M = J\rangle$ must have a magnitude of unity. Its sign is arbitrary but the usual convention is to choose it positive.

The Clebsch-Gordan coefficient exhibits a number of symmetry relations under permutation of its arguments. From the definitions A20 and A21 it is clear that nothing physical is altered if we interchange the order of j_1 and j_2; at most the combined state may change sign owing to the phase conventions we have adopted. Indeed, we find

$$\langle j_2 j_1 m_2 m_1 | JM\rangle = (-)^{J-j_1-j_2} \langle j_1 j_2 m_1 m_2 | JM\rangle \tag{A25}$$

so that

$$|j_2 j_1 JM\rangle = (-)^{J-j_1-j_2} |j_1 j_2 JM\rangle \tag{A26}$$

Similarly, changing the sign of the z-components is equivalent to inverting the direction of the z-axis. This can have no physical significance for an isolated system so we expect a corresponding symmetry relation. We find

$$\langle j_1 j_2 -m_1 -m_2 | J -M\rangle = (-)^{J-j_1-j_2} \langle j_1 j_2 m_1 m_2 | JM\rangle \tag{A27}$$

Other symmetries are described elsewhere (Brink and Satchler, 1968; Messiah, 1962).

As examples, *Tables A1–A3* give simple expressions for the coefficients when

TABLE A1 *Clebsch-Gordan coefficients for $j_2 = 0$ or $J = 0$*

$$\langle j\,0 m 0 | JM\rangle = \delta_{jJ}\,\delta_{mM}$$

$$\langle j_1 j_2 m_1 m_2 | 00\rangle = \delta_{j_1 j_2}\,\delta_{m_1,-m_2}\,\frac{(-)^{j_1-m_1}}{(2j_1+1)^{1/2}}$$

TABLE A2 *Clebsch-Gordan coefficients when $j_2 = 1/2$*

$\langle j_1 \tfrac{1}{2} M-m_2\, m_2 | JM\rangle$

J	$m_2 = 1/2$	$m_2 = -1/2$
$j_1 + \frac{1}{2}$	$\left[\dfrac{j_1 + M + \frac{1}{2}}{2j_1+1}\right]^{1/2}$	$\left[\dfrac{j_1 - M + \frac{1}{2}}{2j_1+1}\right]^{1/2}$
$j_1 - \frac{1}{2}$	$-\left[\dfrac{j_1 - M + \frac{1}{2}}{2j_1+1}\right]^{1/2}$	$\left[\dfrac{j_1 + M + \frac{1}{2}}{2j_1+1}\right]^{1/2}$

TABLE A3 *Clebsch-Gordan coefficients when* $j_2 = 1$

$\langle j_1\, 1\, M - m_2\, m_2 | JM \rangle$

J	$m_2 = 1$	$m_2 = 0$	$m_2 = -1$
$j_1 + 1$	$\left[\frac{(j_1 + M)(j_1 + M + 1)}{(2j_1 + 1)(2j_1 + 2)}\right]^{1/2}$	$\left[\frac{(j_1 - M + 1)(j_1 + M + 1)}{(2j_1 + 1)(j_1 + 1)}\right]^{1/2}$	$\left[\frac{(j_1 - M)(j_1 - M + 1)}{(2j_1 + 1)(2j_1 + 2)}\right]^{1/2}$
j_1	$-\left[\frac{(j_1 + M)(j_1 - M + 1)}{2j_1(j_1 + 1)}\right]^{1/2}$	$\frac{M}{[j_1(j_1 + 1)]^{1/2}}$	$\left[\frac{(j_1 - M)(j_1 + M + 1)}{2j_1(j_1 + 1)}\right]^{1/2}$
$j_1 - 1$	$\left[\frac{(j_1 - M)(j_1 - M + 1)}{2j_1(2j_1 + 1)}\right]^{1/2}$	$-\left[\frac{(j_1 - M)(j_1 + M)}{j_1(2j_1 + 1)}\right]^{1/2}$	$\left[\frac{(j_1 + M + 1)(j_1 + M)}{2j_1(2j_1 + 1)}\right]^{1/2}$

$j_2 = 0, 1/2$ and 1 from which numerical values may be obtained. They should be used in conjunction with the relations A25 and A27. As an exercise, the reader may also use these tables to verify the relations A23 and A24.

Among the various representations of the Clebsch-Gordan coefficient which may be encountered, a popular one is the Wigner $3-j$ symbol. This is renormalised so as to give it a high degree of symmetry (*see* Brink and Satchler, 1968)

$$\begin{pmatrix} a & b & c \\ \alpha & \beta & \gamma \end{pmatrix} = (-)^{a-b-\gamma}(2c+1)^{-1/2}\langle ab\alpha\beta | c - \gamma\rangle \tag{A28}$$

In most systems of physical interest there are more than two component parts and we are faced with the problem of constructing states with good total angular momentum from the angular momenta of these parts. In general there will be more than one way of doing this. For example if we have three angular momenta j_1, j_2 and j_3, there are three ways in which we may couple them to a resultant J by applying twice the coupling relations A20 or A21. One is

$$\mathbf{j}_1 + \mathbf{j}_2 = \mathbf{J}_{12}, \quad \mathbf{j}_3 + \mathbf{J}_{12} = \mathbf{J} \tag{A29}$$

while another is

$$\mathbf{j}_2 + \mathbf{j}_3 = \mathbf{J}_{23}, \quad \mathbf{j}_1 + \mathbf{J}_{23} = \mathbf{J} \tag{A30}$$

Any of the three possibilities is a valid representation but they are not independent. The corresponding eigenfunctions are related by a linear transformation; for example,

$$|(j_1j_2)J_{12}, j_3; JM\rangle = \sum_{J_{23}} |j_1, (j_2j_3)J_{23}; JM\rangle \langle j_1, (j_2j_3)J_{23}; J | (j_1j_2)J_{12}, j_3; J\rangle \tag{A31}$$

The transformation coefficient is real and independent of M; it is used to define the *Racah coefficient W*

$$\langle j_1, (j_2 j_3) J_{23}; J | (j_1 j_2) J_{12}, j_3; J \rangle = [(2J_{12}+1)(2J_{23}+1)]^{1/2} W(j_1 j_2 J j_3; J_{12} J_{23}) \tag{A32}$$

By expanding the left-hand side in accordance with equations A20, the Racah W can be expressed as the contraction (sum over z-components) of a product of four Clebsch–Gordan coefficients. Consequently the Racah coefficients, or the closely related Wigner $6-j$ symbols

$$\begin{Bmatrix} a & b & e \\ d & c & f \end{Bmatrix} = (-1)^{a+b+c+d} W(abcd; ef) \tag{A33}$$

are frequently encountered in any problems involving more than two angular momenta. Their properties are well documented and extensive numerical tables are available.

An example of two coupling schemes like A29 and A30 was encountered in section 3.8.1 when dealing with the collision of two nuclei, each with spin, whose relative orbital angular momentum was non-zero. For further information about these and more complicated angular momentum coupling situations, the reader should consult more specialised books (for example, Brink and Satchler, 1968).

A3 SPHERICAL HARMONICS

Spherical harmonics are solutions of the differential equation A6

$$[\mathbf{L}^2 - \ell(\ell+1)]\, Y_\ell^m(\theta, \phi) = 0 \tag{A34}$$

with integral values for ℓ. The operator $\mathbf{L}^2$ is given explicitly by equation A5. The dependence on ϕ may be factored out

$$Y_\ell^m(\theta, \phi) = \Theta_\ell^m(\theta) \exp(im\phi) \tag{A35}$$

then the eigenvalue equation A7 follows immediately from the explicit form A4 for the operator L_z. We have m an integer with $-\ell \leqslant m \leqslant \ell$. The explicit definition of the Θ_ℓ^m involves an arbitrary choice of phase; the one most commonly used is that of Condon and Shortley (1951)

$$\begin{aligned} \Theta_\ell^m(\theta) &= (-)^m \left[\frac{(2\ell+1)}{4\pi} \frac{(\ell-m)!}{(\ell+m)!}\right]^{1/2} P_\ell^m(\cos\theta) \quad \text{if} \quad m \geqslant 0 \\ &= (-)^m\, \Theta_\ell^{-m} \quad \text{if} \quad m < 0 \end{aligned} \tag{A36}$$

Here $P_\ell^m(\cos\theta)$ is the associated Legendre polynomial (Abramowitz and Stegun, 1970.) The special case with $m = 0$ is the Legendre polynomial, $P_\ell^0(\cos\theta) = P_\ell(\cos\theta)$.

From equation A36 we have that

$$[Y_\ell^m(\theta,\phi)]^* = (-)^m Y_\ell^{-m}(\theta,\phi) \tag{A37}$$

Further, the parity operation (reflection through the origin) replaces (θ, ϕ) by $(\pi - \theta, \phi + \pi)$. Since

$$Y_\ell^m(\pi-\theta,\phi+\pi) = (-)^\ell Y_\ell^m(\theta,\phi) \tag{A38}$$

we see that the spherical harmonics have a definite parity of $(-)^\ell$. They are also orthogonal over the unit sphere and, as defined, normalised so that

$$\int_0^\pi \int_0^{2\pi} Y_{\ell'}^{m'}(\theta,\phi)^* Y_\ell^m(\theta,\phi) \sin\theta \, d\theta \, d\phi = \delta_{\ell\ell'} \delta_{mm'} \tag{A39}$$

where $\delta_{ab} = 1$ if $a = b$ but $\delta_{ab} = 0$ if $a \neq b$. We also have a sum rule

$$\sum_m |Y_\ell^m(\theta,\phi)|^2 = \frac{2\ell+1}{4\pi} \tag{A40}$$

A few examples of low order are

$$Y_0^0 = \left(\frac{1}{4\pi}\right)^{1/2}; \; Y_1^0 = \left(\frac{3}{4\pi}\right)^{1/2} \cos\theta; \; Y_1^{\pm 1} = \mp\left(\frac{3}{8\pi}\right)^{1/2} \sin\theta \exp(\pm i\phi)$$

$$Y_2^0 = \left(\frac{5}{16\pi}\right)^{1/2} (3\cos^2\theta - 1); \; Y_2^{\pm 1} = \mp\left(\frac{15}{8\pi}\right)^{1/2} \cos\theta \sin\theta \exp(\pm i\phi)$$

$$Y_2^{\pm 2} = \left(\frac{15}{32\pi}\right)^{1/2} \sin^2\theta \exp(\pm 2i\phi)$$

The spherical harmonics also satisfy recursion relations (Messiah, 1962; Abramowitz and Stegun, 1970) which enable us to generate values for larger ℓ and m.

The spherical harmonic $Y_\ell^m(\theta,\phi)$ goes through zero $(\ell - m)$ times in the interval $0 < \theta < \pi$; consequently $|Y_\ell^m|^2$ has $(\ell - m)$ minima (= zero) in this interval. This sinusoidal-like behaviour may be seen explicitly from the asymptotic formula for large ℓ

$$P_\ell^m(\cos\theta) \approx (-\ell)^m \left(\frac{2}{\ell\pi\sin\theta}\right)^{1/2} \sin\left[(\ell + \tfrac{1}{2})\theta + (2m+1)\frac{\pi}{4}\right]$$

which holds for $\ell \gg 1$, $m \ll \ell$ and θ not too close to 0 or π.

The $Y_\ell^m(\theta,\phi)$ provide us with a complete set of functions with which we may expand any function of the angles θ and ϕ, as was done in equation 3.34, 4.19, 4.60 or 4.61 for example. The Rayleigh expansion 3.35 of a plane wave is a special case.

When $m = 0$, a spherical harmonic Y_ℓ^0 reduces essentially to a Legendre polynomial P_ℓ (Abramowitz and Stegun, 1970)

$$Y_\ell^0(\theta,\phi) = \left[\frac{2\ell+1}{4\pi}\right] P_\ell(\cos\theta)$$

These P_ℓ constitute a complete orthogonal set which may be used to expand any periodic function f of θ which is symmetric about $\theta = 0, f(-\theta) = f(\theta)$. Since

$$\int_{-1}^{1} P_\ell(x)\, P_{\ell'}(x)\, dx = \frac{2}{2\ell + 1}\, \delta_{\ell\ell'} \qquad \text{(A41)}$$

which is a special case of equation A39, we may write

$$f(\theta) = \sum_\ell a_\ell P_\ell\, (\cos\theta)$$

where

$$a_\ell = \tfrac{1}{2}\,(2\ell + 1) \int_0^\pi f(\theta) P_\ell\, (\cos\theta)\, \sin\theta\, d\theta \qquad \text{(A42)}$$

The results of measurements of an angular distribution or differential cross-sections may be expressed in this form, with the coefficients α_ℓ being chosen to optimise the fit of $f(\theta)$ to the data. Some Legendre polynomials of low order are

$P_0(x) = 1; \qquad P_1(x) = x$

$P_2(x) = \tfrac{1}{2}(3x^2 - 1); \quad P_3(x) = \tfrac{1}{2}(5x^3 - 3x)$

$P_4(x) = \tfrac{1}{8}(35x^4 - 30x^2 + 3)$

For further properties of the spherical harmonics, *see*, for example, Brink and Satchler (1968), also Messiah (1962) and Morse and Feshbach (1953).

A4 EXAMPLE 1: RADIOACTIVE DECAY OF A NUCLEUS

Consider a nucleus, with spin J and z-component M, at rest; the wave function describing its internal state is $\psi_{JM}(\tau)$. Suppose it decays by the emission of a particle with zero spin, such as an α-particle, with orbital angular momentum ℓ. Suppose the daughter nucleus has a spin J' and wave function $\psi_{J'M'}(\tau')$. Then the initial wave function breaks up as follows (compare equation A20)

$$\psi_{JM}(\tau) = \sum_{mM'} \psi_{J'M'}(\tau')\phi_{\ell m}(\tau_\alpha, \mathbf{r})\, \langle J'\ell M'm \,|\, JM\rangle \qquad \text{(A43)}$$

where $m = M - M'$ and $|J' - \ell| \leqslant J \leqslant (J' + \ell)$. Here $\phi_{\ell m}$ describes the α-particle and its motion relative to the daughter nucleus

$$\phi_{\ell m}(\tau_\alpha, \mathbf{r}) = \psi_0(\tau_\alpha)\, u_\ell(r)\, Y_\ell^m\, (\theta, \phi) \qquad \text{(A44)}$$

where $\mathbf{r}$ is the vector joining the centres of mass of the α-particle and daughter nucleus. The probability amplitude that the relative motion will be found with $\ell_z = m$ is just the Clebsch–Gordan coefficient in equation A43. The probability of finding the α-particle at the position (r, θ, ϕ) is $|\phi_{\ell m}|^2$ weighted by the probability of it being in the state $\phi_{\ell m}$. We observe it at some large value of r, moving in the direction with polar angles (θ, ϕ), with a probability given by

$$\left| \sum_{M'm} \psi_{J'M'}\, \psi_0\, u_\ell(r)\, Y_\ell^m(\theta,\phi)\, \langle J'\ell M'm | JM\rangle \right|^2$$

Now we do not observe the internal state of the α-particle or the daughter nucleus, but only the direction of emission, so we must integrate over the internal variables of the α-particle and the daughter nucleus. Since $\int d\tau'\, \psi^*_{J'M'}(\tau') \times \psi_{J'M''}(\tau') = \delta_{M'M''}$, etc., there is no coherence between transitions to different final states with $M' \neq M''$. Further, the radial function $u_\ell(r)$ is a constant at a fixed value of r. Consequently, the angular distribution of the radiation is simply proportional to

$$W_M(\theta) = \sum_{M'} \left| \langle J'\ell M'm | JM\rangle\, Y_\ell^m(\theta,\phi) \right|^2 \tag{A45}$$

where $m = M - M'$. Because the dependence of Y_ℓ^m on ϕ enters only through the factor $e^{im\phi}$, we see that W_M is independent of ϕ; the distribution is symmetric around the z-axis. In the special case that $J' = 0$, so that $\ell = J$ and $m = M$, then the Clebsch–Gordan coefficient is unity and

$$W_M(\theta) = \left| Y_J^M(\theta,\phi) \right|^2 \tag{A46}$$

Another special case occurs when $J = 0$, so that $\ell = J'$, $m = -M'$ and the Clebsch–Gordan coefficient has the value $(-)^{\ell - m}(2\ell+1)^{-1/2}$. Then

$$W_0(\theta) = \frac{1}{2\ell+1} \sum_m \left| Y_\ell^m(\theta,\phi) \right|^2 \tag{A47}$$

$$= 4\pi$$

from equation A40; that is, the angular distribution is constant or *isotropic*. This is a general property of the angular distribution of products from a spin-zero system; it may be shown to be true for $J = 1/2$ also.

Other cases follow by inserting explicit values for the coefficients in equation A45.

If the initial nucleus was not prepared in a single substate M but oriented with a distribution of M values with probabilities p_M, the angular distribution of the decay radiation becomes

$$W(\theta) = \sum_M p_M\, W_M(\theta) \tag{A48}$$

A5 EXAMPLE 2: FORMATION OF A COMPOUND NUCLEUS AND STATISTICAL WEIGHTS

One way of preparing the radioactive nucleus discussed in the previous section is to form it as a compound nucleus in a nuclear reaction. Consider the collision of two nuclei A + a with spins I_A and i_a, respectively. Following equation A21, their wave functions may be combined to form channel-spin functions (compare section 3.8.1)

$$\psi_{I_A M_A}(\tau_A)\,\psi_{i_a m_a}(\tau_a) = \sum_S \psi_{SM_s}(\tau_A, \tau_a)\,\langle I_A i_a M_A m_a | SM_s\rangle \qquad \text{(A49)}$$

where $M_s = M_A + m_a$ and $|I_A - i_a| \leqslant S \leqslant (I_A + i_a)$. Here the Clebsch-Gordan coefficient is the probability amplitude for finding a particular value S of channel spin with z-component M_S when the colliding pair has z-components M_A and m_a. If the incident beam and target are unpolarised, the probability of any given M_A is $(2I_A + 1)^{-1}$ and of any given m_a is $(2i_a + 1)^{-1}$. Consequently the probability of finding a given S and M_S in such a beam is

$$P_{S,M_S} = \frac{1}{(2I_A + 1)(2i_a + 1)} \sum_{M_A m_a} \left|\langle I_A i_a M_A m_a | SM_s\rangle\right|^2 \qquad \text{(A50)}$$

where the sum is constrained to values such that $M_A + m_a = M_S$. Now equation A23 tells us that this sum is just unity so that

$$P_{S,M_S} = \frac{1}{(2I_A + 1)(2i_a + 1)} \qquad \text{(A51)}$$

This is independent of M_S, as would be expected since the two nuclei are not polarised and therefore there is no preferred direction in space. The probability of finding S irrespective of the value of M_s is

$$g(S) = \sum_{M_s} P_{S,M_s} = \frac{2S + 1}{(2I_A + 1)(2i_a + 1)} \qquad \text{(A52)}$$

which is just the statistical weight for channel spin introduced in section 3.8.1.

The total angular momentum J of the system is obtained by combining the channel spin S with the relative orbital angular momentum ℓ

$$\psi_{SM_s}(\tau_A, \tau_a)\,\phi_{\ell m}(\mathbf{r}) = \sum_J \Psi_{JM}\,\langle S\ell M_s m | JM\rangle \qquad \text{(A53)}$$

where $M = M_s + m$ and $|S - \ell| \leqslant J \leqslant (S + \ell)$. Then the spin of any compound nucleus which is formed is limited to one of these J values. Now the Clebsch-Gordan coefficient is the probability amplitude for finding a particular J value in a system with channel spin S, M_s and orbital ℓ, m. Including equation A49, we see that the probability amplitude for finding a particular value of J in a system of two nuclei with M_A and m_a moving with relative angular momentum ℓ, m is just

$$\langle I_A i_a M_A m_a | SM_s\rangle\,\langle S\ell M_s m | JM\rangle \qquad \text{(A54)}$$

In particular, the vector $\mathbf{L}$ is always perpendicular to the direction of motion; if we take this direction (the beam direction in an experiment) as z-axis, then $m = 0$ only and $M = M_s$.

When the incident spins I_A and i_a are randomly oriented, the probability of finding the channel spin S and total angular momentum J is

$$P_{S,J} = \frac{1}{(2I_A + 1)(2i_a + 1)} \sum_{M_A m_a} \left| \langle I_A i_a M_A m_a | SM \rangle \langle S \ell M 0 | JM \rangle \right|^2 \qquad \text{(A55)}$$

where we chose the incident beam direction as z-axis. We saw above that summing the first Glebsch–Gordan coefficient over M_A and m_a (keeping $M = M_A + m_a$ constant) just gives unity, leaving

$$P_{S,J} = \frac{1}{(2I_A + 1)(2i_a + 1)} \sum_{M} \left| \langle S \ell M 0 | JM \rangle \right|^2 \qquad \text{(A56)}$$

Permuting the arguments of the remaining coefficient (Brink and Satchler, 1968) enables us to sum this one also; the result is independent of S (provided $|\ell - S| \leqslant J \leqslant |\ell + S|$) and gives the statistical spin factor of equation 3.86

$$P_{S,J} = \frac{(2J + 1)}{(2I_A + 1)(2i_a + 1)(2\ell + 1)} \qquad \text{(A57)}$$

REFERENCES

Abramowitz, M. and Stegun, I. A. (1970). *Handbook of Mathematical Functions.* New York; Dover Publications

Brink, D. M. and Satchler, G. R. (1968). *Angular Momentum.* 2nd edn. Oxford; Oxford University Press

Condon, E. U. and Shortley, G. H. (1951). *The Theory of Atomic Spectra.* Cambridge; Cambridge University Press

Messiah, A. M. (1962). *Quantum Mechanics I and II.* Amsterdam; North-Holland

Morse, P. M. and Feshbach, H. (1953). *Methods of Theoretical Physics.* New York; McGraw-Hill

Wilkinson, D. H. (1969), ed. *Isospin in Nuclear Physics.* Amsterdam; North-Holland

Appendix B. Transformations between LAB and CM Coordinate Systems

As was discussed in sections 2.2 and 3.1, it is particularly convenient to use a moving coordinate frame in which the centre of mass of two colliding nuclei is at rest. This is called the CMS or centre-of-mass system. A coordinate frame at rest in the laboratory is called the LAB system. We shall only consider non-relativistic kinematics here. (Marmier and Sheldon (1969) give the relativistic case.)

Suppose the target particle A, with mass m_A, is at rest in the LAB and the projectile a, with mass m_a, is incident with velocity $\mathbf{v}_a$. The CMS is moving in the LAB with a velocity

$$\mathbf{V}_{CM} = \frac{m_a}{m_A + m_a} \mathbf{v}_a \tag{B1}$$

hence the projectile has a velocity in the CMS of

$$\mathbf{v}'_a = \mathbf{v}_a - \mathbf{V}_{CM} = \frac{m_A}{m_A + m_a} \mathbf{v}_a \tag{B2}$$

where we use primes to denote quantities measured relative to the CMS. The target has a velocity in the CMS of $\mathbf{v}'_A = -\mathbf{V}_{CM}$. The total momentum of the pair is zero in the CMS so that

$$m_a \mathbf{v}'_a = -m_A \mathbf{v}'_A$$

Thus their speeds are in the ratio

$$\frac{v'_a}{v'_A} = \frac{m_A}{m_a} \tag{B3}$$

The bombarding energy $E = \frac{1}{2} m_a v_a^2$ becomes transformed into

$$\begin{aligned} E &= \tfrac{1}{2} m_a v_a^2 \\ &= \tfrac{1}{2} m_a v_a'^2 + \tfrac{1}{2} m_A v_A'^2 + \tfrac{1}{2} (m_a + m_A) V_{CM}^2 \\ &= \tfrac{1}{2} \mu_\alpha v_a^2 + \tfrac{1}{2} M V_{CM}^2 \\ &= E_\alpha + E_{CM} \end{aligned} \tag{B4}$$

where μ_α is the reduced mass of the pair

$$\mu_\alpha = \frac{m_a m_A}{m_a + m_A} \tag{B5}$$

and M is the total mass, $M = m_a + m_A$. We recognise E_{CM} as the kinetic energy associated with the motion of the centre of mass, while E_α is the kinetic energy of relative motion in the CMS; also

$$E_\alpha = \frac{m_A}{m_a + m_A} E \tag{B6}$$

After a collision, the centre-of-mass motion, hence E_{CM} and V_{CM}, are unchanged. The energy of relative motion E_α will be unchanged if it is an elastic collision, although the *directions* of motion of the two particles will change. If the Q-value of a non-elastic reaction A(a, b)B is $Q_{\alpha\beta}$, the energy of relative motion after the reaction will be

$$E_\beta = E_\alpha + Q_{\alpha\beta} \tag{B7}$$

The reduced mass in the exit channel will be

$$\mu_\beta = \frac{m_b m_B}{m_b + m_B} \tag{B8}$$

In the special case of inelastic scattering, $m_b = m_a$ and $m_B = m_A$ so that $\mu_\beta = \mu_\alpha$. Since the total momentum in the CMS must remain zero, the two residual particles separate in opposite directions with equal but opposite momenta. Hence their speeds in the CMS are related by

$$\frac{v_b'}{v_B'} = \frac{m_B}{m_b} \tag{B9}$$

B1 ELASTIC SCATTERING

After an elastic collision, the speeds of the two particles in the CMS are unchanged (*see* equations B3, B9). This is not true in the LAB because some momentum has been transferred to the previously stationary target. *Figure B1* illustrates the velocity relations after collision. From the sine rule for triangles we have

$$\frac{\sin(\theta_{CM} - \theta_L)}{\sin\theta_L} = \frac{V_{CM}}{v'_a} = x, \text{ say} \qquad \text{(B10)}$$

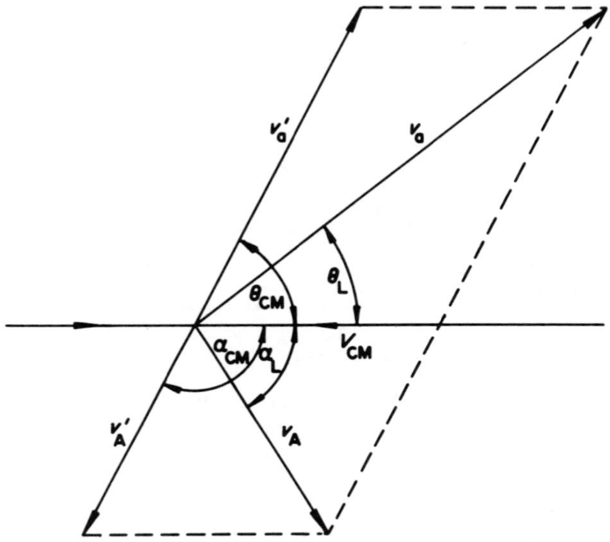

Figure B1 Velocity relationships in the LAB and CM systems for the elastic scattering of two particles

(We follow the popular convention of using subscripts CM and L to denote angles measured in the CMS and LAB, respectively.) Equations B1 and B2 give

$$x = \frac{m_a}{m_A} \qquad \text{(B11)}$$

Equation B10 relates the CMS and LAB angles of scattering of the projectile a. This relation is shown graphically in *Figure B2* for several values of x. When $x \leqslant 1$, θ_L increases monotonically from 0 to π as θ_{CM} increases from 0 to π. For $x > 1$, two values of θ_{CM} contribute to a given value of θ_L and θ_L has a maximum value which is smaller than π. This can be understood physically; $x > 1$ means the projectile is heavier than the target and even a head-on collision will leave the projectile still moving forward. In the CMS this would appear as backward scattering.

The corresponding angles of recoil of the struck particle A (*see Figure B1*) are related by

$$\alpha_{CM} = 2\alpha_L \qquad \text{(B12)}$$

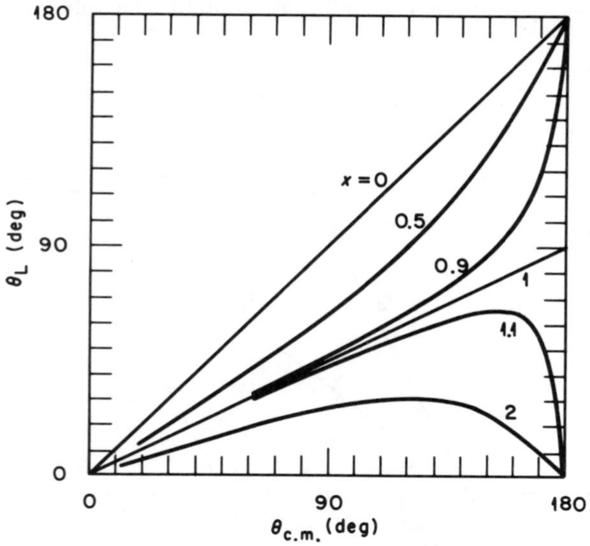

Figure B2 Relationship between scattering angles in the LAB and CM systems. For elastic scattering, x is the ratio of the masses of the two particles, $x = m_a/m_A$. For non-elastic scattering, x is given by equation B20

because $v'_A = V_{CM}$. Further, $\alpha_{CM} = \pi - \theta_{CM}$, so that

$$\alpha_L = \tfrac{1}{2}(\pi - \theta_{CM}) \tag{B13}$$

Another useful relation is obtained by equating components of the momenta perpendicular to and parallel with the beam

$$v'_a \sin\theta_{CM} = v_a \sin\theta_L$$

$$v'_a \cos\theta_{CM} + V_{CM} = v_a \cos\theta_L$$

These yield

$$\tan\theta_L = \frac{\sin\theta_{CM}}{x + \cos\theta_{CM}} \tag{B14}$$

or

$$\cos\theta_L = \frac{x + \cos\theta_{CM}}{(1 + x^2 + 2x\cos\theta_{CM})^{1/2}} \tag{B15}$$

The definition of a cross-section implies that the same number of particles are scattered into the element $d\Omega_L$ of solid angle in the direction (θ_L, ϕ_L) as are scattered into $d\Omega_{CM}$ in the corresponding direction (θ_{CM}, ϕ_{CM}). Thus the differential cross-sections are related by

$$\sigma_L(\theta_L)d\Omega_L = \sigma_{CM}(\theta_{CM})d\Omega_{CM} \tag{B16}$$

Since the transformation between LAB and CMS is symmetric in azimuth about the beam direction, we have $\phi_L = \phi_{CM}, = \phi$ say. Hence we need

$$\frac{\sigma_{CM}}{\sigma_L} = \frac{d\Omega_L}{d\Omega_{CM}} = \frac{d(\cos\theta_L)}{d(\cos\theta_{CM})} \tag{B17}$$

From equation B15 we soon find

$$\frac{d(\cos\theta_L)}{d(\cos\theta_{CM})} = \frac{1 + x\cos\theta_{CM}}{(1 + x^2 + 2x\cos\theta_{CM})^{3/2}} \tag{B18}$$

It is also convenient to have this relation expressed in terms of the LAB angle; it can be shown that

$$\frac{d(\cos\theta_L)}{d(\cos\theta_{CM})} = \frac{(1 - x^2\sin^2\theta_L)^{1/2}}{[x\cos\theta_L + (1 - x^2\sin^2\theta_L)^{1/2}]^2} \tag{B19}$$

B2 NON-ELASTIC COLLISIONS

We shall not derive these results here but leave that as an exercise for the reader. The relations B10 and B14–B19 remain valid if the expression for x is generalised. For the reaction A(a, b)B the expression to use is

$$x = \frac{V_{CM}}{v'_b} = \left[\frac{m_a m_b}{m_A m_B}\frac{E_\alpha}{E_\alpha + Q_{\alpha\beta}}\right]^{1/2} \tag{B20}$$

We note that x is still the ratio of the speed of the centre of mass to the speed of the outgoing particle in the CMS (compare with equation B10). The relation B12 no longer holds because in general we do not have $v'_B = V_{CM}$.

B3 SPECIAL CASES

When $x = 1$, as for the elastic scattering of two particles of equal mass, equation B10 gives $\theta_{CM} = 2\theta_L$ so that θ_L cannot exceed $\frac{1}{2}\pi$ (*see Figure B2*). The CMS and LAB cross-sections are then related by

$$\frac{\sigma_L(\theta_L)}{\sigma_{CM}(\theta_{CM})} = 4\cos\theta_L$$

Consequently, even if the angular distribution is isotropic in the CMS (σ_{CM} = constant, as for the scattering of low-energy neutrons from protons) the angular distribution in the LAB is proportional to cos θ_L. Further, equation B13 shows that for elastic scattering

$$\alpha_L + \theta_L = \frac{\pi}{2}$$

that is, the scattered and recoil particles move at right angles in the LAB.

When $x \ll 1$, we may expand in powers of x. For example

$$\theta_{CM} \approx \theta_L + x \sin \theta_L$$

and if θ is also small

$$\theta_{CM} \approx (1 + x)\, \theta_L$$

Also

$$\frac{\sigma_L(\theta_L)}{\sigma_{CM}(\theta_{CM})} \approx 1 + 2x \cos \theta_L$$

and

$$\frac{\sigma_{CM}(\theta_{CM})}{\sigma_L(\theta_L)} \approx 1 - 2x \cos \theta_{CM}$$

REFERENCE

Marmier, P. and Sheldon, E. (1969). *Physics of Nuclei and Particles*. New York; Academic Press

Appendix C. Some Useful Data

The physical constants were obtained from E. R. Cohen (1976), *Atomic Data and Nuclear Data Tables*, Vol. 18, 587. Note that m = metre, g = gramme, s = second, J = Joule = 10^7 erg, $\pi = 3.14159265$, e = 2.71828183.

C1 PREFIXES

tera (T) = 10^{12}	deci (d) = 10^{-1}	nano (n) = 10^{-9}
giga (G) = 10^{9}	centi (c) = 10^{-2}	pico (p) = 10^{-12}
mega (M) = 10^{6}	milli (m) = 10^{-3}	femto (f) = 10^{-15}
kilo (k) = 10^{3}	micro (μ) = 10^{-6}	atto (a) = 10^{-18}

C2 PHYSICAL CONSTANTS

Speed of light	$c = 2.99792458 \times 10^{8}$ m s^{-1} $\approx 3.00 \times 10^{23}$ fm s^{-1}
Elementary charge	$e = 4.803242 \times 10^{-10}$ esu
	$= 1.602189 \times 10^{-19}$ C
	$e^2 = 1.4400$ MeV fm
Planck's constant	$h = 6.626176 \times 10^{-34}$ J s
	$= 4.13570 \times 10^{-21}$ MeV s
	$\hbar \equiv h/2\pi = 0.65822 \times 10^{-21}$ MeV s
	$\hbar^2 = 41.802$ u MeV fm^2
Fine structure constant	$\alpha \equiv e^2/\hbar c = 7.29735 \times 10^{-3} = 1/137.036$
Avogadro constant	$N_A = 6.022 \times 10^{23}$ mol^{-1}
Boltzmann constant	$k_B = 0.8617 \times 10^{-4}$ eV K^{-1}
Electron volt	eV $= 1.602189 \times 10^{-19}$ J

C3 REST MASSES

atomic mass unit	u	$= 1.660566 \times 10^{-24}$ g = (1/12) mass of neutral ^{12}C atom
	uc^2	= 931.502 MeV
electron	m_e	$= 0.54858 \times 10^{-3}$ u
	m_ec^2	= 0.51100 MeV
muon	m_μ	= 0.1134 u
	$m_\mu c^2$	= 105.7 MeV
pion	$m_{\pi\pm}$	= 0.1499 u
	$m_{\pi\pm}c^2$	= 139.6 MeV
	$m_{\pi 0}$	= 0.1449 u
	$m_{\pi 0}c^2$	= 135.0 MeV
proton	m_p	= 1.007276 u
	m_pc^2	= 938.280 MeV
neutron	m_n	= 1.008665 u
	m_nc^2	= 939.573 MeV
deuteron	m_d	= 2.013553 u
	m_dc^2	= 1875.628 MeV
	binding energy = 2.225 MeV	
α-particle	m_α	= 4.001506 u
	$m_\alpha c^2$	= 3727.409 MeV
	binding energy = 28.30 MeV	

C4 RELATED QUANTITIES

Compton wavelength: electron $\hbar/m_ec = 386.16$ fm
proton $\hbar/m_pc = 0.2103$ fm

Non-relativistic wave number for mass m with energy E

$$k \equiv 2\pi/\lambda = 0.2187\ [m(\text{u})E(\text{MeV})]^{1/2}\ \text{fm}^{-1}$$

Non-relativistic speed for mass m with energy E

$$v \equiv \frac{\hbar k}{m} = 1.389 \times 10^{22}\ [E(\text{MeV})/m(\text{u})]^{1/2}\ \text{fm s}^{-1}$$

Wave number for photon of energy E

$$k \equiv 2\pi/\lambda = 5.068 \times 10^{-3}\ [E(\text{MeV})]\ \text{fm}^{-1}$$

Sommerfeld (Coulomb) parameter for two particles with charges Z_1e and Z_2e, reduced mass m, CM energy E and velocity v

$$n \equiv \frac{Z_1Z_2e^2}{\hbar v} = 0.1575\ Z_1Z_2\ [m(\text{u})/E(\text{MeV})]^{1/2}$$

C5 THE ELEMENTS

Listed are the elements with their chemical symbols and their atomic numbers *Z*. Also given is the mass number *A* of the most abundant naturally occurring isotope. When there is no stable isotope, the *A* for the isotope with the longest known lifetime is given in parentheses. Note that some elements have several stable isotopes; the largest number occur for tin, Sn, which has 10.

Element	Symbol	*Z*	*A*	Element	Symbol	*Z*	*A*
hydrogen	H	1	1	krypton	Kr	36	84
helium	He	2	4	rubidium	Rb	37	85
lithium	Li	3	7	strontium	Sr	38	88
beryllium	Be	4	9	yttrium	Y	39	89
boron	B	5	11	zirconium	Zr	40	90
carbon	C	6	12	niobium	Nb	41	93
nitrogen	N	7	14	molybdenum	Mo	42	98
oxygen	O	8	16	technicium	Tc	43	(97)
fluorine	F	9	19	ruthenium	Ru	44	102
neon	Ne	10	20	rhodium	Rh	45	103
sodium	Na	11	23	palladium	Pd	46	106
magnesium	Mg	12	24	silver	Ag	47	107
aluminium	Al	13	27	cadmium	Cd	48	114
silicon	Si	14	28	indium	In	49	115
phosphorus	P	15	31	tin	Sn	50	120
sulphur	S	16	32	antimony	Sb	51	121
chlorine	Cl	17	35	tellurium	Te	52	130
argon	Ar	18	40	iodine	I	53	127
potassium	K	19	39	xenon	Xe	54	132
calcium	Ca	20	40	caesium	Cs	55	133
scandium	Sc	21	45	barium	Ba	56	138
titanium	Ti	22	48	lanthanum	La	57	139
vanadium	V	23	51	cerium	Ce	58	140
chromium	Cr	24	52	praseodymium	Pr	59	141
manganese	Mn	25	55	neodymium	Nd	60	142
iron	Fe	26	56	promethium	Pm	61	(145)
cobalt	Co	27	59	samarium	Sm	62	152
nickel	Ni	28	58	europium	Eu	63	153
copper	Cu	29	63	gadolinium	Gd	64	158
zinc	Zn	30	64	terbium	Tb	65	159
gallium	Ga	31	69	dysprosium	Dy	66	164
germanium	Ge	32	74	holmium	Ho	67	165
arsenic	As	33	75	erbium	Er	68	166
selenium	Se	34	80	thulium	Tm	69	169
bromine	Br	35	79	ytterbium	Yb	70	174

Element	Symbol	Z	A
lutetium	Lu	71	175
hafnium	Hf	72	180
tantalum	Ta	73	181
tungsten	W	74	184
rhenium	Re	75	187
osmium	Os	76	192
iridium	Ir	77	193
platinum	Pt	78	195
gold	Au	79	197
mercury	Hg	80	202
thallium	Tl	81	205
lead	Pb	82	208
bismuth	Bi	83	209
polonium	Po	84	(210)
astatine	At	85	(210)
radon	Rn	86	(222)
francium	Fr	87	(223)
radium	Ra	88	(226)
actinium	Ac	89	(227)
thorium	Th	90	(232)
protactinium	Pa	91	(231)
uranium	U	92	(238)
neptunium	Np	93	(237)
plutonium	Pu	94	(244)
americium	Am	95	(243)
curium	Cm	96	(247)
berkelium	Bk	97	(247)
californium	Cf	98	(251)
einsteinium	Es	99	(254)
fermium	Fm	100	(253)
mendelevium	Md	101	
nobelium	No	102	(255)
lawrencium	Lw	103	
rutherfordium	Rf	104	(261)
hahnium	Ha	105	

Appendix D. Penetration of Potential Barriers and the Fusion of Very Light Nuclei

The most tightly bound nuclei are those near the middle of the periodic table ($A \sim 50$ to 100, say). Consequently, a sufficiently heavy nucleus may release energy by splitting (fissioning) into two lighter ones (Preston and Bhaduri, 1975). This may occur either spontaneously or after the capture of a neutron, and is the source of energy in a nuclear reactor (as well as the fission or 'atomic' bomb). On the other hand, two lighter nuclei may release energy by combining (fusing) to form a heavier one. Many such fusion reactions have been studied experimentally using beams of heavy ions from accelerators to bombard target nuclei (sections 2.18.12 and 4.12), but only the fusion of two of the lightest nuclei is likely to provide a practical source of energy. The reason is that the repulsive Coulomb force between the two nuclei is proportional to the product of their charges, $Z_1Z_2e^2$, and this Coulomb barrier (section 2.18.8 and *Figure 2.35*) must be overcome before the attractive nuclear forces can initiate the reaction. This fusion process powers the sun and other stars, as well as the fusion, or 'hydrogen', bomb. It is the focus of attention in attempts to produce a controlled thermonuclear reactor. It also provides an interesting example of a practical application of the theory of the tunnelling through potential barriers that is allowed by quantum mechanics (section 3.6).

Deuterium, D, is an attractive fuel to burn because it is readily available from the 'heavy water', D_2O, that is to be found in ordinary water, H_2O. Two deuterium nuclei (deuterons) also have the lowest Coulomb barrier to be overcome, with $Z_1 = Z_2 = 1$. The two reactions of interest are

$$d + d \rightarrow t + p + 4.03 \text{ MeV},$$

$$d + d \rightarrow {}^3\text{He} + n + 3.27 \text{ MeV}.$$

(The capture reaction $d + d \rightarrow {}^4\text{He} + \gamma$ releases 23.9 MeV of energy, but the branching ratio into this channel is extremely small.)

The relative velocity between two deuterons needed to surmount their mutual Coulomb barrier may be obtained by heating a plasma of deuterium to a very high temperature (hence the term *thermonuclear reaction*). The barrier height is of the order of hundreds of keV. Since Boltzmann's constant is $k_B \approx 10^{-4}$ eV K^{-1} (Appendix C), it takes a temperature θ of over a billion (10^9) degrees for the average kinetic energy of thermal motion ($\sim k_B\theta$) to surmount this barrier. Actually, a lower temperature is sufficient (Rolfs and Trautvetter, 1978) because (i) the deuterons may tunnel through the potential barrier (section 3.6) even when their energy is below its top, and (ii) the distribution of velocities in the plasma will provide a considerable proportion of the deutrons with kinetic energies greater than the mean (which would be $\frac{3}{2}k_B\theta$ for a Maxwell distribution). The appropriate conditions for fusion to occur exist in the hot centres of stars, but their achievement on earth, and the containment of the hot plasma, remains a formidable challenge to engineers that is still being addressed.

The quantum mechanical tunnelling through the Coulomb barrier is possible even when the deuterium is not at a high temperature, so that the deuterons have very little thermal energy. This would lead to *cold fusion*. However, its probability is very small (otherwise there would be very little deuterium left on earth!). For example, consider a gas of deuterons at room temperature ($\theta \approx 300$ K). If the distribution of velocities is Maxwellian, the mean kinetic energy of relative motion of two colliding deuterons is $\frac{3}{2}k_B\theta = 0.039$ eV (Clayton, 1968). Then we may use the low energy limit (equation 3.79) for the barrier transmission factor, T. This gives the probability for two deuterons with this mean energy of penetrating their mutual barrier to be $T(E = 0.039 \text{ eV}) \approx 10^{-2176}$, an extremely small number!

However, this is very misleading because there is a distribution of velocities in the gas which we must average over in order to find the overall probability of fusion. The high energy tail of this distribution is very important because the barrier transmission factor, $T(E)$, increases so rapidly with energy, E. What is required in order to evaluate the fusion rate (Clayton, 1968; Rolfs and Trautvetter, 1978) is the average $\langle\sigma(E)v\rangle$, where v is the relative velocity of the pair of deuterons, and then, from equation 3.78, their cross-section for fusion at energy E can be written in the form $\sigma(E) = E^{-1}ST(E)$, with the 'astrophysical S factor' essentially constant (Clayton, 1968). Also, from equation 3.79 we have $\ell n\, T(E) = -bE^{-1/2}$, with b constant. The probability, $P(E)$, of finding a pair with relative energy, E, is proportional to $E^{1/2} \exp(-E/k_B\theta)$ if the distribution is Maxwellian (Clayton, 1968). Then the product $\sigma(E)\, v\, P(E)$ is proportional to $\exp(-E/k_B\theta - b/E^{1/2})$, which is peaked at $E = E_0 = (\frac{1}{2}bk_B\theta)^{2/3}$ with a full width at half-maximum of $\Delta E/E_0 = 4(3E_0/k_B\theta)^{-1/2}$. Usually $E_0 \gg k_B\theta$, so that the peak is sharp. In our case, $b \approx 990$ (eV)$^{1/2}$ and, for example, $E_0 = 3.05\, k_B\theta$ at a temperature of 10^8 K.

However, at a terrestrial room temperature $\theta = 300$ K, the peak moves out much further on the tail of the distribution. Then $E_0 \approx 211\ k_B\theta$ and $\Delta E/E_0 \approx 0.16$. The barrier penetration factor at this energy $E = E_0$ is now $T(E_0) \approx 10^{-183}$, an enormous increase over the value for the mean energy of $\frac{3}{2}k_B\theta$. (Of course, this advantage is partially off-set by the much smaller probability, $P(E_0)$, of finding two deuterons in the gas with this relative energy.) Nonetheless, the fusion probability remains negligibly small.

The probability would be enhanced if some external environment could be found that would 'squeeze' the deuterons closer together and thus assist in overcoming the barrier. For example, the electrons in a deuterium molecule bind the two deuterons and partially shield the Coulomb repulsion between them. It has been estimated (Van Siclen and Jones, 1986) that this may greatly increase the barrier penetration factor, perhaps to greater than 10^{-80}, but still it is small, ensuring that a gas of deuterium remains quite stable under normal conditions.

A further gain is possible by replacing an electron in the deuterium molecule by a muon (Massey *et al.*, 1974). The greater mass of the muon (207 times that of the electron – *see* Appendix C) results in its occupying an orbit with a much smaller radius than that of the electron, thus binding the two deuterons closer together by a factor of about 200. The barrier penetration factor is again estimated to be increased dramatically, perhaps to better than 10^{-4} (Van Siclen and Jones, 1986). This is now sufficiently large for spontaneous fusion to have been observed in muonic deuterium molecules (Massey *et al.*, 1974). Because the muon plays only a transitory role, it has been called *muon-catalysed fusion.* Difficulties in utilising this process include (i) the muons have to be produced independently, and (ii) they only live a short time (2.2×10^{-6} s) before decaying into an electron and two neutrinos.

REFERENCES

Clayton, D. D. (1968). *Principles of Stellar Evolution and Nucleosynthesis*, Chapt 4. New York; McGraw-Hill

Massey H. S. W., Burhop, E. H. S. and Gilbody, H. B. (1974). *Electronic and Ionic Impact Phenomena*, Vol. 5. Oxford; Oxford University Press

Preston, M A. and Bhaduri, R. K. (1975). *Structure of the Nucleus.* Reading, Mass.; Addison-Wesley

Rolfs, C. and Trautvetter, H. P. (1978). *Ann. Rev. Nucl. Part. Sci.* Vol. 28, 115

Van Siclen, C. DeW. and Jones, S. E. (1986). *J. Phys*. Vol. 12, 213

Solutions to Exercises

1.1	(i) 2.4×10^{18} MeV	2.39×10^{-23} fm
	(ii) 1.2×10^{14} MeV	2.39×10^{-19} fm
	(iii) 10.1 MeV	9.02 fm
	4.02×10^{-12} MeV	1.01×10^{-12} MeV
1.2	(i) 70 kW	0.35 g weight
	(ii) 70 kW	0.17 g weight
	(iii) 140 kW	0.17 g weight

1.3 $143.2 \text{ cm s}^{-1} = 5.16 \text{ km hr}^{-1}$ 10.5 cm

1.4 813 MeV 885 kg weight $2.24 \times 10^{27} \times g$

1.5 *See* equation 2.19 $V_c(0) = \frac{3}{2} V_c(R)$

$V_c(0) = 25.3$ MeV $V_c(7) = 16.9$ MeV

$$\overline{V_c(r \leqslant R)} = \tfrac{6}{5} \frac{Z_1 Z_2 e^2}{R} = 20.2 \text{ MeV}$$

1.6 Potentials:

	$r = 2$ fm: 0.72 MeV	5.85×10^{-37} MeV	6.71 MeV
	$r = 1$ fm: 1.44 MeV	1.17×10^{-36} MeV	27.4 MeV

Forces:

	$r = 2$ fm: 0.36 MeV fm^{-1}	2.9×10^{-37} MeV fm^{-1}	8.15 MeV fm^{-1}
	$r = 1$ fm: 1.44 MeV fm^{-1}	1.2×10^{-36} MeV fm^{-1}	47.0 MeV fm^{-1}

1.7 $1.93e$ 6.45×10^{-24} g 7.27 MeV

See Appendix C

1.8 6.07×10^{-14} erg $= 3.79 \times 10^{-2}$ eV $1.35 \times 10^{3} \text{ m s}^{-1}$ 5.625×10^{10} K

1.9 $6.05 \times 10^{33} \text{ dyn cm}^{-2} \approx 3 \times 10^{21} \times K$ (steel)

1.10 $Q = \frac{2}{5} Ze(a^2 - b^2)$ (i) $a/b = 1.35$ (ii) $a/b = 1.373$

1.11 $m = 2m_0$ if $K = m_0c^2$

2.1 $E_A = \dfrac{4M_aM_A}{(M_a + M_A)^2} E \cos^2 \theta_A$ $v_A = \dfrac{(8M_aE)^{1/2}}{(M_a + M_A)} \cos \theta_A$

$M_n \approx 1.16\, M_p$ $E_n \approx 5.7$ MeV

2.2 *See* Appendix B and *Figure B2*

2.3 $E_p = \cos^2 \theta_p$ MeV $\sigma_L(E_p) = 4\, \sigma_{CM} E_p^{1/2}$ $\sigma_L(\theta_p) = 4\sigma_{CM} \cos \theta_p$

2.4 4.029 MeV 9.40 MeV 107.30 MeV

2.5 38.18 MeV 27.57 MeV
$\ell(^{16}O) = 11$ or 12 $\ell(p) = 4$

2.6 359 mb

2.7 $V_c = 21.47$ MeV $V_N = -1.97$ MeV
$F_c = 1.95$ MeV fm^{-1} $F_N = -3.95$ MeV fm^{-1}
$F_c + F_N = 0$ at $r = 11.387$ fm 78°

2.8 $\dfrac{1}{120} \langle r^4 \rangle q^4$

$$F(q) = \frac{3}{(qR)^3} [\sin (qR) - qR \cos (qR)] = \sum_{p=0}^{\infty} (-1)^p \frac{q^{2p}}{(2p+1)!} \langle r^{2p} \rangle$$

$$\langle r^n \rangle = \frac{3}{3+n} R^n$$

Zeros when $\tan x = x$ ($x = qR$): $x_0 \approx 4.5, 7.75, 10.9 \ldots$
$\theta_{min} \approx 19°, 31°, 48°$; $R \approx 3.6, 3.8, 3.4$ fm

2.9 z-axis parallel to **k**; **l** perpendicular to **k**

2.10 $f(\theta) = k^{-1} \sum_{\ell} (2\ell + 1) \exp(i\delta_\ell) \sin \delta_\ell P_\ell (\cos \theta)$ if $\eta_\ell = \exp(2i\delta_\ell)$

$$\frac{d\sigma}{d\Omega} = k^{-2} \left\{\sin^2 \delta_0 + 6\cos(\delta_0 - \delta_1) \sin \delta_0 \sin \delta_1 \cos \theta + 9\sin^2 \delta_1 \cos^2 \theta\right.$$

$$\sigma = \frac{4\pi}{k^2} [\sin^2 \delta_0 + 3 \sin^2 \delta_1]$$

2.11 42 MeV

2.12 2.04% 3.72×10^{-3} K 538 kG

2.13 $d\sigma_{sym} = 4 d\sigma_{unsym}$ (spin-0)
$d\sigma_{sym} = d\sigma_{unsym}$ (spin-$\frac{1}{2}$)

2.14 $\frac{5}{2}$

2.16 $I(\theta) = F_0(\theta) = \sin^2\theta$
$I(\theta) = \frac{2}{3}$ = isotropic

3.3 $\Psi = \chi_1(\mathbf{r})\psi_1(\tau_A) + \chi_2(\mathbf{r})\psi_2(\tau_A)$

$[\nabla^2 - U_{11}(\mathbf{r}) + k_1^2]\,\chi_1(\mathbf{r}) = U_{12}(\mathbf{r})\chi_2(\mathbf{r})$

$[\nabla^2 - U_{22}(\mathbf{r}) + k_2^2]\,\chi_2(\mathbf{r}) = U_{21}(\mathbf{r})\chi_1(\mathbf{r})$

$[\nabla^2 - U_{\text{eff}} + k_1^2]\,\chi_1(\mathbf{r}) = 0$

if $U_{\text{eff}} = U_{11} + \underset{\epsilon\to 0}{\text{Lt}}\; U_{12}\dfrac{1}{\nabla^2 - U_{22} + k_2^2 + i\epsilon}U_{21}$

$\underset{\epsilon\to 0}{\text{Lt}}\dfrac{1}{E - H + i\epsilon} = \mathcal{P}\dfrac{1}{E-H} - i\pi\,\delta(E-H)$; $\mathcal{P} \equiv$ principal value

3.4 $\delta = \tan^{-1}\left[\dfrac{k}{K}\tan KR\right] - kR$ if $K = \left[\dfrac{2m}{\hbar^2}(E-V)\right]^{1/2} \geqslant 0$

$\dfrac{\text{amplitude for } r < R}{\text{amplitude for } r > R} = \left[1 + \left(\dfrac{K_0}{k}\cos KR\right)^2\right]^{-1/2}$ where $K_0 = \left[-\dfrac{2mV}{\hbar^2}\right]^{1/2}$

$\sigma_0 = \dfrac{4\pi}{k^2}\sin^2\delta \to 4\pi R^2\left[\dfrac{\tan K_0 R}{K_0 R} - 1\right]^2$ as $k \to 0$

$\sigma_0 = 0$ if $\tan K_0 R = K_0 R$

See section 3.6.1

3.5 $\left(\dfrac{d\sigma}{d\Omega}\right)_{\text{BA}} = \left(\dfrac{2mV_0R^3}{\hbar^2}\right)^2 g^2(2kR\sin\tfrac{1}{2}\theta)$ if $g(x) = \dfrac{\sin x - x\cos x}{x^3}$

$\left[\exp(i\delta_\ell)\sin\delta_\ell\right]_{\text{BA}} = \dfrac{mV_0kR^3}{\hbar^2}\left[j_\ell^2(kR) + j_{\ell+1}^2(kR) - \dfrac{2\ell+1}{kR}j_\ell(kR)j_{\ell+1}(kR\right.$

$\dfrac{mV_0R^2}{\hbar^2} \ll 1$ if $kR \ll 1$; $\dfrac{mV_0R}{\hbar^2 k} \ll 1$ if $kR \gg 1$

3.6 $\left(\dfrac{d\sigma}{d\Omega}\right)_{\text{BA}} = \left(\dfrac{2mV_0}{\alpha\hbar^2}\right)^2\left(\dfrac{1}{\alpha^2+q^2}\right)^2$ if $q = 2k\sin\tfrac{1}{2}\theta$

Let $\alpha \to 0$, keeping $(V_0/\alpha) = -Z_1Z_2e^2$

3.7 0.59 fm

3.9 $f(\theta) = k^{-1} \sum_{\ell} (2\ell + 1)\, C_{\ell} P_{\ell} (\cos\theta)$

$\sigma_{el} = 4\pi k^{-2} \sum_{\ell} (2\ell + 1) |C_{\ell}|^2$

$\sigma_{abs} = 4\pi k^{-2} \sum_{\ell} (2\ell + 1)\, [Im C_{\ell} - |C_{\ell}|^2]$

$\sigma_{tot} = 4\pi k^{-2} \sum_{\ell} (2\ell + 1)\, Im C_{\ell}$

3.10 $f(\theta) = g(\cos\theta) + h(\cos\theta)\, \boldsymbol{\sigma}\cdot\mathbf{k}\times\mathbf{k}'$ where $\cos\theta = \mathbf{k}\cdot\mathbf{k}'/k^2$
Polarisation vector is parallel to $\mathbf{k}\times\mathbf{k}'$
$f(\theta) = a + b(\mathbf{n}\cdot\boldsymbol{\sigma}_1)(\mathbf{n}\cdot\boldsymbol{\sigma}_2) + c\, \mathbf{n}\cdot(\boldsymbol{\sigma}_1 + \boldsymbol{\sigma}_2) + d\, \mathbf{n}\cdot(\boldsymbol{\sigma}_1 - \boldsymbol{\sigma}_2)$
$\quad + e(\boldsymbol{\ell}\cdot\boldsymbol{\sigma}_1)(\boldsymbol{\ell}\cdot\boldsymbol{\sigma}_2) + g(\mathbf{m}\cdot\boldsymbol{\sigma}_1)(\mathbf{m}\cdot\boldsymbol{\sigma}_2)$
where *a*, *b*, . . . are functions of $\mathbf{k}\cdot\mathbf{k}'$ and $\boldsymbol{\ell}$, **m**, **n** are orthogonal unit vectors:

$\boldsymbol{\ell} = (\mathbf{k} + \mathbf{k}')/2k\cos\tfrac{1}{2}\theta \quad \mathbf{m} = (\mathbf{k} - \mathbf{k}')/2k\sin\tfrac{1}{2}\theta \quad \mathbf{n} = (\mathbf{k}\times\mathbf{k}')/k^2\sin\theta$

$d = 0$ for identical particles

4.1 (i)

If θ_{min} =	35°,	34.6°,	34.15°,	33.75°	
then R =	6.46,	6.51,	6.57,	6.60	fm
Corrected R=	7.26,	7.30,	7.36,	7.38	fm
r_0 =	1.45,	1.44,	1.44,	1.41	fm

(ii) $R = 6.62$ fm corrected $R = 7.69$ fm $\ell = 17.8$

4.3 $\frac{4}{3}\frac{R}{\mu}[\mu^3 - (\mu^2 - 1)^{3/2}]$

4.5 $U(r_p) = J_0\, [\rho(r_p) + \frac{1}{6}\nabla^2\rho(r_p)\langle r^2\rangle + \ldots]$

4.6 $U_i(\mathbf{r}) = \int \rho(\mathbf{r}')\, v_i(\mathbf{r} - \mathbf{r}')\, d\mathbf{r}' \quad i = 0, 1$
$v_0 = \frac{1}{2}(v_{pn} + v_{pp}) \qquad v_1 = \frac{1}{2}(v_{pn} - v_{pp})$
$v_{pn}/v_{pp} = 3 \qquad U_p = -55.3$ MeV $\qquad U_n = -44.7$ MeV

4.8 $W \approx 53 + 25\,(N - Z)/A$ MeV
$W \propto (E + 50)^{-1/2}$

4.9 4.62 fm 0.79 fm 0.068 fm

4.10 At $r = 7.7$ fm (*see* Exercise 4.1)
$U = (2.23 + 0.52i), (2.29 + 0.41i), (2.33 + 0.34i)$ MeV

4.11 $A = (-)^{n-m}\,(k/K)$ with $(Ka + \Delta) = m\pi$ and m integer
$A = (-)^{n-m}$ with $(Ka + \Delta) = (m + \frac{1}{2})\pi$ and m integer

4.12 $\delta_0 = -kR + \tan^{-1}\left[\frac{k}{K}\tan KR\right]$ where $K = \left[\frac{2m}{\hbar^2}(E + V)\right]^{1/2}$

$\sigma_{el} \to 4\pi R^2\left[\frac{\tan K_0 R}{K_0 R} - 1\right]^2$

4.13 $V \approx 51$ MeV
$A \approx 83, 292$
$V_p \approx 57$ ($A = 70$), 60 ($A = 230$)

4.14 $T_{pd} = D(K)F(Q)$
where $D(K) = \int \psi_d^*(s) V_{pn}(s) \exp(-i\mathbf{K}\cdot\mathbf{s})\, ds, \quad \mathbf{K} = \frac{1}{2}\mathbf{k}_d - \mathbf{k}_p$

$F(Q) = \int \phi(\mathbf{r}_n) \exp(-i\mathbf{Q}\cdot\mathbf{r}_n)\, d\mathbf{r}_n, \quad \mathbf{Q} = \mathbf{k}_d - (M_B/M_A)\mathbf{k}_p$

$$C = \left[\frac{\alpha\beta(\alpha+\beta)}{2\pi(\beta-\alpha)^2}\right]^{1/2}$$

$$D(K) = D_0\, \beta^2/(K^2 + \beta^2), \quad D_0 = -\left(\frac{8\pi\epsilon_d^2}{\alpha^3}\right)^{1/2} \left(\frac{\alpha+\beta}{\beta}\right)^{3/2}$$

$$F(Q) = i^{-L} N_L\, [4\pi(2L+1)]^{1/2}\, \frac{R^2}{Q^2 + K_n^2} \left[j_L(QR) h_L^{(1)\prime}(iK_nR) - h_L^{(1)}(iK_nR)\, j_L'(QR)\right]$$

where $f_L' = (df_L/dr)|_{r=R}$

4.15 $L = 3.0, 1.9, 1.1, 0$
$R = 8.36, 7.63, 7.34$, (?); $R \approx 2.0 \times A^{1/3}$
if $j_L(x)$ maximum at $x = 4.5, 3.35, 2.1, 0$ for $L = 3, 2, 1, 0$

4.17 $\Gamma_n = 0.0409\, \Gamma_\gamma$

4.19 *See* Blatt and Weisskopf (1952), page 426; Lynn (1968); Lane and Thomas (1958)

4.21 $\tan\theta_1 = \sin 2\theta_2 \Big/ \left(\frac{m_1}{m_2} - \cos 2\theta_2\right)$

where θ_1 = laboratory angle of projectile of mass m_1 and θ_2 = laboratory angle of recoiling target of mass m_2

4.22 $\mathbf{j}^2 = (\mathbf{i} + \mathbf{l})^2$ commutes with $\mathbf{l}\cdot\mathbf{i}$
$\langle \mathbf{l}\cdot\mathbf{i} \rangle = \frac{1}{2}[j(j+1) - \ell(\ell+1) - i(i+1)]$

4.23 $M = 2$: $W(\theta) = \sin^2\theta$
$M = 0$: $W(\theta) = 1 + 3\cos^2\theta$

Index